Sustainability and Environment: An Integrated Approach

Sustainability and Environment: An Integrated Approach

Edited by Chester Watters

SYRAWOOD
PUBLISHING HOUSE
New York

Published by Syrawood Publishing House,
750 Third Avenue, 9th Floor,
New York, NY 10017, USA
www.syrawoodpublishinghouse.com

Sustainability and Environment: An Integrated Approach
Edited by Chester Watters

International Standard Book Number: 978-1-68286-719-8 (Hardback)

Cataloging-in-Publication Data

Sustainability and environment : an integrated approach / edited by Chester Watters.
   p. cm.
Includes bibliographical references and index.
ISBN 978-1-68286-719-8
1. Sustainability. 2. Environmental protection. 3. Sustainable development. 2. Green technology.
I. Watters, Chester.
HC79.E5 S87 2019
338.927--dc23

TABLE OF CONTENTS

PREFACE

Every book is a source of knowledge and this one is no exception. The idea that led to the conceptualization of this book was the fact that the world is advancing rapidly; which makes it crucial to document the progress in every field. I am aware that a lot of data is already available, yet, there is a lot more to learn. Hence, I accepted the responsibility of editing this book and contributing my knowledge to the community.

Sustainability refers to the process that enables the achievement of technological advancements and the use of resources in a manner that meets human needs in the present day as well as in the future scenario. Sustainability builds on sustainable development, which embraces a combination of various environmental, economic and social factors. Healthy environment and healthy ecosystems are crucial for the sustenance of life on earth. The key strategies to reduce negative human impact on the environment are environmentally-friendly chemical engineering, resource management and environmental protection. Mitigation solutions are also available in effective urban planning, sustainable agriculture, transport, etc. This book unravels the central aspects of this field which will be crucial for the progress of this field in the future. It studies, analyzes and upholds the pillars of environmental sustainability and its utmost significance in modern times. A coherent flow of topics, student-friendly language and extensive use of examples make this book an invaluable source of knowledge.

While editing this book, I had multiple visions for it. Then I finally narrowed down to make every chapter a sole standing text explaining a particular topic, so that they can be used independently. However, the umbrella subject sinews them into a common theme. This makes the book a unique platform of knowledge.

I would like to give the major credit of this book to the experts from every corner of the world, who took the time to share their expertise with us. Also, I owe the completion of this book to the never-ending support of my family, who supported me throughout the project.

Editor

Social Innovation and Sustainable Rural Development: The Case of a Brazilian Agroecology Network

Oscar José Rover [1], Bernardo Corrado de Gennaro [2] and Luigi Roselli [2,*]

[1]   Department of Zootechny and Rural Development, Federal University of Santa Catarina, Florianópolis-SC 88034-001, Brazil; oscar.rover@ufsc.br

[2]   Department of Agricultural and Environmental Science, University of Bari Aldo Moro, 70126 Bari, Italy; bernardocorrado.degennaro@uniba.it

*   Correspondence: luigi.roselli@uniba.it

Academic Editor: Sanzidur Rahman

Abstract: Food is central to human beings and their social life. The growing industrialization of the food system has led to a greater availability of food, along with an increasing risk perception and awareness in consumers. At the same time, there is an increasing resistance from citizens to the dominant model of production and a growing demand for healthy food. As a consequence, an increasing number of social networks have been formed worldwide involving the collaboration between producers and consumers. One of these networks, the Ecovida Agroecology Network, which operates in Southern Brazil, involves farming families, non-governmental organizations, and consumer organizations, together with other social actors. Using a qualitative approach based on participant observation and an analysis of documents, the article examines this network. The theoretical framework used is social innovation, which is commonly recognized as being fundamental in fostering rural development. Results show that Ecovida has instigated innovations that relate to its horizontal and decentralized structure, its participatory certification of organic food, and its dynamic relationship with the markets based on local exchanges and reciprocal relations. Furthermore, such innovation processes have been proven to impact on public sector policies and on the increasing cooperation between the social actors from rural and urban areas.

Keywords: Ecovida Agroecology Network; social innovation; participant observation; Alternative Food Networks; Participatory Guarantee Systems (PGS); sustainable rural development

---

1. Introduction

Rural areas generally lack support in their development, as urban investment normally provides quicker and larger returns. Investment and innovation, however, are required in order to develop any rural or urban area. With limited private funding and public support, social innovation projects (defined in Section 2) are important for the development of rural areas. Social innovation is a key element for any institution, movement, or social network in terms of both organizational and territorial development. However, the development of rural areas also needs the support of public policies.

One way to innovate within rural areas is to transform the organization of the food system into a decentralized social network, acting over a wide geographical area, generating favourable conditions for small-scale farmers to improve their access to the market and to receive differential treatment from public policies (e.g., the institutionalization of participatory certification schemes, structural support to organize local and network markets). This is the working principle of Rede Ecovida de Agroecologia (Eco-life Network for Agroecology, hereinafter referred to as Ecovida), which operates in Southern Brazil and promotes many social innovation initiatives.

This article focus on Ecovida and its dynamics, in order to highlight the role of social innovation in fostering rural development. It is organized as follows. Section 2 reviews the literature on social innovation and its relationship with sustainable rural development. Section 3 explores the Ecovida case study in depth and demonstrates how this network and its initiatives can be interpreted as social innovations. Finally, Section 4 discusses the results and presents the conclusions of the study.

2. The Theoretical Framework: Social Innovation and Rural Development

In his book "The Theory of Economic Development", written in 1912, Schumpeter [1] was the first academic to thoroughly investigate the relationship between development and innovation. His approach shows a clear economic bias and a focus on technology. He also highlights that there is often a cooperative element to an enterprise, which could be interpreted as social innovation, although he does not use this specific term. Following Schumpeter's lead, subsequent research on innovation focused on economic and technological issues within urban contexts. However, in the literature there is still little focus on rural areas and social innovation initiatives, apart from the significant contribution of a few authors [2–5].

Schumpeter's notion of innovation involves a break with the previous development of a given territory. Many of his followers, now called Neoschumpeterians, follow his premises, but interpret the innovations according to their own contexts and historical timelines. As they study innovations in terms of their technological elements, they introduced the notion of "technological trajectory" defined as "a process that generates technological changes over time, indicating how those innovations interact with each other and how they follow a given development pattern" [6] (p. 258). Institutional theories on development have a similar notion but without a technological specification, which they call "path-dependence", referring to "random and remote facts in time (which) have a clear influence on institutional evolution" [7] (p. 114). These approaches underline the importance of history in understanding social and innovative processes.

In the last few years, as the global preoccupation with environmental sustainability has grown, as also confirmed by the 2015 Paris Agreement on Global Warming [8], the interest in innovations that promote sustainable development have been increasingly valued, with greater emphasis on governance initiatives based on the involvement of communities. Neumeier [3] states that a lack of social innovation restrains vitality and the further development of rural areas. In his opinion, social innovation is the foundation of sustainable rural development. Bock [5] highlights the importance of the "global financial crisis, which produced massive public budget cuts" to stimulate the prominence of "self-determination, self-help and self-reliance as components of social innovation" [5] (pp. 555–556). She emphasizes "the problematic side of social innovation (...) as promoted like a solution in a context where the development base is also weakened as a result of policy interventions" [5] (p. 559).

In many cases, social innovation is perceived as an invention, as a change linked to the economy or technology, or as a result of organizational innovation. Social innovation, however, goes beyond these meanings. Neumeier [3] (p. 55) shows that social innovation happens when there is a change in attitudes, behaviour, or perception within a group who enters a network in which people work around common interests, setting up new paths to collaborative action within, or beyond, the borders of the group. According to this author, although innovation occurs within a specific social group, this improvement should be understood within the broader context of where the social innovation occurs.

From another viewpoint, Murray, Grice, and Mulgan [2] (p. 3) consider that social innovation means new products, services, and organization models which meet social needs, create new relationships and collaborations, and provide new possibilities for action for the society in which these changes occur. Additionally, Caroli et al. [4] interpret social innovation as processes that transcend the social, itself, regarding the production of goods and services as a way of overcoming social and environmental problems, as well as market failures. Bock [5] (p. 561), together with the previous authors, place social needs at the core of social innovation. On the other hand, Neumeier [3] suggests that, although needs are important for social innovation, it is also possible to interpret social innovation beyond the focus on needs. According to Aléssio and Rover [9] (p. 116), social innovation depends on "organizational dynamics (which) may form differently, according to local contexts in each territory, allied to technological processes adjusted to local routes, more or less dependent of each territory's history".

Neumeier [3] suggests that social innovations happen as social practices and not as technical artefacts, which depend on group work connected to social networks and the availability of social capital [10,11]. Such social practices need to be stimulated in order to lead to innovation, by both the internal and external agents of the initiatives. In this way social innovations develop in terms of collaborative action, representing novelty in the subjective perception of the individuals involved and in their attitudes and behaviour. Their practical implementation is connected to a vision of superiority as a solution to existing methods, with its main focus on the construction of social skills and assets, and not on the fulfilment of social needs. Thus, according to Neumeier, the material results are merely additional, as the focus of social innovations is not materialistic.

Murray, Grice, and Mulgan [2] highlight six stages of social innovation, which they see as non-mandatory, as well as non-linear: (1) inspiration and social context; (2) proposals and ideas; (3) building prototypes and experiments; (4) confirming the innovation; (5) organizing and promoting the growth and expansion of the experiment; and (6) changes in the reference system which, as an ultimate goal of social change, involves structural changes within the context where the innovation occurs. Conversely, Neumeier [3] presents a more academic proposal, focused on identifying the presence and organization of social innovation in different rural areas. His conceptual starting point is that social innovation represents the behaviour and perception changes within a group. Consequently, Neumeier limits his study to the strict social character of innovation. He believes that there are three key stages of social innovation: (a) problematization, which is triggered by an initial impulse that stimulates social actors to act; (b) drawing attention, engaging, and increasing the interest in innovation of the social group/s involved; and (c) coordinating participants in implementing new behaviours, during which a dynamic co-learning process develops.

To date, empirical studies on social innovation are still limited, both in urban and rural areas. Since social innovation in rural contexts is much more important for the promotion of development, due to their higher socioeconomic vulnerability, the main object of this study is to analyse the specific processes of social innovation promoted by Ecovida in the rural areas of Southern Brazil aimed at promoting a sustainable rural development. Based on the theoretical approach proposed by Neumeier [3] and Aléssio and Rover [9], we analyse the driving forces behind the participation of the social actors; the dynamics of the actor-network underlying its composition and organizational changes: the importance of non-social elements in the decisions taken by social actors. We also investigate the dependence and relationship of the general organizational dynamics of Ecovida in each micro-region, as well as Ecovida's relationship with sector policies.

3. The Case of Rede Ecovida de Agroecologia

3.1. Methodological Approach

The method chosen to examine the role of social innovation in fostering rural development is the case study approach. This qualitative method studies the characteristics of a particular entity or phenomenon, and is helpful for exploring a complex research area about which little is known. The case study we focused on is the Rede Ecovida de Agroecologia (Ecovida). The research is based on combined data collection tools: participant observation and an analysis of documents. Participant observation has occurred throughout the network's existence, through the direct participation of one of the authors at Ecovida from its foundation. The analysis of documents focused on Ecovida's official documents and scientific papers [12–24] in order to describe the features and the dynamics of the organization.

In this section, the qualitative data gathered are synthesized and interpreted to describe the Ecovida experience, from its foundation up to the present, using the theoretical scheme proposed by Neumeier [3] and Aléssio and Rover [9].

3.2. History, Organization and Social Innovation

Ecovida was created in November 1998 in the state of Santa Catarina. The initial external problematization was caused by the state government of Santa Catarina which had, earlier in the year, launched a program for the regulation and certification of organic products in the state. In order to resist this initiative, a group of non-governmental organizations (NGOs) joined together to build a non-state alternative to this policy. In 1999, Ecovida had already reached out to the entire southern region of Brazil (Figure 1), a region in which the NGO network, known as the Rede Tecnologias Alternativas (Alternative Technology Network), was already operative. In this year, the first meeting of farmer groups (FGs), NGOs, and other interested organizations took place, having changed the network's name to "Rede Ecovida de Agroecologia".

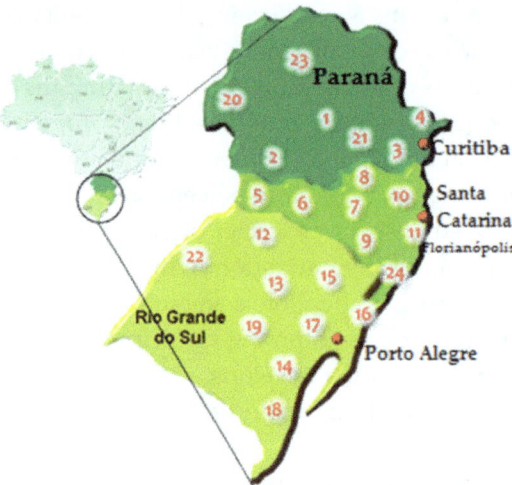

Figure 1. Map of Brazil with the geographical reach of Rede Ecovida and its regional centres. Source: [14].

The establishment of Ecovida highlights some important issues regarding the process of social innovation. Its original mission, which was clearly external and governmental, led some organizations to join in opposition to a centralized certification model which did not respect the history of the pioneer farmers and organizations who were already adopting organic methods in the area. Since its foundation, this network has been promoting a participatory certification scheme with its own label integrating the social actors involved in the network. More than a certification label for the products, the process involved agro-ecological production, joint working, and mutual social learning.

Ecovida's social innovation has always focused on benefitting agroecological family farmers and their organizations. Groups of family farmers have been set up, particularly in regions with supporting NGOs (or through joint efforts when such groups already existed). These groups are the organizational core of Ecovida. To become part of the network, farmers need to belong to a farmer group (FG) and each group is linked to a regional centre (RC), which acts as the main functional body of its organizational dynamics.

Figure 1 shows the location of the RCs in 2015, which had extended to a large area seventeen years after its foundation. Table 1 presents the historic evolution of the number of members and organizations connected to Ecovida. It started with 343 farming families, 35 farmer groups, and four NGOs in 1999, and by 2016 there were 4500 farming families, 300 farmer groups, 30 NGOs, and 28 RCs. Its growth is tied to the social capital it built up and to the increasing demand for organic food both in Brazil and internationally [25]. The growth in consumer organizations connected to Ecovida is in line with the growth in demand, which proves the potential of this network to involve organizations beyond the rural areas.

Table 1. Evolution of the Ecovida network.

Year	Farmer Families	Farmer Groups	Municipalities	NGOs	Consumer Organizations	Regional Centres	Traders	Agro-Industries
1999	343	35	N/A	4	N/A	N/A	N/A	N/A
2005	2438	272	180	28	06	21	N/A	N/A
2007	2700	290	205	35	08	24	N/A	N/A
2009 *	3000	300	220	35	08	25	N/A	N/A
2011 **	2444	213	178	35	21	17	39	113
2016	4500	300	170	30	20	28	N/A	N/A

Produced by the authors. Sources: [20–22]. NGOs: non-governmental organizations. * For 2009 it is difficult to assess the precise data from the whole of Rede Ecovida [15–21]; ** According to the author [22] only 17 Regional Centres answered his survey, despite 24 confirming their existence. This means that there was no decrease in the Ecovida numbers for this specific year.

Ecovida is a horizontal and decentralized network, and takes non-binding decisions, which are accepted by the majority of its members but which, in some cases, are not fully integrated at a local level. Different levels of progress depend on the capacity of each RC's organization. In some regions, Ecovida has strong organizational connections, which are not so developed in others. The fragility of several RCs is demonstrated by the lack of information provided by Ecovida, by not knowing exactly how many family farmers and organizations are associated with it.

By opposing the hegemonic and centralized development model, many endeavours have focused on the micro, specific, and immediate, despite the fact that many organizations have a strategic vision for how this model could be transformed [26]. Starting with the construction of local alternatives to the hegemonic paradigm, networks, such as Ecovida, came together to coordinate a number of organizations with common social transformation projects. Due to the micro-regional and/or sectorial activities of many social organizations, networking was regarded as a solution to increase their effectiveness. Especially from the 1990s onwards, as they brought alternative and significant experiences together, many local organizations formed social networks, such as Ecovida, in order to promote a common action. Many networks with similar profiles have been set up in Southern Brazil, such as NGOs, solidarity economies, family agro-industries, popular enterprises, or credit structures [13]. These could be considered as part of the so-called Alternative Food Networks (AFNs) [27,28]. In fact, according to Renting et al. [27], AFN is "a broad embracing term to cover newly emerging networks of producers, consumers, and other actors that embody alternatives to the more standardised industrial mode of food supply".

Ecovida requires a member to participate in an organization (farmers' group, association, cooperative, etc.) in order to take part in its network. A farmer can participate in the Ecovida network only after joining a FG. This is a key condition for obtaining the organic certification of his/her produce. It is a decentralized network whose managerial decisions are taken at each RC, of which the FGs constitute the core engine. Despite its national coordination bodies, the network's decision-making is multi-directional as its role is secondary compared to FGs and RCs. These latter organs make decisions and select the organizational paths to follow.

The current organization of Rede Ecovida, as shown in Table 2, is the result of a historic path of each territory involved and their local organizational dynamics. Until the 1980s, the members of Rede TA-Sul carried out intense political and organizational training with family farmers from their own micro-regions. From the 1990s onwards, many of the NGOs of Rede TA-Sul directed their work more towards the productive sector through a method which would later be known as agroecology. This was an alternative to an agriculture, which depended on industrial inputs, on the market, and was based on the premises of the Green Revolution (it refers to the process of modernization of agriculture, started in the 1960s. It consists in the development of modern or high-yield crop varieties, particularly of corn, rice and wheat, released to farmers in Latin America and Asia, associated with the intense use of capital, improved seeds, chemical inputs, and moto-mechanization). As productive practices advanced, it was also necessary to establish trading locations which enabled producers to make a profit from agroecological production. Pedagogically speaking, it is possible to assign four stages to the organizational dynamics of the territories of Ecovida up to the present (Table 3).

Table 2. Decision and organizational bodies of Ecovida.

Organizational Bodies	Decision and Organization
Plenary Meeting	Every two years with all those who are part of the Network. It works as a General Assembly where strategic guidelines are decided.
General Assembly of RCs	With two members of each regional centre (RC) with two meetings per year.
Coordination Meetings	Where decisions from the Plenary Meetings and General Assembly of RC's are put into action, working at two levels: (a) General Coordination; (b) Coordination in each State (Paraná, Santa Catarina and Rio Grande do Sul).
Regional Centres	Spread across micro-regions in southern Brazil, these assume specific organizational strategies with respect to their own specific situations, led by the principles and guidelines of Ecovida. They are made up of: (i) Coordination; (ii) Secretariat; (iii) Financial Sector; (iv) Ethics Committees *; (v) Groups.

Source: Based on Perez-Cassarino [22] and Ecovida [14]. * Ethics committees guarantee the application of principles and rules of Ecovida by all of its members. They exist in the groups, RCs, and throughout the network.

Table 3. Historical steps in the organization dynamics of family farmers in the Ecovida territories.

Year	Stage	
1970s/1980s	Stage 1: Political organization of family farmers and setting-up of social and political organizations	
1990s	Stage 2: Food production based on alternative farming (agroecology), adding value to local agro-biodiversity	Transition from the 1990s to the 2000s matches the creation of Ecovida, characterized by a global increase in organic street markets and the need to offer trade opportunities to develop agroecological production with the establishment of trading posts
2000s	Stage 3: Trading produce as well as fostering a national proposal for participatory certification of organic productions	
2010s	Stage 4: Strengthening of elements from previous stages, adding more focus on trade and production and less on valuing political and organizational training	

Source: Stages 1 and 2 adapted from Arns et al. [13] and Arl [12].

In the 1990s, the spread of street markets for organic produce required norms that guaranteed their quality, also requiring Ecovida-related groups to be involved in the market development of their practices. In 1998 these organizations, with clear pro-social ideals, then created Rede Ecovida de Certificação Participativa (Ecovida Network for Participatory Certification) in order to guarantee the organic quality of their products. As they understood that this process would limit their approach to such a specific element such as certification, they changed their designation to Rede Ecovida de Agroecologia (Eco-life Network for Agroecology). This change is a social innovation itself. In fact, the network brings together two new elements: the Participatory Guarantee System (PGS) and its social setup of the market.

3.3. Functioning of Participatory Guarantee Systems (PGS)

It is not possible to distinguish the higher quality of organic products just by their aesthetic appearance compared with products from conventional agriculture where chemically-synthesized products (e.g., pesticides, fertilizers) are used. This makes the organic quality cue a credence attribute for which the market success relies on the trust between producers and consumers. Brazil thus created three forms of guarantee systems for organic quality assurance [29]:

(1)   Third-party certification, operated by an independent company, subject to Conformity Assessment Bodies (in Portuguese: Organismos de Avaliação da Conformidade, or OAC);

(2)   Participatory Guarantee Systems, operated by a Participatory Body for Conformity Assessment (in Portuguese: Organismos Participativos de Avaliação da Conformidade, or OPAC), under which Ecovida operates; and

(3)   Social Control Organizations (in Portuguese: Organização de Controle Social, or OCS), operated by local organizations, intended to be used only to sell products according to direct marketing strategies.

Each of these accredited guarantee systems enable the trade of produce across different regions. However, third-party certification permits the use of the label of the Brazilian System for the Evaluation of Organic Compliance, thus enabling the international sale of the produce because it conforms to international rules. Participatory Guarantee Systems also allow the use of such labels, but the sales are limited to within Brazil. As for the accreditation by a Social Control Organization, this can only be used by family farmers and only permits them to sell directly at local consumption points.

Now let us turn to the work Ecovida carries out with participatory certifications. Ecovida has played a key role in the formulation of this kind of guarantee system under Brazilian law. In order to promote participated certification, as a social innovation, recognition by law was required. Due to its experience, Ecovida worked together with the federal government and played a central role in the shaping of this certification scheme. Social innovation, in Ecovida, does not represent the state's withdrawal, as there is still a link between the state and civic organizations. In this case there is no "at distance of state support" [5] (p. 569), but cooperation to establish the norms and operate the participatory certification.

Participatory certification involves the exchange of knowledge and shared learning between its members, as well as a specific inspection by crossed monitoring. To perform this control, farmers within the same network or organization, but from different groups, visit and monitor other producers, with special focus on any cases of non-compliance. Non-compliance of rules is very important because it harms the whole organization, affecting the Ecovida label itself. This method of cross-monitoring can be particularly effective for organic production and trade, as it stimulates the exchange of knowledge, seeds, and other elements that make up its organizational capital. As for Ecovida, Frison, and Rover [17] show how expertise and experiences are exchanged within each group and among different groups: in terms of handling techniques, control methods, as well as the general knowledge of organic production. There is also a broader awareness of the Brazilian legislative bodies regarding the efficiency of this certification method.

The success of the two systems for organic quality assurance that do not require a third-party involvement (OPAC and OCS) is supported by the wide adhesion of Brazilian farmers (Figure 2).

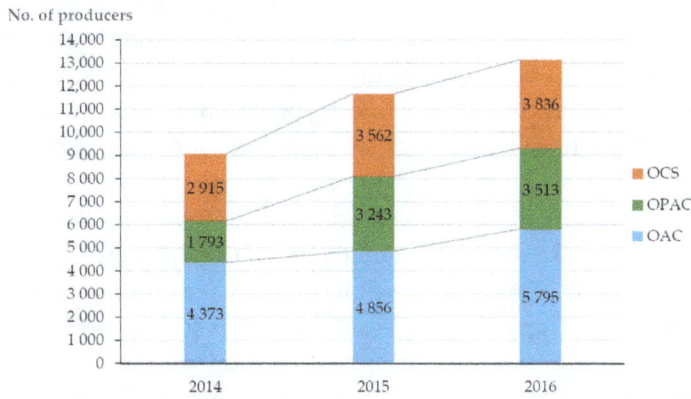

Figure 2. Number of organic producers in Brazil according to the organic accreditation system. OAC: Third-party certification; OPAC: Participatory Guarantee Systems; OCS: Social Control Organizations. Source: Brazilian National Register of Organic Producers [30].

In 2006, 90,497 farms across Brazil declared that they were producing agroecologically, but only 5106 were certified as organic [31]. Ten years later, in 2016, the organic certification system included 13,144 registered farms [23], most of which are certified under OPAC and OCS systems. From 2006–2016, the number of certified producers grew by 144%. In the last three years, from 2014 to 2016, the total growth rate was almost 45%, but the highest increase was recorded for producers involved in the OPAC scheme (+96%) (Figure 2). The high costs and bureaucracy involved in the process of accreditation explain the difference in the numbers between the self-declared organic producers and those actually certified. Additionally, the lower accreditation costs and bureaucracy explain the success of the OPAC and OCS accreditation systems. Ecovida represents nearly 70% of Brazilian producers certified by OPAC, although its geographical operations are limited to Southern Brazil, where it represents nearly 100% of the participatory certification farmers.

Ecovida certification is first produced within each FG as the network's relationship is with the group and not with each individual farmer. In this way the group is certified and not the producers themselves, although legislation requires each producer's product to be labelled. The main inspection within this certification model is then carried out by a neighbouring producer who does not want any misconduct from his/her fellow producers to negatively impact on his/her own production or label. In addition to the inspection of producers, RCs include an ethics committee, whose members take the responsibility for documenting and coordinating the certification procedures. Another task of the RC is the promotion of knowledge exchange between farmers from different groups.

Although there are significant positive elements, the Ecovida participatory certification has two important weaknesses: (i) its potential markets are restricted because this kind of certification only permits commercialization in Brazil; and (ii) there is a risk that one "free rider" member of the network may perform an illegal action, thus creating a problem for the Ecovida label.

Participatory Guarantee Systems, such as that promoted by Ecovida, can be regarded as a form of social innovation as they promote a change in attitudes, behaviour, and perceptions among its participants. They, thus, create new consolidated paths of collective action as well as representing a new model for the certification process of organic produce, in collaboration with the Federal Government of Brazil.

3.4. Social Set-Up of the Markets

Ecovida plays an active part in the Brazilian social movement of Solidarity Economy, aimed at "making farmers and consumers come closer" [16]. Its concept of agroecology originally gave priority to biodiverse production systems, and resists trading with wholesale or retail outlets. Trading has, in fact, become a barrier to the expansion of production and the management of Ecovida's producers, as the production grew in all RCs and the markets' capacity to absorb their produce was restricted [23]. This limitation has led organizations to set up new trade routes under a philosophy of fair trade, for example the Southern Circuit of Food Circulation. This increased the number of local farmers' markets in some regions as well as providing a trading post (a Box) to gather and distribute food from the rural areas and organizations in a metropolitan area (Florianópolis). Of the many initiatives that exist, we will now briefly describe the three most important ones: the Trade Circuit; local and regional markets; and so-called Organic Trading Boxes. In describing these initiatives, we provide a general idea of the key elements that make them social innovations from a commercial perspective.

The Trade Circuit started in 2006 and is based in some RCs which act as meeting and distribution points for the traded products. These wholesale points, which sometimes operate retailing activities, also complement the supply of organic produce in each region with those from other RCs.

Figure 3 shows the evolution of the Trade Circuit between 2008 and 2016. It shows the expansion of the Circuit to other regions, and the greater integration and creation of new routes within different regions. This Circuit follows a set of principles that give it a distinctive character from other commercial systems. According to Magnanti [19], among these principles, the following need highlighting: (a) production must necessarily be from family farm agriculture, produced in diverse systems that guarantee the self-supply of the producing family; (b) selling organizations must also acquire products, as their primary goal is not monetary exchange, but to guarantee the diversity of available goods in each region; (c) the criteria for the establishment of a price list must be assessed regularly in order to guarantee that the work of farmer families is fairly paid and that products are accessible to consumers.

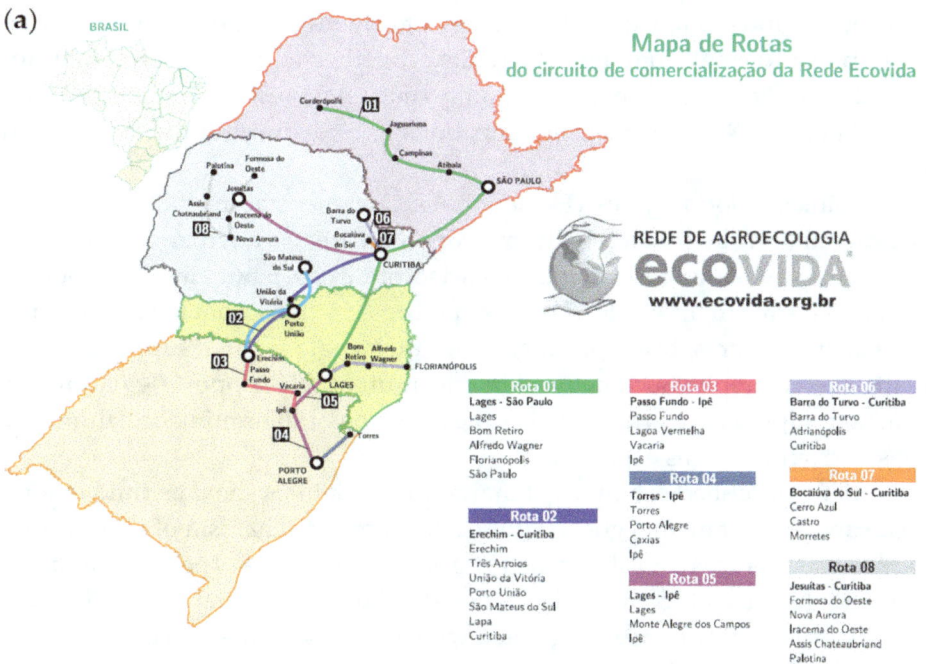

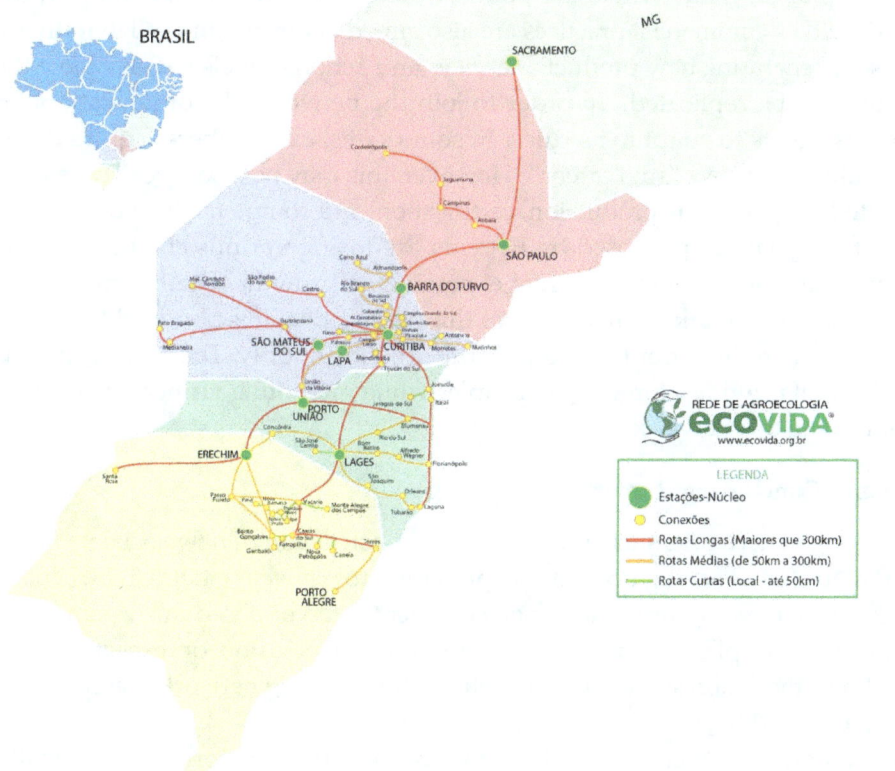

Figure 3. Route map of Ecovida Network's Trade Circuit in 2008 (**a**) and 2016 (**b**). Source: [15–19].

Local and regional markets are the primary and most important commercial strategy of Ecovida. "There are more than 100 free fairs for ecological produce and other means of trade in the entire territory of action of Ecovida" [14]. Its vision of agroecology aims to generate productive processes that sustain ecosystems and organizational dynamics, which guarantee the social reproduction of family farmers. Markets are socially equipped to fit this approach quite well. Despite being small-scale sale points, they are active and demonstrate the strategy of bringing producers and consumers together through direct trade and small commercial circuits. Together with these fairs, sales to institutions (such as local authorities for school meals, community centres) represent a commercial form that allows direct trade, enabling the RCs and their organizations to trade most of the farmers' produce in all its diversity.

The Box Orgânicos Florianópolis (Florianópolis Organic Trading Box), located in the Santa Catarina state capital and surrounded by its metropolitan area, is another commercial outpost that serves the innovative dynamics of Ecovida. It started out in 2013 but already supported the Trade Circuit, as well as the street markets. RC farmers and organizations from the area of this Box, as well as from other Ecovida centres, bring products from their regions, thus supporting each other and diversifying their food supply. Such networking, its mutual self-supporting nature, and the larger and wider-reaching supply network, as well as the diffusion of their influence, through cooperation, are gains that stand out from this experience.

The three briefly described examples of marketing initiatives show an innovation process that combines social and environmental goals with the economic mechanism of supply and demand, in order to produce scale returns and strengthen commercial logistics. The social and environmental ethics of Ecovida interact with commercial demands and create social innovation targeted at the economic viability of the family farmers associated with it. The "socially-built character of the economic regulations" [32] (p. 217), by which Ecovida is ruled, balances trading processes and endogenous market strategies.

In these commercial innovations of Ecovida, the different levels of state intervention have been important. For instance, many wholesale points were organized with the support of the Federal Government. Ecovida's commercial practices are also aimed at developing collaborative action within the network itself, generating new products, services and action models which enable these potential social innovations to be replicated. In order to join the network, the organizational dynamics of Ecovida oblige its agents to adapt to its rules. In some cases, ex-members acquire experience in the market and decide to pursue a "solo career". However, this may be also a positive element of its role as a laboratory to learn about the production, certification, and commercialization of organic products.

However, the dynamics proposed by Ecovida are facing various challenges, namely how to maintain its organizational principles while experiencing growth, whether in terms of territorial coverage or number of network members. From this perspective, it is also vital to activate additional ways to access the market in order to cope with the growing supply. The broadening of commercial practices should fulfil Ecovida's founding principles while promoting farmers' groups, which are one of the key elements in this network.

4. Discussion and Concluding Remarks

When working in networks such as Ecovida, the construction of new social projects requires the negotiation of interests and setting up common principles for collective guidelines. This results from a historical process connected to rural grassroots movements who opposed the hegemonic development model by proposing principles such as decentralization, social inclusion, or respect for local ecosystems. This process follows the vision shared by many alternative farming networks based on the philosophy of food democracy [18,33,34].

Ecovida is an organizational network that innovates and transforms the agricultural systems in which it takes part. It has become the main production and organic certification network in Brazil. The need to open up markets to farmers and organizations has stimulated important trading innovations, such as those described above. The network resists conventional production methods with its top-down or directive organizational systems and trade mechanisms which set producers and consumers apart, while decisions regarding human food production, processing, and distribution have become increasingly centralized. Ecovida uses socially available tools, empowers and promotes new institutional structures, and broadens the opportunities for systemic transformation through diversified processes. The social innovation dynamic of Ecovida is also highlighted by its specific history resulting from production and organizational strategies and from the relationship with its own territorial context (Table 4).

Table 4. Main components in the social innovation and collaborative dynamics of Ecovida.

Analytical Elements of Social Innovation	Characteristics of Ecovida
1. Motivations for the participation of its actors	Ideological engagement by transforming the model of rural development, based on agroecology and biodiversity. Exchange of knowledge, seeds, information on methods and techniques, among other things.
2. Composition and dynamics of actor-network	A decentralized and multi-directional network where decision processes occur simultaneously at different levels. Open and transparent communication through a mailing list without any management filters.
3. Influence of non-social elements on decisions of social actors	Agrobiodiversity of production systems is prioritized. The need for agrobiodiverse trade stimulates the organization and engagement in specific markets.
4. Dependence and relationship with specific territorial processes and general policies	It influences public regulations and adapts itself to the norms of participatory certification of organic produce. It has been supported by infrastructure to improve its market dynamics, mainly by the Federal Government. It has its own regulations with general principles and guidelines, following the norms of national organic produce legislation. RCs base their actions on the general principles and rules of the network but keep a strong autonomy, working as the central functional organs of Ecovida.

Produced by the authors, based on the theoretical approach proposed by Neumeier [3] and Aléssio and Rover [9].

Just as social innovation is considered as a strategic pillar of development in rural areas, the diversification of ecosystems has been highlighted as the central constituent in redesigning sustainable agro-ecosystems [35]. When combined, diversification (through agroecology) and social innovation are two factors that have helped in the creation and growth of Rede Ecovida.

Ecovida represents a collective alliance to contribute to the growth of agroecology as a means of production and as a model of sustainable rural development. Belonging to Ecovida is, on the one hand, a territorial issue, as the network has deep roots in the effective and spatial dynamics of society. In addition, it also has an ideological aspect, for it is not only interested in produce, but also proposes new means of production and new rural development patterns. Arl [12] regards Ecovida as part of a social project which, at the same time, takes the reality of each ecosystem and territory into consideration.

Food trade, as with any other trade, has its own exchange circuit with tensions between socially-regulated paths, as well as diversions motivated by competition [36]. Ecovida, therefore, promotes diversions which correspond to a socio-economic strategy beyond market competition, in social conflict with conventional production, innovating towards an agro-food democracy. Quoting Appadurai [36] (p. 45), "diversions only have a meaning if associated with the paths from which deviation took place". In the conflict with and in opposition to conventional production, Ecovida has innovated by creating its own means of participatory certification and trade. It has, thus, innovated by promoting a deviation from the conventional methods of production, certification, and trade, creating alternatives that can be applied to other socio-productive contexts. It does this in conjunction with other social actors, of which the Federal Government of Brazil has been key in the last two decades.

Altieri and Nicholls [35] highlight that for organic agriculture to be ecologically and socially sustainable, a social organization is required that is embedded with the values of ecology and sustainability. The choice of direct and local sales as the primary mechanism of distribution, followed by the exchange of products between RCs, through their Trade Circuit, strengthens both the territorial and organizational bonds within Ecovida, as well as between this organization and its partners. Its mechanisms of distribution enable farmers to grow a wider variety of produce. This gives them the opportunity for a less specialized production with less dependence on larger production scales. This can then facilitate a wider diversity in production and a better ecosystem, which is key to the ecological resilience of these agro-ecosystems.

The public spirit characterizing Ecovida creates new opportunities to produce and distribute organic food. In addition, organic production generates environmental services and public goods (e.g., biodiversity conservation, climate change mitigation, maintaining soil functionality, agricultural and rural landscape, rural vitality) [37]. Therefore, building actions and policies to favour the production, distribution, and consumption of organic food is of major importance, particularly when promoted by the citizens themselves. Ecovida provides essential social innovation which helps in the planning of other social and political initiatives, as well as public policies towards a truly sustainable agriculture and a better food democracy.

Acknowledgments: The authors would like to thank the two anonymous reviewers for their useful comments, needless to say that any shortcomings are the responsibilities of the authors alone. In addition, we are grateful to the Coordination for the Improvement of Higher Education Personnel (CAPES)-Brazil.

Author Contributions: Oscar José Rover, Bernardo Corrado De Gennaro and Luigi Roselli conceived, designed and performed the research; Oscar José Rover, Bernardo Corrado De Gennaro and Luigi Roselli analysed the data and wrote the paper.

Conflicts of Interest: The authors declare no conflict of interest.

References

1. Schumpeter, J.A. *The Theory of Economic Development: An Inquiry into Profits, Capital, Credit, Interest and the Business Cycle*; Opie, R., Ed.; Transaction Publishers: Piscataway, NJ, USA; London, UK, 2008.
2. Murray, R.; Caulier, J.; Mulgan, G.G. Il Libro Bianco Sulla Innovazione Sociale. Available online: http://www.societing.org/wp-content/uploads/Open-Book.pdf (accessed on 9 September 2016).
3. Neumeier, S. Why do social innovations in rural development matter and should they be considered more seriously in rural development research?—Proposal for a stronger focus on social innovations in rural development research. *Sociol. Rural.* **2012**, *52*, 48–69. [CrossRef]
4. Caroli, M.G. *Modelli ed Esperienze di Innovazione Sociale in Italia: Secondo Rapporto Sull'Innovazione Sociale*; International Center for Research on Social Innovation (CERIIS): Milano, Italia, 2015. (In Italian)
5. Bock, B.B. Rural marginalisation and the role of social innovation; A turn towards nexogenous development and rural reconnection. *Sociol. Rural.* **2016**. [CrossRef]
6. Furtado, A. Opções tecnológicas e desenvolvimento do terceiro mundo. In *Desenvolvimento e Natureza: Estudos para Uma Sociedade Sustentável*, 3rd ed.; Cavalcanti, C., Ed.; Cortez/Fundação Joaquim Nabuco: São Paulo, Brazil, 2001; pp. 256–275.
7. Arend, M.; Cario, S.A.F.; Enderle, R. Instituições, inovações e desenvolvimento econômico. In *Pesquisa & Debate. Revista do Programa de Estudos Pós-Graduados em Economia Política*; Departamento de Economia da Pontifícia Universidade Católica de São Paulo: São Paulo, Brasil, 2012; Volume 23, pp. 110–133.
8. United Nations. Framework Convention of Climate Change. Paris Agreement. Available online: http://unfccc.int/resource/docs/2015/cop21/eng/l09r01.pdf (accessed on 9 September 2016).
9. Aléssio, B.C.; Rover, O.J. O desenvolvimento regional como processo de encadeamento de dinâmicas organizativas e trajetórias tecnológicas: O caso da região oeste catarinense. *Redes* **2014**, *19*, 113–129. (In Portuguese) [CrossRef]
10. Casieri, A.; Nazzaro, C.; Roselli, L. Trust building and social capital as development policy tools in rural areas. An empirical analysis: The case of the LAG CDNISAT. *New Medit.* **2010**, *9*, 24–30.

11. Casieri, A.; De Gennaro, B.; Medicamento, U. Framework of economic institutions and governance of relationships inside a territorial supply chain: The case of organic olive oil in the sierra de segura (Andalusia). *Cah. Agric.* **2008**, *17*, 537–541.

12. Arl, V. *Caderno de Formação 01*; Rede Ecovida de Agroecologia: Lapa, Brazil, 2007.

13. Arns, C.E.; Zuniga, G.; Rover, O.J. *Relatório Sobre um Projeto de Desenvolvimento Sustentável e Alternativo para a Região Sul do Brasil com Base em Reflexões e Práticas de Organizações do Campo Popular*; Coordenação dos Fóruns da Região Sul do Brasil: Curitiba, Brazil, 2002.

14. ECOVIDA ONLINE. Available online: http://www.ecovida.org.br (accessed on 18 November 2015).

15. Darolt, M.; Grando, G.; Almeida, F. *Cartilha: Circuito Sul de Circulação e Comercialização de Alimentos Ecológicos*; Rede Ecovida de Agroecologia: Curitiba, Brazil, 2016.

16. ECOVIDA ONLINE. Available online: http://www.ecovida.org.br/?sc=SA002&stp=STP0002 (accessed on 18 February 2011).

17. Frison, E.; Rover, O.J. Entraves para a certificação orgânica do leite numa central cooperativa de agricultores familiares do oeste catarinense. *Revista Brasileira de Agroecologia* **2014**, *9*, 70–83. (In Portuguese)

18. Lamine, C.; Darolt, M.; Brandenburg, A. The Civic and Social Dimensions of Food Production and Distribution in Alternative Food Networks in France and Southern Brazil. *Int. J. Sociol. Agric. Food* **2012**, *19*, 383–401.

19. Magnanti, N.J. Circuito Sul de Circulação de Alimentos da Rede Ecovida de Agroecologia. *Agriculturas* **2008**, *5*, 26–29. (In Portuguese)

20. Meirelles, L. A Rede Ecovida de Agroecologia hoje. In Proceedings of Palestra Proferida Durante o VII Encontro Ampliado da Rede Ecovida de Agroecologia, Ipê, Brasil, 13–15 November 2009.

21. Meirelles, L. A Rede Ecovida de Agroecologia. Carta Maior. Maio de 2016. Available online: http://cartamaior.com.br/?/Editoria/Meio-Ambiente/A-Rede-Ecovida-de-Agroecologia/3/36143 (accessed on 27 September 2016).

22. Perez-Cassarino, J. A Construcao Social de Mecanismos Alternativos de Mercados no Ambito da Rede Ecovida de Agroecologia. Ph.D. Thesis, Universidade Federal do Paraná, Curitiba, Brazil, 2012.

23. Rover, O.J. Agroecologia, mercado e inovação social: O caso da Rede Ecovida de Agroecologia. *Ciências Sociais Unisinos* **2011**, *47*, 56–63. (In Portuguese) [CrossRef]

24. Rover, O.J.; LAMPA, F.M. Rede Ecovida de Agroecologia: Articulando trocas mercantis com mecanismos de reciprocidade. *Revista Agriculturas* **2013**, *10*, 22–25.

25. Willer, H.; Lenoud, J. *The World of Organic Agriculture: Statistics and Emerging Trends 2014*; Research Institute of Organic Agriculture (FiBL): Frick, Switzerland; IFOAM—Organics International: Bonn, Germany, 2014.

26. Villasante, T.R. *Redes e Alternativas: Estratégias e Estilos Criativos na Complexidade Social*; Vozes: Petrópolis, Brazil, 2002.

27. Renting, H.; Marsden, T.K.; Banks, J. Understanding alternative food networks: Exploring the role of short food supply chains in rural development. *Environ. Plan. A* **2003**, *35*, 393–411. [CrossRef]

28. Aguglia, L. La filiera corta: Una opportunità per agricoltori e consumatori. *Agriregionieuropa* **2009**, *17*, 16–20. (In Italian)

29. Sacchi, G.; Caputo, V.; Nayga, R.M. Alternative labeling programs and purchasing behavior toward organic foods: The case of the participatory guarantee systems in Brazil. *Sustainability* **2015**, *7*, 7397–7416. [CrossRef]

30. Ministério da Agricultura, Pecuária e Abastecimento–MAPA, Cadastro Nacional de Produtores Orgânicos. Available online: http://www.agricultura.gov.br (accessed on 19 July 2016).

31. Instituto Brasileiro de Geografia e Estatística (IBGE). Censo Agropecuário. Available online: http://www.ibge.gov.br/ (accessed on 27 September 2016).

32. Zaoual, H. *Nova Economia das Iniciativas Locais: Uma Introdução ao Pensamento Pós-Global*; Coppe/UFRJ-Instituto Alberto Luiz Coimbra de Pós-Graduação e Pesquisa de Engenharia: Rio de Janeiro, Brazil, 2006.

33. Pascucci, S. Governance structure, perception, and innovation in credence food transactions: The role of food community networks. *Int. J. Food Syst. Dyn.* **2010**, *1*, 224–236.

34. Renting, H.; Schermer, M.; Rossi, A. Building food democracy: Exploring civic food networks and newly emerging forms of food citizenship. *Int. J. Sociol. Agric. Food* **2012**, *19*, 289–307.

35. Altieri, M.A.; Nicholls, C.I. Agroecology and the Search for a Truly Sustainable Agriculture. Available online: http://www.agroeco.org/doc/agroecology-engl-PNUMA.pdf (accessed on 27 September 2016).

36.   Appadurai, A. Introduction: Commodities and the politics of value. In *The Social Life of Things: Commodities in Cultural Perspective*; Cambridge University Press: New York, NY, USA, 1988; pp. 3–63.

37.   Food and Agriculture Organization of the United Nations (FAO). *The State of Food and Agriculture: Innovation in Family Farming*; FAO: Rome, Italy, 2014.

Exploring Socio-Technical Features of Green Interior Design of Residential Buildings: Indicators, Interdependence and Embeddedness

Yan Ning [1,*], Yadi Li [1], Shuangshuang Yang [1] and Chuanjing Ju [2]

[1]   Department of Construction and Real Estate, Southeast University, Nanjing 210096, China; liyd_seu@126.com (Y.L.); yangss95@163.com (S.Y.)
[2]   Department of Business Administration, Southeast University, Nanjing 210096, China; 101012004@seu.edu.cn
*   Correspondence: ningyan@seu.edu.cn

Academic Editor: Umberto Berardi

Abstract: This research aims to develop indicators for assessing green interior design of new residential buildings in China, grounded in the socio-technical systems approach. The research was carried out through a critical literature review and two focus group studies. The results show that the boundaries of green interior design were identified with respect to three dimensions, namely performance, methodology and stakeholders. The socio-technical systems approach argues for the recognition of the interdependence between the systems elements and the feature of embeddedness. The interdependence of the systems elements exists within each of these three dimensions and across them. It is also found that the socio-technical systems of green interior design are embedded in the social, regulatory and geographic context. Taking interior design of residential buildings as the empirical setting, this study contributes to the literature of green building assessment by presenting a socio-technical systems approach.

Keywords: environmental impact assessment; green interior design; green building; socio-technical systems; embeddedness; China

1. Introduction

Green development has become the national strategy for economy development and the topmost governmental agenda in China. The Chinese government initiated five principles for national development in the fifth Plenary Session of the 18th the Communist Party of China Central Committee in 2015. These are innovation, coordination, green, openness and sharing. According to the BP Statistical Review of World Energy, in 2012, China accounted for 21.9% of total worldwide primary energy consumption. The building sector consumes about 27.0% of the country's total energy. The building sector has accounted for approximately 43% of China's total energy consumption from the life-cycle perspective [1]. Thus, achieving green in the building sector will significantly contribute to a reduction in overall use of carbon and energy.

The Chinese Government has announced a series of action targets and roadmaps for achieving green buildings. For example, the Ministry of Finance (MOF) and the Ministry of Housing and Urban-Rural Development (MOHURD) jointly announced the 'Implementation Plan for Accelerating Green Building Development' that set the targets for creating 1 billion m^2 of new green building areas by 2015, constituting at least 30% of new building areas by 2020. In 2014, the Central Government and State Council initiated the "New National Urbanization Plan 2014–2020" in which 50% of new building are projected to reach the green building standards by 2020. The 'Evaluation Standard for Green

Building' (ESGB) was published and updated in 2006 and 2014 respectively [2]. All these regulatory contexts motivate the entire industry to strive for a green building paradigm shift.

Increasing studies also shade light on energy/carbon reduction in the building sector in China [3,4], either for commercial [5,6] or for residential buildings [7]. Studies also focused on specific stages of green building delivery, for instance design [8], construction [9] and retrofit [10]. Policies for addressing problems with respect to green building delivery were also extensively examined [4], in terms of challenges and opportunities [3,11,12].

However, the interior design of new residential buildings is rarely examined in China. Although pollution control and indoor environmental standards associated with interior design of residential buildings have been sparsely addressed, there is a lack of established tools for assessing their environmental impacts. This gap in knowledge is significant given that the interior design and construction constitute a considerable market share in the construction sector, and green interior design is of vital importance to the green building delivery.

This study aims to develop indicators for assessing green interior design of new residential buildings. This study argues for a socio-technical systems approach [3], which emphasizes that green interior design is characterized by systems and embeddedness features. The systems feature embraces the interdependence among the systems element [13,14] and the embeddedness feature implies that green interior design is embedded into the social, regulatory and geographic context. This study focuses on new residential buildings as they have distinct features from other building types (e.g., office and hotels) and existing residential buildings.

The paper is organized as follows. Section 2 presents a literature review of green interior design and socio-technical systems approach, followed by a conceptual framework for defining the boundaries of green interior design (Section 3). Section 4 reports on the research method of focus group studies. The key results of the indicators for assessing green interior design and discussion of the socio-technical systems features are presented in Section 5. Conclusions and recommendations for future studies are shown at the end.

2. Literature Review

2.1. Green Interior Design of Residential Buildings

People spend 90% of their time indoors [15]. However, it is found that levels of indoor pollutants are usually two to five times higher than outdoor levels [16], which could be detrimental to the health and well-being of occupants [17]. Thus, improving the indoor environment is of great importance to their well-being [18]. However, in classical interior design, designers often prioritize on meeting the aesthetic and functional needs of the clients, rendering environmental issues less important [19].

Prior studies have fallen short of providing a cohesive description of the boundaries of green interior design of residential buildings thus far. In a simple manner, studies argued that green interior design intends to cover a wider scope than the classical approach. These include material, aesthetic qualities, environmental and health impacts, availability, ease of instalment and maintenance, and life-cycle cost [20]. Kang and Guerin [18] defined environmentally sustainable interior design practice as three aspects: global sustainable interior design, interior materials, and quality indoor environments.

In the green building standards for new construction or renovation, e.g., Leadership in Energy and Environmental Design (LEED), and the Building Research Establishment Environmental Assessment Method (BREEM), parts of the assessment credits are associated with interior design. In addition, six aspects of indoor environment are assessed in ISO 16813:2006. These are indoor air quality, thermal comfort, acoustical comfort, visual comfort, energy efficiency and HVAC system controls. These are applicable to environment design for new construction and the retrofit of existing buildings. However, it mainly deals with the indoor environment, referring less to space performance and material savings.

In addition, the LEED for Homes Design and Construction (LEED BD + C: homes and multi-family low-rise; LEED BD + C: Multifamily Midrise) specified the requirements on location and transportation, sustainable sites, water efficiency, energy and atmosphere, materials and resources, indoor environmental quality, innovation and regional priorities. It defined broader requirements for achieving green interior design, such as credits assigned to location and transportation, sustainable sites and outdoor water use. These aspects might not be applicable to the China's context as these aspects are dealt with by architectural designers and fixed prior to the interior design. Similarly, the local assessment tool ESGB comprises energy savings, land savings, water savings, material savings, environment protection, and building functional requirements during the complete building life cycle. Notwithstanding these evaluation tools, there is a lack of well-established tools, specifically with respect to assessing green interior design of new residential buildings.

Research has examined green interior design of offices and commercial buildings [21]. The LEED and BCA (Building and Construction Authority) Green Mark initiated assessment tools for offices and commercial interiors. The BCA published the "BCA Green Mark for Office Interior" [22]. It comprises energy efficiency, water efficiency, sustainable management and operation, indoor environmental quality, other green features [22]. LEED for commercial and institutional interiors addressed sustainable sites, energy and atmosphere, materials and resources, and indoor environmental quality [23]. The BREEAM UK non-domestic building refurbishment and fit-out schemes have four assessment parts. These are building fabric and structure, core services, local services and interior design. Refurbishment and fit-out projects can be assessed against one or all of the four parts, or any combination [24]. However, the indicators developed in these tools, while providing valuable inspiration for assessing green interior design of residential buildings, would not be applicable to the residential building in China.

2.2. Socio-Technical Systems Approach

Socio-technical systems are referred to as "a somewhat abstract, functional sense as the linkages between elements necessary to fulfill societal functions" [25] (p. 898). The socio-technical systems approach not only focuses on achieving interior design at the design and construction stages, but also on functionality at the occupancy phase.

The systems approach highlights that system elements are tightly interrelated and interdependent with each other [25]. Open systems are another important feature of the socio-technical systems approach. Although sustainable building is increasingly recognized as involving complex socio-technical systems [3,14], studies rarely investigate their features. Drawing on the analytical framework [25,26], green interior design is considered to have complex socio-technical systems. Aside from the indicators for green interior design being developed, this study examined the socio-technical systems features of green interior design of residential buildings.

2.3. Empirical Context of Green Interior Design in China

Given the open system feature of the socio-technical systems, the economic, social and regulatory contexts of green interior design of residential buildings in China are elaborated in detail. These contextual factors together shape the development of green interior design of residential buildings.

2.3.1. Economic Context of Green Interior Design

Due to with rapid urbanization, the decorating industry has huge development potential. The capital of the decorating industry reached 3160 billion CNY in 2014, as compared to 1180 billion CNY in 2005 (see Figure 1). Residential decoration constitutes almost half of this capital. The growth rate in these years was kept stable around 10% [27]. This indicates there is a great market potential in interior design and construction.

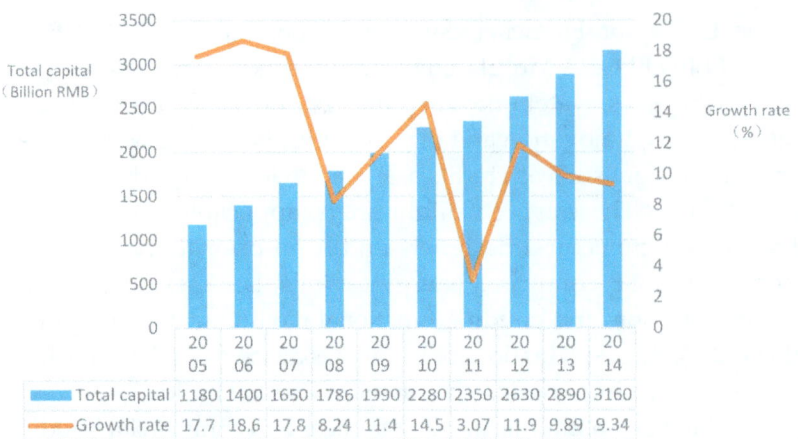

Figure 1. Total capital of the decoration industry and growth rate. Source: MOHURD [27].

Two significant drivers of this growth could be identified. The first is the political incentive of fine decoration (whereby the developer undertakes the decoration work) in the building sector. In the last decade, the majority of residential buildings were handed over without fine decoration; the interior design and construction is entirely left to the buyers. However, increasing studies found that this delivery system resulted in a huge amount of waste and environmental problems [28]. As a consequence, the policy now is re-oriented to incentivize fine decoration. According to MOF and MOHURD [29], all new residential buildings are suggested to be handed over with fine decoration already complete. The second driver is the booming market of the refurbishment of existing residential buildings. The first mass construction of residential buildings in China took place in the late 1990s. It was gradually observed that these buildings underwent varying degrees of refurbishment.

2.3.2. Social Context of Green Interior Design

Contractor registration heads in China comprise three categories, namely general contractors, specialist contractors and labor subcontractors. Construction firms that undertake decoration work belong to one type of these specialists. Their work scope covers decoration work and directly related supporting works [30]. There were around 140,000 firms that undertook decoration work in 2014.

The old registration system was transformed from a three-class grade to two-class grade (i.e., first and second Class) in 2014 [30]. Grades are classified in accordance with financial capability, personals and track record. Firms in the class one have no limits in tender amount, whereas those in class two are limited to a contract amount below 20 million CNY. For design firms, there are three grades (see Table 1).

Table 1. Design and construction firm categories in residential decoration.

Firm Types	Grades	Registration Criteria
Design firm	Class one Class two Class three	Financial capability and track record Personals Technology and management systems
Construction firm	Class one Class two	Financial capability Personals Track record

Source: MOHURD [30,31].

2.3.3. Regulatory Context of Green Interior Design

Regulations of relevance to interior design fall into three levels, namely national, industrial and provincial levels. The former two are applicable to all regions, whereas the latter refers to local regulations coming into effect in a specific province. A review of the existing regulations, codes and rules in China was carried out. The national and industrial regulations cover one aspect and multiple aspects are presented in Supplementary Materials Annex 1 and 2 respectively. Provincial-level regulations are presented in the similar manner (see Annex 3 and 4). From these tables, three patterns could be observed.

First, existing rules and regulations are closely associated with interior design. Thus, the development of green interior design should be compatible with the existing regulations. Specifically, green interior design should comply with the civil building regulations. This is because buildings are classified into two types in China, namely industrial and civil building (see Figure 2); residential buildings belong to the latter category. In any case, rules and regulations are highly localized. Developing indicators for green interior design thus requires a contextualized plan.

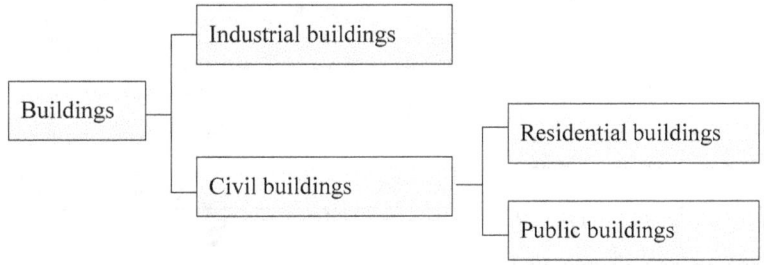

Figure 2. Building type classification in China. Source: adapted from Li & Yao [3].

Second, a lack of well-established assessment tool for green interior design is observable at national, industrial or provincial levels. Such a deficiency might severely hinder the green transformation of the interior design and construction for residential buildings.

Third, green interior design practices face complex and intricate regulation systems. Ye et al. [32] found that contents in most standards largely overlap; some mandatory provisions in local standards are unnecessary. Besides, green interior design could be controlled and monitored by multiple governmental entities. Thus, to promote such practices through regulations would require coordination among multiple governmental entities. Lack of such coordination would impose great obstacles to practice uptake and implementation.

3. Conceptual Framework of Green Interior Design for Residential Buildings

This study adopted the framework of Ju, Ning and Pan [33] to initially define green interior design of residential buildings. It deals with three dimensions, which are performance, methodology and stakeholders (see Table 2). The former two present the technical aspect, whereas the latter portrays the social aspect.

(1) The performance of green interior design comprises five aspects. These are effective space utilization, healthy indoor environment, energy saving, water conservation and material saving. These five aspects were initially identified from the extant literature and further verified through focus groups;

(2) The methodology deals with the temporal and spatial dimensions. The temporal dimension includes workflow and material flow. Work flow covers design, material selection, construction, maintenance and end-of-life. Although this study deals with green design, it is recognized that design solutions have great impacts on construction, maintenance stages. The material dimension deals with the material flow of the cradle-to-grave life cycle. The spatial dimension describes the location of physical subjects. It has to properly deal with the component of interior design, compatibility with architectural and mechanical, electrical and plumbing (MEP) design and outdoor environment;

(3)   The stakeholder dimension refers to the actors who play a role in achieving green interior. These include developers, contractor, designers, suppliers, end-users, the government and industrial organizations.

Table 2. Boundaries of green interior design of residential buildings.

Systems	Dimension		Indicator
Technical systems	Performance		Space performance, indoor environmental quality, energy efficiency, water conservation and material savings
	Methodology	Temporal dimension: workflow	Design, material selection, construction, operation, maintenance and end-of-life
		Temporal dimension: material flow	Raw material extraction, Transportation from extraction site to factory, manufacturing, transportation from factory to building site, construction installation, operation, renovation, deconstruction and recycling/landfill site
		Spatial dimension	Components of interior design, architectural design, MEP design and outdoor environment
Social systems	Stakeholders		Developers, contractors, designers, suppliers, end-users, government and industrial organizations

Source: Adapted from Pan and Ning [14] and Ju et al. [33]. MEP: mechanical, electrical and plumbing.

4. Research Methods

4.1. Focus Group Studies

This study adopted focus groups to develop indicators for and explore the socio-technical systems feature of green interior design of residential buildings in China. Focus groups are useful for exploring a particular topic [34]. The purpose of the first-round focus group was to derive the indicators, and verify five categories in the conceptual framework as well as interdependence and embeddedness features. The second focus group aimed to validate the indicators and socio-technical systems features obtained in the first round.

Participants were selected using purposive sampling. Participants from a wide range of organization types were targeted. In the end, seven participants were invited in the first round. Detailed background information is shown in Table 3. In the second round, another two participants from environmental assessment firms and construction firms were invited, together with six participants from the first round. The reason for inviting the same participants in two rounds is because they would help to closely validate the framework derived from the first round. Each focus group lasted for three hours.

Table 3. Profiles of focus group participants.

Participants	Organization	Designation	Round 1	Round 2
1	Government organization	Director	Yes	Yes
2	Government organization	Deputy director	Yes	Yes
3	Governmental organization	Officer	Yes	Yes
4	Environmental assessment firm	Director	Yes	Yes
5	Research institute for building science	Director	Yes	Yes
6	Construction group	Vice general manager	Yes	Yes
7	Academia	Associate professor	Yes	-
8	Environmental assessment firm	Engineer	-	Yes
9	Construction firm	Chief Executive Officer	-	Yes
Total			7	8
Duration			3 h	3 h

Following the suggestion of Cyr [34], at the start of first focus group, participants were requested to: (1) comment on proposed definition (Table 2); (2) suggest specific indicators under each aspect and comment on the appropriateness of the five aspects; (3) discuss why the aspects and associated indicators are essential for interior design; and (4) comment on the socio-technical systems features.

4.2. Data Analysis

Data analysis was carried out following the rules of systematic combining approach [35]. Two types of unit of analysis were adopted, namely individual and interaction levels [34]. The individual level unit of analysis was used to triangulate the proposed indicators and socio-technical systems feature; the interactive unit of analysis was appropriate for exploring the indicator development and socio-technical systems features.

Themes (i.e., the five categories) and indicators identified from the literature review were verified from the focus groups, relying on labels that could represent similar descriptions across different participants. In the end, indicators of green interior design were identified. Emerged aspects were compared with the existing findings. Through going back and forth between framework, data sources and analysis, this step fulfilled the match between theory and data in systematic combining [35].

5. Results and Discussion

5.1. Importance of Green Interior Design and Socio-Technical Perspective

Participants agreed that developing indicators for assessing green interior design of residential buildings is of vital importance for green building delivery in China. They summarized three reasons. Firstly, the regulatory context of considering green building delivery as a national strategy in China has been widely accepted. Developing a standard for green interior design could well fit with the national policy vision. In addition, a green building paradigm shift requires concerted efforts from all parties throughout the building cycle. The interior design is an essential stage.

Secondly, in practice, there exists a considerable market demand for delivering green interior. Participants commented that end-users in China expressed enormous concern on the environmental pollution caused by the interior decoration. Thus, there is a strong demand for green interior design. However, the market still failed to fully meet the end-users' requirements.

Thirdly, participants acknowledged the importance of the systems approach in addressing green interior design. Existing standards and regulations prescribe some aspects of interior works, for example, the indoor air quality and energy efficiency. However, no tools are available for articulating green interior design in a systems approach.

5.2. Aspects of Green Interior Design

Participants agreed with the presentation of the five key aspects of green interior design and suggested specific indicators under each aspect (see Table 4).

Table 4. Indicators for green interior design for residential buildings.

Aspects	Indicators	Description
Space performance	Efficient use of space (SP1)	• Properly-configured functions (SP11) • Multiple functions-oriented (SP12)
	Adaptive use (SP2)	• Adaptive use and consideration of potential future needs (SP21) • Space flexibility and adjustable when new requirements arise (SP22)
	Compatibility with architectural and MEP design (SP3)	• Compatibility with the architecture design (SP31) • Compatibility with the MEP design and configuration (SP32)

Table 4. *Cont.*

Aspects	Indicators	Description
Indoor environmental quality	Acoustic (IEQ1)	• Sealing of gaps around windows and doors, openings of high sound conduction (IEQ11) • Use of materials for increasing sound absorption and insulation (IEQ12) • Reduction of vibration noise arising from water flows in pipes (IEQ13)
	Lighting (IEQ2)	• Maximization of the use of natural daylight through openings without impairing the structure (IEQ21) • Use of light-colored interiors that reflect light from windows or skylights (IEQ22) • Use of high performance artificial lights and appropriate configuration (IEQ23) • Avoid using materials with high surface reflectance (IEQ24)
	Thermal comfort (IEQ3)	• Increase air tightness through air barriers around windows and doors (IEQ31) • Use of passive technologies, shading, reflection, absorption devices (IEQ32) • Humidity control (IEQ33)
	Indoor air quality (IEQ4)	• Proper mechanical flushing of indoor pollutant sources (IEQ41) • Air purification filters to prevent outdoor pollution (IEQ42) • Improving air circulation (IEQ43) • Prevent interior pollution migration (IEQ44) • Selection of low-pollutant materials (IEQ45) • Air quality monitoring systems (IEQ46) • Removal of sources of water or moisture (IEQ47)
Energy efficiency	Envelope (EE1)	• Energy-saving windows and door treatments (EE11) • Use of insulation in interior walls (EE12) • Choice of appropriate shading devices (EE13)
	Lighting and daylight (EE2)	• Selection of high performance lighting and control devices (EE21) • Selection of lighting supported by renewable energies (EE22) • Implementation of a flexible lighting control systems with plug and play components such as wall controls, sensors, and dimming ballasts (EE23) • Smart controls such as occupancy sensors and daylight dimming (EE24)
Water conservation	Water conservation (WC)	• Selection of water-efficient appliances, fixtures and fittings (WC1) • Installation of devices to monitor water leakage (WC2) • Reduction in the volumes of sewage (WC3) • Water usage monitoring (WC4) • Recycling of domestic wastewater (WC5)
Material-saving	Ease of maintenance (MS1)	• Selection of high performance decoration materials and products (MS11) • Ease of maintenance finishes, materials and products (MS12)
	Environmental friendly materials (MS2)	• Use materials salvaged from waste (MS21) • Selection of recyclable materials (MS22) • Selection of localized materials (MS43)
	Buildability (MS3)	• Plan material use (MS31) • Use of standard sizes of materials and products (MS32) • The technical interface (MS33) • Selection of industrial modules produced off-site (MS34)
	Life-cycle cost optimization (MS4)	• Consideration of different component service lives in order to achieve lowest life-cycle cost (MS41)

5.2.1. Space Performance (SP)

Participants indicated that the major concern of end-users is to maximize the interior space use. To enhance space performance, interior design should embrace occupants' behaviors and requirements. Three indicators were developed for guiding effective space utilization. These are proper space planning (SP1), adaptive use (SP2) and compatibility with the architecture and MEP design (SP3).

Efficient use of space (SP1). The interior space should be properly configured (SP11), enabling a smooth activity flow. When planning the room spaces, multiple function purposes (SP12) should be taken into account, which would help to maximize space use for different functions. To achieve this, it is important to investigate the true requirements of users.

Adaptive use (SP2) indicates the consideration of future needs (SP21) and ensures space flexibility which presents adjustability when new requirements arise (SP22). Flexible designs aim to meet occupants' unforeseen requirements. Along with the rapid technological change as well as possible alteration of the function and workflows in the room, occupants may need more flexible and adaptable interiors to accommodate these unforeseen changes. One example is to maximize the user's control of the environment, for instance mobile furniture, or building utilities that are reconfigurable and expandable.

Compatibility with the architecture and MEP design (SP3). As interior design is fully based on existing architectural design (SP31) and develops in tandem with the MEP design (SP32), it is important to ensure design elements are compatible with each other. The interior design needs to fully make use of the existing conditions imposed by the architectural design. The design team should be familiar with the base architecture in order to achieve unified scale and compatibility.

5.2.2. Indoor Environmental Quality

The results show that achieving indoor environmental quality (IEQ) was manifested by acoustic (IEQ1), lighting (IEQ2), thermal comfort (IEQ3) and indoor air quality (IEQ4).

Acoustic performance (IEQ1). Three strategies were proposed to enhance acoustic performance. These are sealing gaps around windows, doors and openings (IEQ11), selecting materials with high sound absorption and insulation (IEQ12) and reducing noise vibrations arising from water flows in pipes (IEQ13). Proper selection of the absorptive surfaces would help to eliminate noise disturbance.

Lighting performance (IEQ2). To improve indoor lighting performance, passive strategies are helpful, such as maximizing the use of natural daylight through openings (IEQ21). This should also improve the daylight use efficiency through light-colored interiors that reflect light from windows or skylight (IEQ22). This is consistent with prior studies that conclude that internal reflectance of materials and finishes affect daylighting [36]. It might be necessary to avoid using materials of high surface reflectance (IEQ24). In addition, it is important to select high performance artificial lights and deploy appropriate configurations (IEQ23).

Thermal comfort (IEQ3). To achieve thermal comfort, several strategies could be adopted, such as increasing air tightness through air barriers around windows and doors (IEQ31). It is also recommended to adopt various effective passive technologies, such as shadings, reflections and absorption devices (IEQ32). Lastly, humidity control technologies could be adopted (IEQ33) in certain period in Jiangsu Province.

Indoor air quality (IEQ4). Participants commented that indoor air quality is the most serious concern in interior design and construction. The topmost strategy is to select low pollution materials (IEQ45), such as green labelling materials. Selecting low pollution materials will reduce the pollutants brought into the building. The second strategy is to improve air circulation (IEQ43), and adopt proper mechanical flushing of indoor pollutant sources (IEQ41). Quite often homes that are poorly ventilated will have high levels of biological contaminants arising from mould growth on damp surfaces. It is important to prevent interior pollution migration (IEQ44), for example preventing cooking smoke migrating from the chicken.

As haze is currently a serious concern in the north China, air purification filters to prevent outdoor pollution (IEQ42) and air quality monitoring systems (IEQ46) are considered to be a solution. The last strategy is to remove sources of water or moisture that encourage fungal growth (IEQ47). This needs to avoid external and internal leaks and adopt proper humidity control devices.

5.2.3. Energy Efficiency (EE)

For interior design, energy efficiency could be achieved by improving envelope (EE1) and lighting and daylight (EE2). It is worth noting that the selection of air-conditioner and water heater is often decided by the end-users rather than the developer or the interior designers in China. Thus these two energy consumption sources were excluded.

Improve envelope (EE1). It is common to adopt energy saving window and door treatments (EE11). Another strategy is to use high-insulation interior walls (EE12). Exterior insulation walls are excluded here because they are often included in the main structure construction work package rather than the interior work. It is also recommended to use internal shading devices (EE13).

Lighting and daylight (EE2). High performance lighting and control devices (EE21) are suggested, to uptake lighting supported by renewable energies (EE22), and to implement a flexible lighting control system with plug and play components (e.g., wall controls, sensors, and dimming ballasts) (EE23). Various smart controls such as occupancy sensors and daylight dimming are preferable (EE24).

5.2.4. Water Conservation (WC)

Through the focus group, five strategies were proposed to reduce water consumption. Two are of relevance to technological aspects, such as selecting water-efficient appliances, fixtures and fittings (WC1), and reducing the volumes of sewage (WC3). Two strategies are particular to monitoring water usage (WC4) and water leakage (WC2). The last strategy is to recycle domestic wasted water (WC5).

5.2.5. Material-Saving

In order to save materials, four strategies would be helpful. These are use of materials with ease of maintenance (MS1), selection of environmentally-friendly materials (MS2), increase in buildability (MS3) and optimization of life-cycle cost (MS4).

Ease of maintenance (MS1). To save materials, it is necessary to design for ease of maintenance. A first suggestion is to select materials that are of low maintenance (MS11). Using easily maintained finishes will be critical. Another useful method is to select high performance fittings and products (MS12). Extra consideration should be given to products used in heavy-use areas and specific functional areas.

Environmentally-friendly materials (MS2). The environmentally-friendly materials indicate that there is life-cycle optimization, with respect to raw materials, manufacturing, transportation, installation, use and disposal or reuse. It is suggested to use materials salvaged from wastes (MS21) and use recyclable (MS22) and local materials (MS23). Using materials containing recycled content is also preferable. As the carbon labeling is currently on promotion in China, carbon labeling materials are suggested.

Buildability (MS3) is another important assessment criterion from the life-cycle perspective. In order to enhance buildability, it is helpful to use materials and products of standard sizes (MS32) and properly deal with the technical interface between products assembling (MS33). One important strategy is to promote industrial modules produced off-site (MS34). This would largely reduce the on-site waste. Lastly, it is better to draw a material use plan in advance (MS31). This would largely reduce waste and rework.

Life-cycle costing (MS4). It is also recognized that short-term solutions, although less expensive, do not necessarily produce cost savings in the long run. Therefore, a life-cycle costing approach should be materialized in the green interior design. One important strategy is to systematically consider the life span of different components in order to optimize life-cycle cost (MS41). It is important to prepare a fine match of different finishes and products [37].

5.3. Socio-Technical Feature of Green Interior Design

Participants reached a consensus about the socio-technical systems feature of green interior design. These are two-layer system feature and embeddedness feature (see Figure 3).

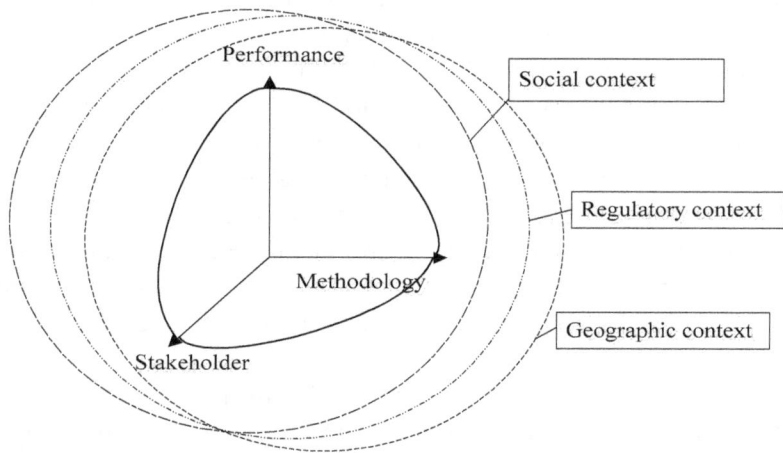

Figure 3. The socio-technical systems features of green interior design.

5.3.1. Two-Layer Systems Features

Systems feature implies the necessity of examining interdependence among elements of the socio-technical systems. Three dimensions of green interior design, namely performance, methodology and stakeholders, are multifaceted and interwoven with each other. The former two present the technical aspect, whereas the latter has the feature of social systems. To achieve green interior design, it is important to properly deal with the interdependence of the socio-technical systems.

In the first layer, interdependence exists between performance, methodology and stakeholders. It is recognized that achieving green interior design is highly technical and requires technological advancement and various passive design solutions. However, mere reliance on the technologies is often found to be limited, which requires the complements from the social aspect. The social aspects refers to the collective endeavors of the key stakeholders. Participants commented that parties' interests are often not well aligned. This would further impede the technology adoption and implementation in green interior design and construction.

In the second layer, interdependence exists within each dimension. With regards to the temporal dimension, the green design solutions will influence the buildability and maintenance in the later stages. Feedbacks collected from the construction and maintenance will inform the design decision making. Within the performance dimension, it is found that independence exists among five aspects. For example, materials show multiple functions (e.g., buildability, maintenance, life-cycle cost and environmental impacts), which will influence the five aspects of interior design to a varying extent. In addition, some finishes may provide satisfactory durability, yet have limitations in the ease of maintenance. This leads to conflicts in the durability and level of maintenance.

5.3.2. Embeddedness Feature of the Socio-Technical Systems

Embeddedness, considering green interior design as an open system, is another feature of the socio-technical systems of green interior design. Embeddedness feature could be interpreted in relation to the social, regulatory and geographic context.

(1) Social context of green interior design

Firstly, the interior design reflects the needs of the occupants who have inherent habit and preferences. Technical specifications thus need to be customized to the occupants' requirements. This is consistent with Li & Yao [3] who argued that it is important to understand building users' demands, expectations and behavior/lifestyle and this requires socio-technical knowledge. Thus, designers' skills for understanding the needs of end-users are desirable. Interior designers' behavioral intentions associated with the green interior design is determined by their attitude, subjective norms, and perceived behavioral control [38]. This indicates that it is hard to isolate the technical interior design solutions from the social settings.

In addition, persuading the client to accept green ideas is often difficult. This is consistent with the results from the survey of green building technology adoption in China which found that stakeholders' reluctance to use was the largest barrier [12]. High upfront investment and inertia might be possible reasons [39]. Thus, it is necessary to inform end-users on the long term value of green interior design.

Secondly, the green interior design practices are socially contextualized in the project setting. Although it is recognized that designer's knowledge of green interior design is of great importance to the implementation, they may not always put it into practice [40]. Project scheduling pressure might be one obstacle.

Thirdly, apart from designers and end-users, achieving green interior design requires the participation of other parties. For instance, interior design requires coordination with structural, mechanical, and electrical engineers. This entails an integrated approach. Key to achieving an integrated approach is open communication and early involvement [41,42]. These two strategies would help to avoid common errors, mistakes and rework.

Lastly, it is essential to investigate the cultural beliefs and customs. Many nationalities and religions attach significance to certain colors, patterns and materials. For instance, most Western cultures consider black the color of mourning. Eastern/Oriental cultures associate white with mourning.

(2) Regulatory context of green interior design

The articulation of the socio-technical systems of green interior design varies from one context to another due to regulatory differences. The regulatory context is manifested through two aspects. Firstly, green interior design must comply with current regulations. Secondly, it is recognized that the achieving green interior design involves co-option of various standards. For example, in the US, green interior design may be of relevance to selecting ENERGY STAR appliances and low-flow plumbing fixtures [23]. In Singapore, adoption of water efficient fittings covered under the Water Efficiency Labelling Scheme [22] is encouraged.

Participants expressed their concern of regulatory deficiencies for motivating practitioners to adopt green interior design in China. They compared the achievement of green interior works to green building in terms of political impacts. They commented that practitioners now have great motivation to invest in green building solutions because the ESGB exists for legitimating such behaviors and the government also provides momentary incentives. However, for the green interior works, no such policies could be used to recognize the green efforts. This could deter adoption and diffusion.

(3) Geographic context of green interior design

Weather conditions are one facet of the geographic context. China has five climate zones. These are Severe Cold, Cold, Hot Summer and Cold Winter, Hot Summer and Warm Winter, and Mild Zone [3]. This study examined interior design in the Jiangsu Province in the middle of China where the weather conditions is Hot Summer and Cold Winter. Purification filters would be suggested in the green interior because the outdoor air quality suffers due to severe haze pollution. But, in the South China, there may be high levels of humidity. This would require use of an air dryer.

Local climatic conditions are important criterion when selecting materials and finishes. Special maintenance requirements would be required for heavy snow or rain, very arid or humid climates, unusual soil conditions and sand and high level of sun exposure.

6. Conclusions

This study aimed to develop indicators for and explore the socio-technical systems of green interior design of residential buildings in China, grounded in the socio-technical systems approach. The study was carried out through a combination of a critical literature review and two focus group studies.

One important result is the conceptual framework for defining the boundaries of green interior design for residential buildings. Consistent with prior studies [33], it deals with three dimensions, namely performance, methodology and stakeholders. The performance dimension comprises space performance, indoor environmental quality, energy efficiency, water conservation and material saving. The methodology dimension deals with the temporal (i.e., work and material flow) and spatial dimensions. The stakeholder dimension refers to the actors who have a role in achieving green interior design. This framework provides a system understanding of the boundary of green interior design for new residential buildings.

Another finding is that this study verified proposed five aspects of performance (i.e., space performance, indoor environmental quality, energy efficiency, water conservation and material saving, see Table 3) and developed indicators for each aspect. The systems framework was verified to be valid.

The last finding is the identification of the socio-technical systems features of green interior design. Although prior studies argued for a socio-technical systems approach in green building delivery [3], this is rarely examined in a systems approach. Distinct from prior systems approaches that examine indicators in isolation, this systems approach focuses on three dimensions (e.g., performance, methodology and value) and emphasizes the interdependence between the systems elements. Interdependence exists both within and across each aspect. Crucial to the green interior design is their feature of being embedded into the social, geographic and regulatory context.

Taking interior design of new residential buildings in China as the empirical setting, this study contributes to the knowledge by presenting two features of socio-technical systems approach, namely the interdependence and embeddedness. The practical implication is that practitioners and policy makers should recognize the socio-technical systems feature of the green interior and take the one-fits-all green strategies with caution. This is because these strategies often take the end-users to be passive recipients, whereas the socio-technical system features argue for active and collective participation of the key stakeholders. In addition, practitioners and policy makers could customize the indicators developed in this study to their specific projects with considerations of the social, regulatory and geographic contexts.

This study only examined the green interior design in China, grounded in the socio-technical systems approach. To reinforce the socio-technical features of green interior design, comparative studies among different nations are recommended. Additional research design (e.g., case studies, interview, questionnaire survey) could be employed to further validate the key findings.

Another recommendation for future studies is to adopt larger scale surveys to further gauge the extent to which green interior design has been implemented in various geographic contexts. Despite the development of indicators for assessing green interior design, parameters for each indicator are worth further in-depth examination. For example, although internal shading devices at windows (EE13) are recommended to increase energy efficiency, the specific parameters of the shading devices are still not known. Further studies in this regard are thus recommended.

Acknowledgments: This research was supported by the National Science Foundation of China (71502032; 71602031), and the Priority Academic Program Development of Jiangsu Higher Education Institutions. Reviewer's constructive comments are highly appreciated.

Author Contributions: Yan Ning and Chuanjing Ju conceived and designed the research. Yadi Li and Shuangshuang Yang collected and analyzed the data and contributed the policy review.

Conflicts of Interest: The authors declare no conflict of interest.

References

1. Zhang, Y.; He, C.Q.; Tang, B.J.; Wei, Y.-M. China's energy consumption in the building sector: A life cycle approach. *Energy Build* **2015**, *94*, 240–251. [CrossRef]
2. Ministry of Housing and Urban-Rural Development (MOHURD). *Evaluation Standard for Green Building*; MOHURD: Beijing, China, 2014.
3. Li, B.; Yao, R. Building energy efficiency for sustainable development in China: Challenges and opportunities. *Build. Res. Inf.* **2012**, *40*, 417–431. [CrossRef]
4. Zhang, Y.; Wang, Y. Barriers' and policies' analysis of China's building energy efficiency. *Energy Policy* **2013**, *62*, 768–773. [CrossRef]
5. Chen, H.; Lee, W.L. Energy assessment of office buildings in China using LEED 2.2 and BEAM Plus 1.1. *Energy Build.* **2013**, *63*, 129–137. [CrossRef]
6. Jiang, P.; Keith, T.N. Opportunities for low carbon sustainability in large commercial buildings in China. *Energy Policy* **2009**, *37*, 4949–4958. [CrossRef]
7. Gong, X.; Akashi, Y.; Sumiyoshi, D. Optimization of passive design measures for residential buildings in different Chinese areas. *Build. Environ.* **2012**, *58*, 46–57. [CrossRef]
8. Yu, J.; Yang, C.; Tian, L. Low-energy envelope design of residential building in hot summer and cold winter zone in China. *Energy Build.* **2008**, *40*, 1536–1546. [CrossRef]
9. Shi, Q.; Zuo, J.; Huang, R.; Huang, J.; Pullen, S. Identifying the critical factors for green construction—An empirical study in China. *Habitat Int.* **2013**, *40*, 1–8. [CrossRef]
10. Xu, P.; Chan, E.H.W. ANP model for sustainable Building Energy Efficiency Retrofit (BEER) using Energy Performance Contracting (EPC) for hotel buildings in China. *Habitat Int.* **2013**, *37*, 104–112. [CrossRef]
11. Cai, W.G.; Wu, Y.; Zhong, Y.; Ren, H. China building energy consumption: Situation, challenges and corresponding measures. *Energy Policy* **2009**, *37*, 2054–2059. [CrossRef]
12. Du, P.; Zheng, L.Q.; Xie, B.C.; Mahalingam, A. Barriers to the adoption of energy-saving technologies in the building sector: A survey study of Jing-jin-tang, China. *Energy Policy* **2014**, *75*, 206–216. [CrossRef]
13. Morrison-Saunders, A.; Pope, J. Conceptualizing and managing trade-offs in sustainability assessment. *Environ. Impact Assess. Rev.* **2013**, *38*, 54–63. [CrossRef]
14. Pan, W.; Ning, Y. The dialectics of sustainable building. *Habitat Int.* **2015**, *48*, 55–64. [CrossRef]
15. Environmental Protection Agency (EPA). Question about Your Community: Indoor Air. 2013. Available online: http://www.epa.gov/region1/communities/indoorair.html (accessed on 20 March 2016).
16. Environmental Protection Agency (EPA). Buildings and Their Impact on the Environment: A Statistical Summary. 2009. Available online: http://www.epa.gov/greenbuilding/pubs/gbstats.pdf (accessed on 20 March 2016).
17. Kubba, S. Chapter 5—Design Strategies and the Green Design Process. In *LEED Practices, Certification, and Accreditation Handbook*; Butterworth-Heinemann: Boston, MA, USA, 2010.
18. Kang, M.Y.; Guerin, D.A. The state of environmentally sustainable interior design practice. *Am. J. Environ. Sci.* **2009**, *5*, 179–186. [CrossRef]
19. Mazarella, F. Interior Design. 2011. Available online: http://www.wbdg.org/design/dd_interiordsgn.php (accessed on 20 March 2016).
20. Hayles, C.S. Environmentally sustainable interior design: A snapshot of current supply of and demand for green, sustainable or Fair Trade products for interior design practice. *Int. J. Sustain. Built Environ.* **2015**, *4*, 100–108. [CrossRef]

21. Fadeyi, M.O.; Taha, R. Provision of Environmentally Responsible Interior Design Solutions: Case Study of an Office Building. *J. Archit. Eng.* **2012**, *19*, 58–70. [CrossRef]
22. Building and Construction Authority (BCA). *BCA Green Mark for Office Interior*; version 1.1; BCA: Singapore, 2012.
23. Leadership in Energy and Environmental Design (LEED). *LEED Reference Guide for Green Interior Design and Construction for the Design, Construction and Renovation of Commercial and Institutional Interiors Projects*; U.S. Green Building Council: Washington, DC, USA, 2009.
24. Building Research Establishment Environmental Assessment Method (BREEM). *BREEAM International Refurbishment and Fit-out*; BREEAM: London, UK, 2015.
25. Geels, F.W. From sectoral systems of innovation to socio-technical systems: Insights about dynamics and change from sociology and institutional theory. *Res. Policy* **2004**, *33*, 897–920. [CrossRef]
26. Murphy, J.T. Human geography and socio-technical transition studies: Promising intersections. *Environ. Innov. Soc. Transit.* **2015**, *17*, 73–91. [CrossRef]
27. Ministry of Housing and Urban-Rural Development (MOHURD). *Statistic Yearbook of Chinese Docoration Industry*; MOHURD: Beijing, China, 2015.
28. Li, Z.; Kong, S. Mass production of fine decoration. *Constr. Econ.* **2013**, *3*, 66–69.
29. Ministry of Finance (MOF); Ministry of Housing and Urban-Rural Development (MOHURD). *Notices on Implementation Suggestions for Accelerating the Promotion of Green Buildings*; MOF; MOHURD: Beijing, China, 2012.
30. Ministry of Housing and Urban-Rural Development (MOHURD). *Contractor Registration Head Standard*; MOHURD: Beijing, China, 2015.
31. Ministry of Housing and Urban-Rural Development (MOHURD). *Registration Standards for Engineering Firms*; MOHURD: Beijing, China, 2007.
32. Ye, L.; Cheng, Z.; Wang, Q.; Lin, H.; Lin, C.; Liu, B. Developments of green building standards in China. *Renew. Energy* **2015**, *73*, 115–122. [CrossRef]
33. Ju, C.; Ning, Y.; Pan, W. A review of interdependence of sustainable building. *Environ. Impact Assess. Rev.* **2016**, *56*, 120–127. [CrossRef]
34. Cyr, J. The Pitfalls and Promise of Focus Groups as a Data Collection Method. *Sociol. Methods Res.* **2014**, *45*, 231–259. [CrossRef]
35. Dubois, A.; Gadde, L.E. Systematic combining: an abductive approach to case research. *J. Bus. Res.* **2002**, *55*, 553–560. [CrossRef]
36. Yeh, A.G.; Yuen, B. Introduction. In *High-Rise Living in Asian Cities*; Springer: Dordrecht, The Netherlands, 2011; pp. 1–8.
37. Grant, A.; Ries, R. Impact of building service life models on life cycle assessment. *Build. Res. Inf.* **2013**, *41*, 168–186. [CrossRef]
38. Lee, E.; Allen, A.; Kim, B. Interior design practitioner motivations for specifying sustainable materials: Applying the theory of planned behavior to residential design. *J. Inter. Des.* **2013**, *38*, 1–16. [CrossRef]
39. Jensen, O.M. Consumer inertia to energy saving. In *ECEEE 2005 Summer Study*; Panel 6; ECEEE: Stockholm, Sweden, 2005; pp. 1327–1334.
40. Bacon, L. Interior Designer's Attitudes toward Sustainable Interior Design Practices and Barriers Encountered When Using Sustainable Interior Design Practices. 2011. Available online: http://digitalcommons.unl.edu/archthesis/104/ (accessed on 20 March 2016).
41. Häkkinen, T.; Belloni, K. Barriers and drivers for sustainable building. *Build. Res. Inf.* **2011**, *39*, 239–255. [CrossRef]
42. Mollaoglu-Korkmaz, S.; Swarup, L.; Riley, D. Delivering sustainable, high-performance buildings: Influence of project delivery methods on integration and project outcomes. *J. Manag. Eng.* **2011**, *29*, 71–78. [CrossRef]

3

Discerning and Addressing Environmental Failures in Policy Scenarios Using Planning Support System (PSS) Technologies

Brian Deal [1] and Haozhi Pan [2,*]

[1] Department of Landscape Architecture, University of Illinois, Champaign, IL 61820, USA; deal@illinois.edu
[2] Department of Urban and Regional Planning, University of Illinois, Champaign, IL 61820, USA
* Correspondence: hpan8@illinois.edu

Academic Editor: Tan Yigitcanlar

Abstract: The environmental consequences of planning decisions are often undervalued. This can result from a number of potential causes: (a) there might be a lack of adequate information to correctly assess environmental consequences; (b) stakeholders might discount the spatial and temporal impacts; (c) a failure to understand the dynamic interactions between socio-ecological systems including secondary and tertiary response mechanisms; or (d) the gravity of the status quo, i.e., blindly following a traditional discourse. In this paper, we argue that a Planning Support System (PSS) that enhances an assessment of environmental impacts and is integral to a community or regional planning process can help reveal the true environmental implications of scenario planning decisions, and thus improve communal planning and decision-making. We demonstrate our ideas through our experiences developing and deploying one such PSS—the Land-use Evolution and impact Assessment Model (LEAM) Planning Support System. University of Illinois researchers have worked directly with government planning officials and community stakeholders to analyze alternate future development scenarios and improve the planning process through a participatory, iterative process of visioning, model tuning, simulation, and discussion. The resulting information enables an evaluation of alternative policy or investment choices and their potential environmental implications that can change the way communities both generate and use plans.

Keywords: environmental market failure; environmental planning; planning support systems; knowledge communication; plan making

1. Introduction

It has been suggested that among other attributes, good urban planning might function as a publicly-driven counterweight to potential market failures [1]. Similarly, as a specific planning disciplinary area, environmental planning might be viewed as a tool to counteract market forces that tend to result in environmentally unsustainable developmental patterns. For example, in a self-optimizing free market, no one will pay to avoid environmental externalities, especially those that are difficult to pin down or happen at stochastic spatial or temporal scales—unless there are planning or regulatory frameworks in place. These types of frameworks usually depend on environmental assessment processes and/or tools to uncover the environmental impacts of development or investment decisions and use planning-related measures to offset them [2,3]. In a typical planning process however, these assessments are usually poorly derived, poorly understood, and poorly applied. This can result from a number of potential issues. For example, there might be a lack of adequate information to correctly assess environmental consequences, or stakeholders might discount the spatial and temporal implications of the impacts. There may also be a failure to understand the dynamic and complex interactions between socio-ecological systems including secondary and tertiary response mechanisms. Finally, it might be that the gravity of the status quo, i.e., blindly following a traditional discourse, is the easiest and most compelling route for an un-informed group of stakeholders.

In this paper, we argue that if interpretations of environmental impacts are not comprehensively derived and sufficiently exposed to stakeholders, the true costs of development and investment decisions remain masked and will result in massive inefficiencies and ultimately environmental failure. We make this point by looking at environmental market failures within a planning context and the potential for planning support systems to counteract these failures. We do this by first connecting the literature on environmental market failures to planning practices (Section 2) in order to understand how these failures are typically manifested in planning-related processes. In Section 3, we suggest three steps that might be incorporated into the planning process in order to counteract these failures. These steps include: (1) a better understanding of the environmental issue; (2) providing more useful information to stakeholders; and (3) using planning processes to tackle spillover effects. In Section 4, we propose that the use of Planning Support Systems (PSS) can help support these steps. To illustrate the ideas in real-world planning practices, we introduce the implementation of the University of Illinois' Land-use Evolution and impact Assessment Model (LEAM) PSS in three application and deployment cases: Peoria, IL, St. Louis, MO, and McHenry County, IL. In each case, we demonstrate how the LEAM PSS was used to inform specific planning decisions in response to potential environmental systems degradation. In Section 5, we conclude the paper with a general discussion and conclusions on our ideas and lessons learned, including a brief discussion on the future of PSS technology in environmental assessment.

2. The Realization of Environmental Market Failures in Planning

In the literature on environmental market failures, environmental impacts are often considered undervalued; typically described as an externality or side effect instead of a primary or secondary cost [4] The consistent undervaluation of environmental costs often results in suboptimal decisions for individuals, organizations, or markets as a whole [4,5]. The root causes of these environmental market failures are difficult to pin down, although they are typically attributed to information imperfection, global externalities that do not affect local benefits, and/or complicated coordination of technology and ecological innovation. Andrew [5] points out that when market participants have sparse or uneven information, sub-optimal decisions are likely to ensue. This suggests that if information on environmental impacts are not known, market participants in routine development decisions are likely to generate environmental market failures [5–7].

Brueckner [7] considers planning as a potential remedy to the problem. We agree. We generally argue however, that if the environmental impacts of development are not *sufficiently* estimated, trusted, and/or objectively derived, planning will be unable to overcome the negative environmental impacts of market-driven forces. We also recognize that environmentally driven planning approaches can lead to undesirable and unintended environmental failures if done in a poor or incomplete way [7–9]. In the following, we discuss four potential ways in which urban planning typically fails to realize the true environmental implications of development or investment decisions (as noted above): (1) a lack of adequate information; (2) stakeholder spatial or temporal discounting; (3) a failure to understand the dynamic interactions between socio-ecological systems; or (4) the gravity of the status quo.

2.1. A Lack of Adequate Information to Correctly Assess Environmental Consequences

In the literature on market failures, *imperfect information* or information inefficiency are potential causes of failure. These are usually a product of asymmetric or incomplete access to information [5]. This means that certain types of information may be inaccessible to one party but not another, producing an asymmetrical transaction. An often used example is the used car market, where a seller of a used vehicle has much better information on the car's condition than a potential buyer. In this case the failure is due to the asymmetry of the information available to each side of the transaction causing a potential over-valuation of the vehicle's worth by the buyer. In planning and development projects, imperfect information is usually the result of information availability rather than asymmetry. This occurs when a complex development issue is difficult to assess or understand so that decisions are made based on incomplete information. This is especially true in the case of environmental impacts, where the issues may not be well defined, or even understood by the stakeholders or planners involved in the process—giving rise to suboptimal decision making and ultimately the host of environmental failures we currently see in our urban systems.

He et al. [10] point out that a persistent failure to deliver environmentally sustainable development solutions is closely linked to the separation of environmental assessment from the typical urban planning process. We suggest that insufficient adoption of environmental assessment is only part of the picture. A more essential problem is that most environmental assessment does not take full account of the environmental impacts of development. So that even when it is considered a part of a planning and development decision process, the delivery of incomplete information on impacts can lead to highly probable failure. For example, in a typical environmental assessment, a commercial development that engulfs a patch of agricultural land might associate the implications of the transaction only in terms of the primary impacts produced by the development (site-related impacts) or the loss of agricultural lands, which can be more regional in scope. If however, the development is proximal to ecologically sensitive areas, the development could have enormous ecosystem service impacts. This typifies hidden environmental costs that are often neglected by traditional environmental assessments.

2.2. A Failure to Understand the Dynamic Interactions between Socio-Ecological Systems including Secondary and Tertiary Response Mechanisms

Holmberg and Karl-Henrik [9] proposes that the non-linearity of markets often confound plans intended for sustainable development. For example, in the Laguna West master planned community in Sacramento County, CA, new urbanist and sustainable design principles were used as a basis for the development. Despite rigorous planning and analysis however, when the community began opening, the market for units in the community rose at a startling rate and subsequent phases subverted the planned economic diversification and many sustainability-oriented goals [11]. Such market-driven complexities can easily lead to a host of unexpected failures and shortcomings.

The general complexity of urban systems further undermines our ability to evaluate the real environmental costs of planning decisions. Urban planning problems are often referred to as "wicked problems" that are "inherently different from the problems that scientists and engineers deal with" [12]. Science and engineering-based approaches however, often deal with only primary impacts and fall way short of robust outcomes when applied to such wicked complexity. The non-linearity of future urban development patterns, and dependencies among developmental decisions often overwhelm the analytical capacities of typical environmental assessments, especially (most of) those that use static assumptions [13]. This difficulty in assessing secondary and tertiary (latent) impacts leads to incomplete information and confusion, which ultimately leads to environmental failure.

2.3. Stakeholder Discounting of Spatial and Temporal Environmental Impacts

Secondary and tertiary environmental costs typically spill over long spatial distances and trickle into distant futures. These long distances and time lines can lead to the problem of discounting. Discounting occurs when a value is subjectively lowered because it is removed or distant in terms of time, space, or socio-cultural relationships. Discounting is an important concept when discussing environmental sustainability (and climate change). Essentially the argument is that we should not discount the value of environmental resources (i.e., a healthy climate) for future generations because it will be just as valuable to them in 50 years as it is to us today. However, Fall [14] argues that we cannot predict how future generations will value environmental resources and that environmental crisis could alter the discount rates and slow the rate of environmental damage. Weitzman [15] presents a deductive argument for valuing environmental resources with a low discount rate for long-term planning projects showing that they are theoretically likely to produce the best return on investment. Hoel and Sterner [16] take this argument a step further and argue that the rising scarcity of environmental goods will also increase the relative cost and have direct effects of the discount rates. The idea of devaluing costs that are removed from a particular context in terms of space or time is a central problem in planning because it strikes at planning's very purpose: to adequately account for the costs associated with policy and investment strategies (i.e., *plans*) made within that context.

Berke and Conroy [6] argue that although plans are always linked to global concerns, "local plans should acknowledge that communities function within the context of global (and regional) environmental, economic, and social systems". Empirical data suggest that those links do not always lead to actions; that stakeholders tend to overlook and discount the spatial and temporal environmental costs of their plans. They further point out that the high discount rates that undervalue environmental costs with spatial and temporal distance help create plans with severe environmental impact spillovers. We argue that stakeholder discounting plays a large role in this phenomenon.

2.4. The Gravity of the Status Quo, i.e., Blindly Following a Traditional, Growth-Oriented Discourse

Some developers, businesses, and even some communities profit by maintaining the historical, developmental pattern status quo. Some argue that these patterns have typically been put in place at a time of 'unawareness' of their environmental implications [6,9]. Patterns of sprawl development for example, were started before their social, health and environmental ramifications were well known. In other words, some historic patterns may no longer be favorable when their associated environmental costs are assigned. Another useful example is energy production; much of our fossil-based energy infrastructure was developed before the actual costs of carbon emissions were known [5]. Now that we have a better sense of their true costs however, some of the infrastructure used in the energy industry is no longer viable. Similarly, urban development patterns continue to follow the sprawling, growth-oriented path initiated during the 1950s that make it increasingly difficult to bring our cities within what is considered ecologically sustainable and equitable [9].

In the sustainable design literature, Strategic Sustainable Development (SSD) is a proposed method that might tackle the complexity and uncertainties associated with sustainable developmental policies and thereby challenge the status quo [10]. SSD incorporates a set of techniques including life cycle analysis, indicator development, natural capital accounting, forecasting, emissions analysis, and backcasting to provide more accurate environmental assessment to current developmental trends and policy scenarios.

In summary, we suggest that in an urban planning context, environmental market failures might be attributed to a lack of knowledge of environmental impacts, a lack of understanding of secondary or tertiary impacts, and decision interactions, poor communication between spatially spread stakeholders, and a poor understanding of alternatives from current developmental paths. In the following section, we consider approaches to addressing these shortcomings through the use of spatially explicit Planning Support Systems (PSS). We posit that PSSs can help engage environmental decision-making in planning by providing more critical and useful information on the environmental implications of development decisions thereby helping stakeholders avoid planning-related environmental failures.

3. Improving Planning Decisions with PSSs

The use of PSSs is a relatively new addition to a long-running and persistent discourse on the role of technology in urban planning and policy-making [17,18]. One of the first comprehensive looks at planning PSS was assembled by Brail and Klosterman in 2001 [19]. Brail later writes more specifically on large-scale urban modeling systems [17]. Geertman and Stillwell followed, documenting best practices in PSS technology [18] with an update on the state of art in 2009 [20]. More recently, the emphasis has shifted to the management of information needs [21–23], use-based systems [24,25], and web-based strategies of information retrieval and delivery [23,26]. More recent conceptualizations also include Michael Batty's "Smart City" [24], described as "a fusion of ideas about how information and communications technologies might improve the functioning of cities". The idea revolves around the need to coordinate and integrate technologies that have synergies in operation but have been developed as separate entities—much like what now might be considered a PSS.

One specific element typical of some spatially explicit planning Support Systems (PSSs), is a Land-Use Change (LUC) modeling core. These models are usually built with an intention to help planners and decision makers better understand spatial data, the dynamics and complexity of urban development patterns, and in some cases assess the impacts that changing urban patterns have on environmental, economic, and social systems. Although it has been more than 30 years since Cellular Automata (CA) technologies were tested in the development of these spatially explicit modeling cores [25], and more than 20 years since urban planners begin to adopt and adapt LUC models to develop first generation PSSs [27], these models have only recently become adaptable enough to be operationally used in support of planning.

3.1. Reveailing Environmental Costs through PSSs

Pioneers have now pushed PSS LUC model prototypes out of research labs and have begun to package the LUC models into operational applications with visual and online user interfaces. The Landuse Evolution and impact Assessment Model [28], WhatIf [29], and UrbanSim [30] are among three of the best operational examples in the US. These PSSs aim to enable non-experts with an ability to input data, operate, localize, and test their PSS LUC models with future potential scenarios, in some instances through web-based portals with simple mouse clicks. Some of these models are close to realizing this ideal (LEAM included). In our opinion however, in order to fully take advantage of the enormous benefits of PSSs within a structured planning process, a connection to knowledgeable modelers and planners is needed. These experts can: (a) facilitate the process and the technology needed to guide communities in applying the tools efficiently and effectively; (b) ensure the quality of PSSs inputs and outputs; (c) explain the implications of modeled outcomes (to non-technical audiences); and (d) they can help suggest implementation strategies in backcasting [9] and scenario analysis exercises. Generally, however, we believe these tools are fundamentally important for countering environmental market failures.

As previously noted, many hidden environmental costs are the result of a weakness in uncovering secondary (or tertiary) environmental impacts. This is compounded by an inability to understand the dynamics and interactions in and between impacts. LUC and PSS technologies are an effective means for addressing these issues. Spatially explicit PSSs bring together LUC models in an information delivery system (visualization interfaces) to provide planners and communities with critical knowledge of various dynamic systems and interactions that can facilitate more effective communicative planning approaches [19]. Infusing critical (environmental) knowledge to stakeholders and making (larger scaled) information available can engage a larger stakeholder group, that is capable and interested in understanding common environmental spillovers. The ready availability and democratization of information can also coerce communities and individuals to address the environmental externalities of their actions. Environmental impact information infused into public discussion through PSS technologies can also enable a more effecting weighing of the trade-offs that might emerge between economic benefits and environmental costs.

In addition, some planning and analysis techniques are greatly improved by PSS technologies [26,31,32]. The ability to easily perform backcasting analysis for example, can help plan-making processes achieve outside-the-norm envisioned, preferred futures rather than depending on a projection from the status quo perspective. These kinds of techniques are important for breaking historic, unsustainable, path dependent developmental patterns.

Given the potential of PSSs to aid urban planning in uncovering hidden environmental costs, the question becomes how these PSS tools become ubiquitous to real-life planning practices. The University of Illinois and the Landuse Evolution and impact Assessment Modeling Laboratory has had some success in operationalizing and implementing their Planning Support System in a variety of regions worldwide, most notably and successfully in the Midwestern US. In the following sections, we describe several examples of the LEAM operationalization process. These examples help to reveal some substantive ways in which the LEAM PSS has impacted the planning and policymaking and in the process helped to alleviate environmental market failures. First, through use-based PSS development and implementation. Next, by helping to counteract potential market distortions through useful information development and dissemination. Finally, through an ability to integrate seamlessly and positively affect the process of environmental assessment in the process of plan making. We also propose a future vision of PSS technologies that evolve from user-driven to user-awareness. We suggest this will enable them to be even more effective in practical environmental planning implementation.

3.2. *The LEAM PSS*

The need in planning and policy making to answer both 'what-if' and 'so-what' questions is fundamental to the LEAM PSS framework. The PSS consists of two major organizational parts: (1) a Land-Use Change (LUC) model—defined by a dynamic set of sub-model drivers that describe the local causality of change and enable an ability to test and play out potential 'what-if' scenarios; and (2) impact assessment models that facilitate interpretation and analysis of the modeled future land-use changes depending on local interest and applicability—these help to assess 'so-what' questions and explicate the potential implications of a modeled scenario.

LEAM LUC model utilizes a hybrid Cellular Automata (CA) approach. Similar to CA, LEAM utilizes a structured lattice surface (cells) with state-change conditions that evolve over time. The lattice is shaped by biophysical factors (such as hydrology, soil, geology and land form), and socioeconomic factors (employment, household structure, administrative boundaries, and planning areas). These factors, when taken in combination, provide a contoured lattice with high and low spots that represent each cell's probability of potential change in land use. Probabilities are predicated on local interactions (e.g., the accessibility of the cell to a given attractor), global interactions (e.g., the state of the regional economy), and other mechanisms of causation (e.g., social forces). Specific rules can be applied and tested. Controlling the constraints in the rule set can be used to produce diverse sets of planning scenarios. Unlike other large-scale efforts, LEAM works at a finely scaled resolution (30×30 m) that includes cell-based micro models. This enables loosely and tightly coupled linking with other models that might operate at a different spatial scale, including regional macro socioeconomic models and transportation infrastructure and demand. The effect is that a wider range of potential 'what-if' scenario sets can be tested and assessed [26] LEAM has been both tightly and loosely coupled with other models that operate at various spatial and temporal scales including: economic forecasting models [33], bi-directional travel demand models [34]; ecosystem service models [34]; water quality models [35]; water quantity models [36]; and social cost models [37]. Coupling these models with LEAM dynamics and making the information useful through a 'use based' implementation process has helped decision makers make sense of the complex interactions between urban change and environmental systems.

4. The LEAM Use-Based PSS Implementation

In the LEAM implementation process, the LUC model evolves as an iterative process of data collection, model building, dialogue, visualization, and general presentation and access. Local planners, policy makers and stakeholders (convened by local planning entities and identified broadly as possible) provide feedback and input about the local salience and value of any given simulation. This feedback is gathered regularly and begins at project inception. It is used to more effectively capture the local condition, to provide a better local version of the tool and to inform local stakeholders about the tool and its uses. This form of *use-driven* modeling and system development, which takes place in very public forums, is what most distinguishes the LEAM approach [22]. The feedback and local dialogue elements are critical in the creation of useful PSS tools especially in terms of overcoming market distortions. Constant internal and external review and interaction are critical to informing both the modeler and the local stakeholders of modeled changes, improvements and scenario outcomes. Presenting this use-driven approach in publically accessible PSS visualization portals helps provide another layer of feedback and interaction. Consensus building is performed and achieved using typical planning procedures (see [29] for a more detailed discussion on use-based modeling and consensus building).

In applying this use-based model process we have found that LEAM can influence decision-making through various pathways [29]. In the following, we describe some specific pathways and their effect on the plan-making process in past LEAM applications. The three cases presented below: Peoria, IL, St. Louis, MO, and McHenry County, IL represent three ways in which the LEAM PSS has made significant impacts on the practice of planning. These include: counteracting distortions, facilitating dialogue, and integrated plan making for challenging the status quo.

4.1. Counteracting Distortions in Peoria, IL

Planning decisions take place over extremely long periods of time—sometimes involving different generations, over large distances. This raises questions of inter-generational and geographic equity. As noted previously, this is compounded by the fact that planning decisions have complicated spatial interactions and environmental impacts that are often secondary, or even tertiary. Environmental costs may accrue in one form, to one generation, in one part of the geography (community, state, nation, world); while the benefits accrue in a different form, for a different generation, in another part. Delivering objective, unbiased and apolitical information can help counteract these types of phenomena that can emerge in typical public planning processes [38].

Many of these issues are the result of an under-estimation of environmental impacts that are sometimes the result of information distortions. Information distortions are usually the product of local knowledge that is deeply situated in the web of accepted norms, meanings, and beliefs. If incorrect (distortions of actual causal relationships), their locally embedded nature makes information distortions difficult to overcome with traditional planning communication processes [39]. If left uncorrected, they can lead to problematic conclusions in public discussions. One example can be seen in local and personal discounting discussed above.

In our work, we argue that providing an ability to objectively test and evaluate current and future conditions can be a powerful tool for counteracting potential local distortions and poorly considered discounting that leads to costly future consequences. In an early LEAM application in Peoria, IL we helped provide those objective arguments for local and regional planners in a simple example.

In the early 2000s, the three counties surrounding Peoria, Illinois (Woodford, Tazwell, and Peoria) were witnessing significant conversion of very fertile and productive agricultural land to residential and commercial land-use. There was a distinct sense of unease about this trend, although there was no specific analytical proof for its existence. Woodford County in particular, was concerned about its agricultural heritage. The county outlined several strategies for preserving agricultural land. One particular strategy required a change in the county zoning ordinance that would require 40-acre minimum lot size on current agricultural lands. At roughly the same time in a regional planning exercise, a number of simulations of future land-use change were being run for the tri-county region and reviewed and critiqued in public workshops [31]. These simulations established the extent and spatial distribution of future growth in a business-as-usual scenario described through maps and depictions of the impacts of this growth. Other scenarios explored included higher and lower growth rates and various public investments and policy ideas, including the proposed ordinance change in Woodford County. Figure 1 shows expected land-use outcomes in a 'business-as-usual' scenario. After our simulation, we used the land cover data provided by the county's planning board to assess loss of agricultural and ecological lands associated with future growth using 30-by-30-m resolution.

To discuss the simulation results, local planners held two big meetings and several smaller focused group meetings. Stakeholders involved in those public discussions included staff from Illinois Department of Natural Resources, citizens, non-government organizations (NGOs), local planners, and government entities. Public discussions on the 40-acre zoning requirement scenario revealed that the ordinance change would reduce consumption of agricultural land as intended but would also bring with it unexpected regional consequences. Compared to the business-as-usual scenario, the amount of agricultural land lost to development over thirty years with the proposed ordinance change dropped from 10,000 acres to 7000 acres. This was expected. Unexpectedly, when the proposed ordinance was included in a simulation, new development that would have been located in the agricultural area moved to environmentally sensitive bluffs along the Illinois River, resulting in the loss of 12,500 acres of forestland (Figure 1). This revelation changed perceptions of the proposed ordinance and the ordinance was put on hold until a ravine overlay district focused on protecting the river bluffs was put in place.

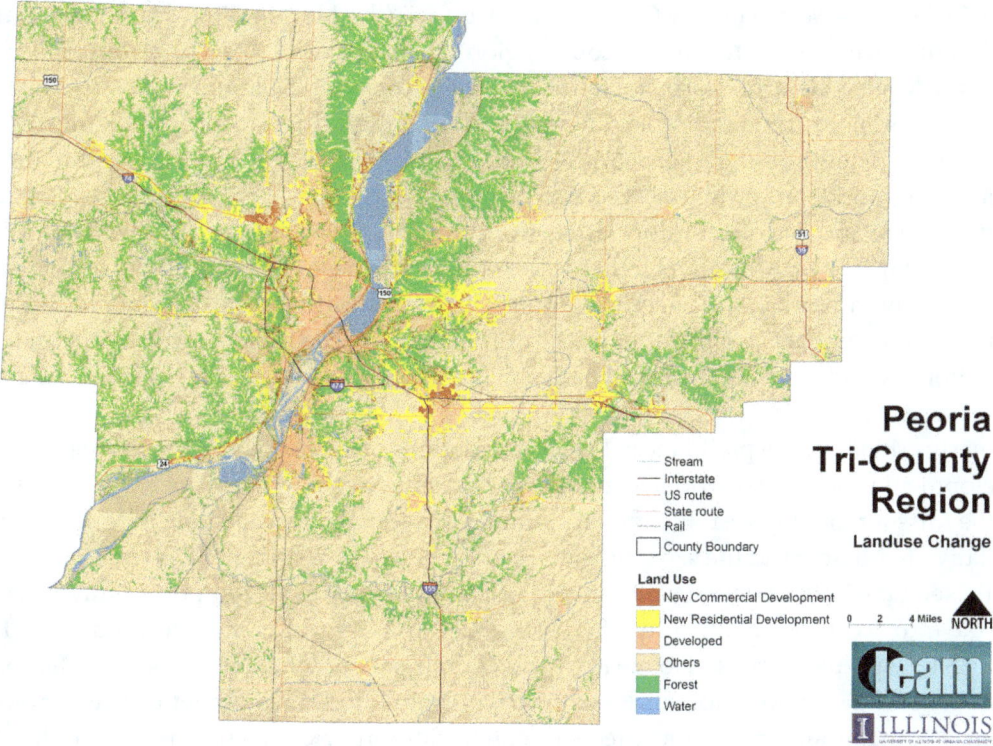

Figure 1. Land use outcomes in the Tri-County Region of Peoria in a 'business-as-usual' scenario. Yellow areas represent new residential developments; dark red patches are commercial. The arrow refers to the desirable bluff areas along the river.

Initially, any negative consequences of the proposed ordinance appear to have been discounted. Once people saw the simulations and understood the consequences of the proposed ordinance, this discount rate was substantially decreased to the point that protecting the bluffs became a higher priority.

It appears from the above case that explicating and elaborating on the future consequences of various public policy and investment choices may alter the extent to which these consequences are discounted by stakeholders in public deliberations. When future consequences are vaguely known and ill-defined, they are easily discounted. When potential consequences are represented in tangible and objective ways however, they make the familiar unfamiliar, they challenge habitual ways of thinking, and question what appears evident and taken for granted [40]. In short, they can counteract normative distortions.

4.2. Facilitating Dialogue with the St. Louis Blueprint Model

PSSs can assist planners in convening local stakeholders to discuss and validate environmental assessment results and in the process arouse a *regional* consciousness of the potential spatial and temporal spillover of environmental impacts. Geertman [41] points out that PSS tools can enhance participatory planning processes, because "a greater degree of access to relevant information will lead to the consideration of a greater number of alternative scenarios—which in turn will result in a better informed public debate". One project that exemplifies this idea is the application of LEAM to the two-state, eight-county region around St. Louis, Missouri (MO) [32,42]. In this project, we coupled various other models with LEAM to analyze the potential impacts of the land-use change results. Two in particular were a four step transportation model (that utilized over 2200 Transportation Analysis Zones, TAZ) to calculate travel time changes for each scenario, and a regional economic input-output model (conducted on a household level) that provided demand for space and assessed economic implications.

In 2003, the East–West Gateway Coordinating Council (the Metropolitan Planning Organization and Council of Governments for the St. Louis region), began to use LEAM (in a version later called the Blueprint Model) as a platform for encouraging a regional dialogue on issues of economic development, social equity and environmental sustainability. Instead of initializing the process with a lengthy model-building exercise, the initial focus was set on quickly producing a set of simulations. This quick-start process served two purposes: to quickly begin the process of engagement and build interest; and to collect information from the local stakeholders on the state of the local condition for adapting LEAM model to fit local conditions. These early simulations were subjected to public scrutiny in workshops, meetings and other public forums. Participants in these forums provided valuable insights into the dynamics of urban LUC in the region and a direction for future modeling efforts. Conducted on an annual basis, they also provided an excellent platform for dialogue among participants.

One early critique of the preliminary LEAM simulations presented was aimed at the way in which new development was being distributed across the two sides of the Mississippi River—the Illinois on the east, the Missouri on the west. Preliminary simulations showed considerable new developments in Illinois relative to Missouri; at the same time, the central business district (CBD) is in Missouri and has historically seen the bulk of new development. These simulations utilized posted travel speeds and did not take into account the difficulty of crossing the Mississippi River from Illinois into the CBD. When congested speeds were used to measure travel time (taking into account how traffic congestion makes portions of the region more or less attractive), simulated development shifted from Illinois to Missouri. A major factor was the effects of congestion on bridges and the approaches to them (bridges represent severe choke-points with very little opportunity for alternative routing). In the regional dialogue, this outcome highlighted the critical role played by bridges in the distribution of new development across the region.

The construction of a new Mississippi River bridge had been the subject of planning studies, preliminary design and environmental impact analysis for over 20 years in the region. A concerted civic and political effort to secure earmarked federal funding was only partially successful. The resulting funding shortfall called into question the original bridge proposal and how it would be implemented. Alternatives considered included covering the shortfall with a toll and constructing less expensive alternatives such as enhancing the capacity of an existing bridge; there was no regional consensus on the way forward. Facing a stalemate on the issue, the regional planning organization, the East–West Gateway Council of Governments (EWGateway) took the lead and sought to inject an analytical basis into the regional debate. In order to do this, however, it became crucial to go beyond traditional cost-benefit analyses and to jointly simulate and analyze future transportation and land-use consequences of the different choices.

Numerous simulations were created by coupling LEAM with a regional econometric input-output model (LEAMecon) and a regional travel demand model (TransEval). The land-use, economic and transportation outcomes in the simulations, and those of a baseline 'No-Build' simulation, were the basis for comparisons. Aggregate differences appear to be slight: building the bridge appears to slightly increase development in Illinois (Madison and St. Clair counties), while slightly decreasing development in Missouri (St. Louis and Jefferson counties). Interestingly, imposing a toll increased land development in far northwestern Missouri (St. Charles County). Figure 2 displays differences in LUC between the Full Build and No Build simulations at a finer resolution; red cells see more growth in the Full Build simulation, green cells see more growth in the No Build simulation. The map presents a more complex set of differences and suggests that aggregating to the county level masks greater change: while building the bridge facilitates greater land development in the Illinois side of the region and takes away from development on the Missouri side of the river, there are significant differences in development at the local level.

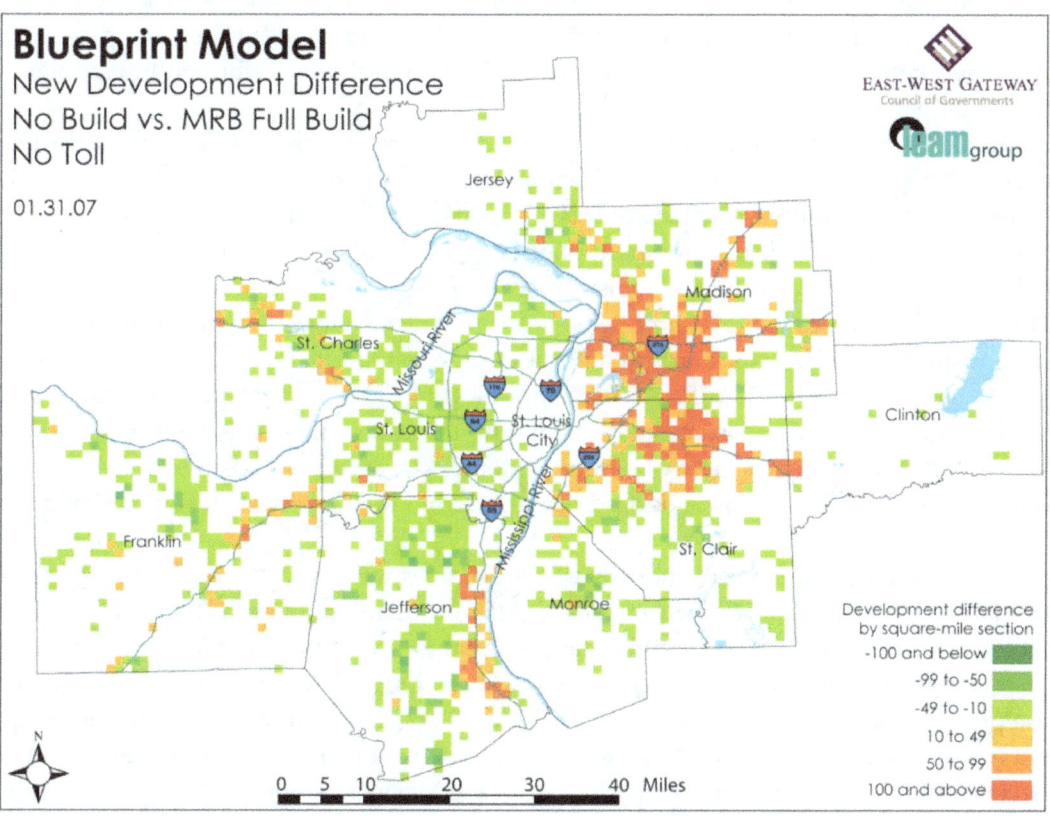

Figure 2. Differences between the Mississippi River Bridge Full Build (**red**) and No Build (**green**) simulations in the St. Louis Blueprint Model.

EWGateway convened three big public meetings and tens of smaller meetings to discuss the simulation results. As might be expected, discussions around these simulations were quite intense. Outcomes appeared counter-intuitive; imposing a toll on the bridge increased total travel times in the region. Working through the complex interactions suggested a striking explanation; the toll was diverting traffic to the other bridges across the river that do not impose a toll, increasing congestion on these bridges, and increasing travel times. This explanation brought into question the wisdom of using a toll to cover the budget shortfall. There were other insights generated: patterns of land use are likely to change if additional river crossings are built, therefore cooperative land-use policies and controls must be put in place in these areas to manage these impacts. Ultimately, however, only slight differences were uncovered even though the magnitude of the investment required for each of the scenarios tested was very different. This suggests that perhaps the lowest cost alternative is preferable, but it also suggests that demand-side tactics, such as investing in a better regional jobs-housing balance, might be more cost effective.

In this case, the LEAM PSS assisted planners in convening local stakeholders to discuss and validate each bridge scenario and their potential implications. The user-based process clearly facilitated a regional dialogue and aroused a regional consciousness of the potential spatial and temporal spillover effects. The process also helped challenged conventional thinking about the fundamental needs and benefits of the proposed investment.

4.3. Planning for a Deviation from Current Developmental Path in McHenry County, IL

In the previous discussion, including environmental assessment in planning processes and decision-making informed a public dialog so that participants had a better appreciation for the future environmental consequences of their public policy and public investment choices [33]. These planning processes however, are essentially forward looking exercises. Many similar processes that do not use PSS technologies rely heavily on projecting existing conditions into the future. They generally fail to capture changing paradigms or emergent behaviors. This often results in the continuation of existing developmental paths.

The use of PSS tools enable a broad range of multi-directional analysis that might be useful in analyzing or planning for structural change—including those needed to address a host of environmental market failures. The idea of *backcasting* using PSS technologies for example, has been shown effective in sustainable development planning [43]. Deal et al. [44] propose that backcasting from a desirable future state using PSS tools enables planners to step outside current developmental trends to test ideas and reexamine assumptions. Where a forecast projects an image of the future based on a current situation, a backcast starts at a point in time in the future and draws a developmental path back to the current condition. This is useful for plotting a path that responds to "how do we get there" kinds of questions from future states that might not emerge from existing trends.

A LEAM application in McHenry County, IL demonstrates a PSS-led backcasting exercise that helped the county understand how a desired deviation from their current developmental path might be achieved.

McHenry County, IL defines the northwest edge of the seven counties that make up the Chicago metropolitan region. It is approximately 60 miles northwest of downtown Chicago. It has a population of 318,000. Its location and unique natural features create a quality-of-life that is attractive to many. Since 1990, the county's population has grown 40%, averaging 2.3% growth annually. The previous land-use plan for the county was the McHenry County Land Use Plan 2010, which was compiled in 1993 and updated in 2000. However, McHenry County Regional Planning Commission (RPC) deemed this plan increasingly irrelevant and began to compile the McHenry County 2030 Comprehensive Plan in 2007 [26,45].

In 2007, with the help of the State of Illinois Department of Natural Recourses, the LEAM Laboratory began to build a PSS for the County. The original work was designed to assess the potential future implications that urban land use changes would have on the natural resources in the County. It was soon put to use to inform the discourse and test some of potential distortions emerging from their 2030 process [45]. The PSS development process first established a 'reference' or 'business-as-usual' scenario as a baseline for assessing the impacts of various land use policies being discussed. The reference scenario simulated LUC if current growth pattern trends continue to 2030. Other model scenarios were then compared to the reference scenario in order to understand the impact that the tested policies might have on various important county assets. LEAM was coupled with other impact models (as described above) to assess: land-use, water demand, water quality, wetlands, natural areas, agricultural uses, and groundwater protection (the list was determined by county stakeholders). Of particular interest to the County was the loss of important agricultural and ecological lands associated with future growth. These were evaluated linking LEAM to the Land Evaluation and Site Assessment (LESA) modeling framework from the Illinois Department of Agriculture [46].

The McHenry County Planning board convened more than 10 public meetings made up of a range of public interests, stakeholder groups, government employees and officials, and planners. The process revealed an early, major concern with the projected population forecast for 2030 that was derived using LEAMecon. This was the bellwether issue that underlay a larger conflict on the future of the

County between pro and anti-growth advocates; one group of residents hoping to continue the past development trends and environmental groups urging protection of environmentally sensitive and agricultural lands. After lengthy discussions LEAM simulated a range of potential scenarios identified by the County Planning Board (18 prime scenarios) of future land-use patterns. The RPC identified various preferred outcome scenarios [26,45]. One such scenario, the Compact Contiguous Growth (CCG) composite scenario is shown in Figure 3. The difference between the reference scenario and the CCG composite scenario is on the left. Areas in blue receive more development in the CCG composite scenario; areas in red receive more development in the Reference scenario. A notable difference is seen in the southwest portion of the County. This is due to a limit that the CCG composite scenario places on the amount of vacant land available in that part of the county. The right shows an urban growth boundary that was also considered as part of the scenario analysis.

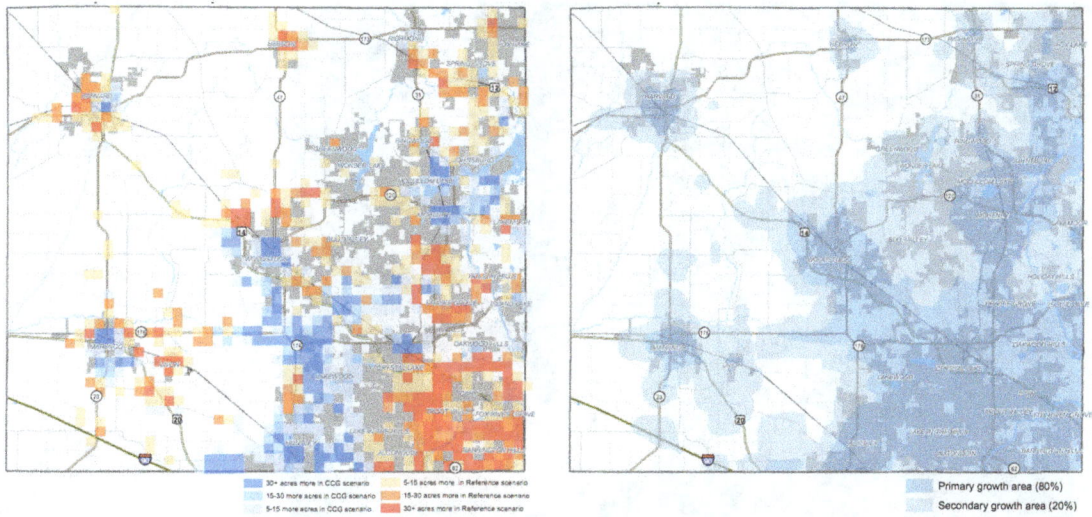

Figure 3. McHenry County Compact Contiguous Growth (CCG) scenario growth projection to 2030 compared with a reference scenario (left). Comparison using 1 mile × 1 mile section scaling. Red cells are unique to the reference scenario; blue cells represent changes as a result of the CCG scenario. The right image shows as urban growth boundary that was also a part of the CCG scenario analysis.

The spatially and temporally explicit information generated by the LEAM PSS helped the McHenry County Regional Plan Commission (RPC) objectively assess the impacts of their proposed policies. For example, one scenario identified spatial locations where farmland and ecologically sensitive areas were identified as at risk [47]. With LEAM information, the RPC could specify where to set up ecological/agriculture preservation districts to prevent a disruption of the critical areas.

The process represents a typical PSS backcasting exercise. First, a desirable outcome was established—in this case, minimizing agricultural and ecological land losses. Backcasting how to achieve this desirable outcome required an analysis of the complex interactions between a host of variables, so that many multiple model iterations were examined in order to understand exactly what this preservation meant to county stakeholders. Once these were unraveled, a coherent set of policy levers were developed for how these outcomes for the future might be achieved. In this case agriculture productivity was closely tied to the introduction of a new (and also desired) transportation investment (a new interchange conflicted with highly productive agricultural lands). The LEAM PSS provided useful spatially explicit information on where, when, and how to plan for an alternative, desirable future that was just outside current development patterns.

The LEAM PSS enabled a 'continuous planning' process to take shape in the county. The spatial and aspatial data created for the PSS allows the community to continuously interact with the data and models associated with the plan (a visualized interactive tool is shown in Figure 4). It is now a living comprehensive plan where critical questions can be examined, progress on critical issues can be updated and communicated, and success or failure can be determined and re-assessed. We suggest this type of tool and process are critical for challenging existing (and unsustainable) growth and development practices and challenging the status quo.

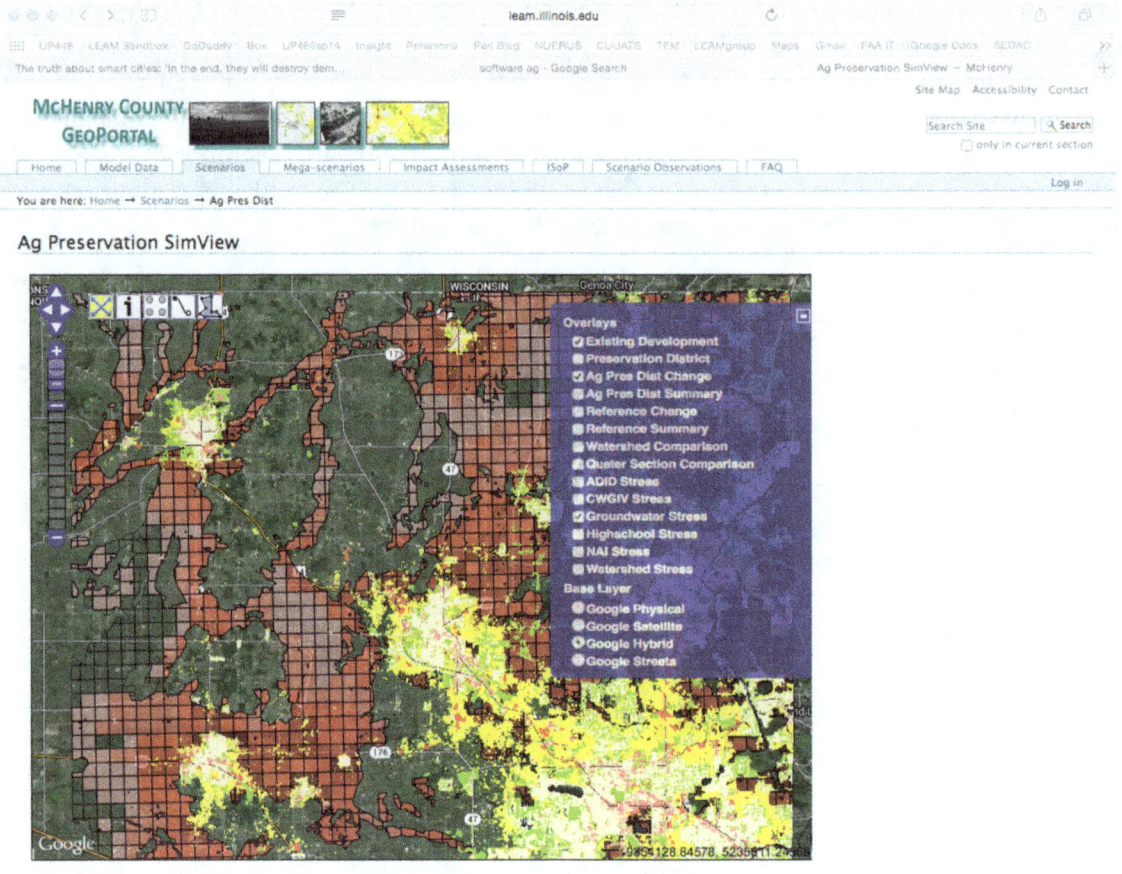

Figure 4. The LEAM PSS's 'GeoPortal' for McHenry County. The center is the map viewer window with overlay results viewed on an interactive google map. On the right is a legend for the displayed maps. The viewer displays a forecasted agricultural preservation scenario and its associated groundwater stress. Yellow and bright right cells demonstrate future growth scenario; darker red cells (big cells) represent groundwater stress associated with the growth.

5. Discussion and Conclusions

In this paper we posit that if environmental impacts are not sufficiently exposed in the planning process, environmental market failures will likely materialize. We note that this can take several forms including: a lack of adequate information, stakeholder discounting, a failure to understand complex interactions, or by following a traditional, growth-oriented path. We argue that the use of dynamic and complex systems models that are able to effectively derive a truer picture of the environmental impacts of a given scenario plan can help offset some of this failure. We then demonstrate how the operational deployment of the LEAM model and PSS in real world planning settings supports this notion. We described several communities that were able to avoid decision traps that could bring about otherwise unseen environmental impacts.

In Peoria, IL we showed that providing useful information on the future consequences of planning decisions help alter stakeholder discounting. From our experience, we conclude that when potential consequences are represented in tangible and objective ways they can challenge habitual ways of thinking and help counteract normative distortions.

In the St Louis, MO metro area, the LEAM PSS assisted planners in convening local stakeholders and facilitating a regional dialogue. This helped raise a regional awareness of the potential spatial and temporal spillover effects and challenged conventional thinking on the proposed investment.

In McHenry County, IL the LEAM PSS enabled a county wide dialogue in a 'continuous planning' process where the data and models associated with the plan can be continuously accessed, examined, and communicated so that success or failure of each proposed policy can be determined and re-assessed. We conclude this is critical for challenging existing (and unsustainable) planning practices.

The following are some additional observations and lessons learned from our experiences in PSS deployment and environmental assessment.

5.1. The Process of Constructing Useful Planning Support Models Is More Important Than the Modeled Results

One important lesson learned is that in order to foster local thought experiments and discussion, it may not be necessary to wait until model results are highly calibrated and fine-tuned. In fact, we argue that planning and decision *impact* can be more important than model accuracy. Preliminary results that can be presented early in the planning process and are easily modified can serve to ignite regional dialogue. For example, the presentation of both favorable and unfavorable scenarios in terms of environmental impacts would be useful in helping a community understand and articulate their preferences. This articulation can help stakeholders and planners communicate with each other as well as steer policy toward achieving important sustainability goals that are sometimes difficult to express. This does not absolve planners from providing accurate depictions—especially when challenging accepted norms however. At some point in the process reliable and trust worthy information is required. But it does suggest that even 'quick and dirty' results can be very useful when planners, stakeholder representatives, and decision-makers work cooperatively to interpret them through the lens of expertise and local perspective.

5.2. The Physical Representation of Information Matters

In operationalizing LEAM, we have been extremely careful about how we represent and display the numerous graphic and visual representations that are created. We take great pains to explain the logic of these representations to stakeholders involved. While we have to this point, limited ourselves to maps and graphs, we have developed a variety of ways to represent spatial and temporal data [42]—roughly adhering to the six I's of Senbel and Church [48]—information, inspiration, ideation, inclusion, integration, and independence. While we recognize that any cartographic representation is necessarily selective truth-telling, we also recognize that visual representations are critical for affecting stakeholder understanding and for unraveling the problem of discounting the future.

5.3. User-Driven, Use-Driven, and User-Awareness in PSS Deployment

User-driven models (such as WhatIf? and UrbanSIM) are packaged into software that can be run with some basic skills sets by planners or policy makers. *Use-driven* models such as LEAM are created for specific projects at specific instances, and sometimes built to answer specific questions. Use-based models involve modelers, contextual experts, and planners to help build operate and explain the models in the planning process. Stakeholders and citizens are also critical to their development and deployment. They provide important, but difficult to obtain local, historical knowledge, insights about local dynamics, and intimate knowledge of local social constructs. They also provide the specific goals and visions for the future that are so essential to the planning and modeling process.

Use-driven and user-driven PSS models are not mutually exclusive however, component pieces of both are useful. For example, user-based interfaces allow non-experts to directly modify model factors according to their understandings of their locality and generate their own results. Modelers should be on hand to interpret those results and facilitate public discussions based on different model results.

User-aware PSSs are a next-generation process that combines use/user-driven PSS models. User-aware PSS can adjust its user-interface to different users (modeling experts or layman users) and communicate diverse user-inputs with each other in the system. This can make future PSS operationalization more democratic and interactive.

5.4. Post-Plan Involvement Is Also Important for Operationalizing a PSS

Currently, PSS-generated information is mainly utilized during the pre-implementation phase of the planning process. Although monitoring can be an important component of plan implementation (that sometimes can determine the success or failure of a plan), due to time constraints it is not often feasible for planners to collect new data, look back and analyze previous plans. A next generation PSS might automate this process so that "actual" emerging developments and real-time environmental impacts can be included in development of scenarios and plans. As the time and effort for post-plan monitoring are conserved, planners and modelers can afford additional time with the communities to ensure that implementation can more closely follow the plan intent. This can also be useful in helping stakeholders to understand the dynamic interactions between socio-ecological systems and socio-physical decisions.

5.5. The Future of PSS Technology in Environmental Assessment

We see a general need for information and planning systems which current PSS technologies provide do not meet; especially in the area of environmental assessment. We need smart planning support systems that possess basic self-awareness with respect to their data and users. These should be capable of learning, of spatial and temporal reasoning, of understanding system integration rules, and be able to adequately address environmental impacts from a broad perspective. More specifically, and in the spirit of the above, these technologies should be capable of producing reasonable and adequate environmental impact information that can reduce stakeholder spatial and temporal discounting and challenge the traditional and unsustainable status quo.

Author Contributions: Both authors (Brian Deal and Haozhi Pan) contributed to the paper by collecting cases and writing the content.

Conflicts of Interest: The authors declare no conflict of interest.

References

1. Klostermann, R. Argument for and against Planning. *Town Plan. Rev.* **1985**, *56*, 5–20. [CrossRef]
2. Randolph, J. *Environmental Land Use Planning and Management*; Island Press: Washington, DC, USA, 2004.
3. Marsh, W.M. *Landscape Planning: Environmental Applications*; John Wiley and Sons: Hoboken, NJ, USA, 2005.
4. Jaffe, A.B.; Newell, R.G.; Stavinsc, R.N. A tale of two market failures: Technology and environmental policy. *Ecol. Econ.* **2005**, *54*, 164–173. [CrossRef]
5. Andrew, B. Market failure, government failure and externalities in climate change mitigation: The case for a carbon tax. *Public Adm. Dev.* **2008**, *28*, 393–401. [CrossRef]
6. Berke, P.R.; Conroy, M.M. Are we planning for sustainable development? An evaluation of 30 comprehensive plans. *J. Am. Plan. Assoc.* **2000**, *66*, 21–33. [CrossRef]
7. Brueckner, J.K. Urban sprawl: Diagnosis and remedies. *Int. Reg. Sci. Rev.* **2000**, *23*, 160–171. [CrossRef]
8. Naess, P. Urban planning and sustainable development. *Eur. Plan. Stud.* **2001**, *9*, 503–524. [CrossRef]
9. Holmberg, J.; Karl-Henrik, R. Backcasting from non-overlapping sustainability principles—A framework for strategic planning. *Int. J. Sustain. Dev. World Ecol.* **2000**, *7*, 291–308. [CrossRef]
10. He, J.; Bao, C.-K.; Shu, T.-F.; Yun, X.-X.; Jiang, D.; Brwon, L. Framework for integration of urban planning, strategic environmental assessment and ecological planning for urban sustainability within the context of China. *Environ. Impact Assess. Rev.* **2011**, *31*, 549–560. [CrossRef]
11. Katz, P.; Scully, V.J.; Bressi, T.W. *The New Urbanism: Toward an Architecture of Community*; McGraw-Hill: New York, NY, USA, 1994; Volume 10.
12. Rittel, H.W.J.; Webber, M.W. Dilemmas in a General Theory of Planning. *Policy Sci.* **1973**, *4*, 155–169. [CrossRef]

13. Barredo, J.I.; Kasanko, M.; McCormick, N.; Lavalle, C. Modelling Dynamic Spatial Processes: Simulation of Urban Future Scenarios through Cellular Automata. *Landsc. Urban Plan.* **2003**, *64*, 145–160. [CrossRef]

14. Fall, E.H. *The Worst-Case Scenario and Discounting the Very Long Term*; HAL: Bangalore, India, 2006.

15. Weitzman, M.L. Why the far-distant future should be discounted at its lowest possible rate. *J. Environ. Econ. Manag.* **1998**, *36*, 201–208. [CrossRef]

16. Hoel, M.; Sterner, T. Discounting and relative prices. *Clim. Chang.* **2007**, *84*, 265–280. [CrossRef]

17. Brail, R.K. *Planning Support Systems for Cities and Regions*; Lincoln Institute of Land Policy: Cambridge, MA, USA, 2008.

18. Geertman, S.; Stillwell, J. *Planning Support Systems in Practice*; Springer: New York, NY, USA, 2003.

19. Brail, R.K.; Klostermann, R.E. *Planning Support Systems: Integrating Geographic Information Systems, Models, and Visualization Tools*; ESRI Press: Redlands, CA, USA, 2001.

20. Geertman, S.; Stillwell, J. *Planning Support Systems: An Introduction*; Springer: Berlin, Germany, 2009.

21. Brömmelstroet, M.T. Equip the warrior instead of manning the equipment: Land use and transport planning support in the Netherlands. *J. Transp. Land Use* **2010**, *3*, 25–41.

22. Deal, B. *Sustainable Land-Use Planning: The Integration of Process and Technology*; VDM Verlag: Saarbrücken, Germany, 2008.

23. Budthimedhee, K.; Li, J.; George, R. ePlanning: A Snapshot of the Literature on Using the World Wide Web in Urban Planning. *J. Plan. Lit.* **2002**, *17*, 227–246. [CrossRef]

24. Batty, M. *Smart Cities and Big Data: How We Can Make Cities More Resilient*; Keynote Address at the Joint AESOP-ACSP Congress: Dublin, Ireland, 2013.

25. White, R.; Engelen, G. Urban systems dynamics and cellular automata: Fractal structures between order and chaos. *Chaos Solitons Fractals* **1994**, *5*, 563–583. [CrossRef]

26. Deal, B.; Pallathucheril, V. A Use-Driven Approach to Large-Scale Urban Modelling and Planning Support. In *Planning Support Systems Best Practice and New Methods*; Springer: Dordrecht, The Netherlands, 2009; Volume 95, pp. 29–51.

27. Geertman, S.; Stillwell, J. Planning support systems: An inventory of current practice. *Comput. Environ. Urban Syst.* **2004**, *28*, 291–310. [CrossRef]

28. LEAM Lab. LEAM Portal. Available online: http://www.leam.illinois.edu/ (accessed on 22 December 2016).

29. What if? Inc. Welcome to What if?TM 2.0 2015. Available online: http://www.whatifinc.biz/ (accessed on 22 December 2016).

30. UrbanSim. Available online: http://www.urbansim.com/urbansim/ (accessed on 22 December 2016).

31. Deal, B.; Pallathucheril, V. The Land Evolution and Impact Assessment Model (LEAM): Will it play in Peoria? In Proceedings of the 8th International Conference on Computers in Urban Planning and Urban Management, Sendai, Japan, 27–29 May 2003.

32. Deal, B.; Pallathucheril, V. Simulating Regional Futures: The Land-use Evolution and impact Assessment Model (LEAM). In *Planning Support Systems for Cities and Regions*; Brail, R., Ed.; Lincoln Institute for Land Policy: Cambridge, MA, USA, 2008; pp. 61–84.

33. Pallathucheril, V.; Deal, B. Communicative Action and the Challenge of Discounting. In Proceedings of the 53rd Annual Conference of the Association of Collegiate Schools of Planning, Cincinnati, OH, USA, 1–4 November 2012.

34. Deal, B.; Kim, J.H.; Hewings, G.J.; Kim, Y.W. Complex urban systems integration: the LEAM experiences in coupling economic, land use, and transportation models in Chicago, IL. In *Employment Location in Cities and Regions*; Springer: Berlin, Germany, 2013; pp. 107–131.

35. Choi, W.; Deal, B.M. Assessing hydrological impact of potential land use change through hydrological and land use change modeling for the Kishwaukee River basin (USA). *J. Environ. Manag.* **2008**, *88*, 1119–1130. [CrossRef] [PubMed]

36. Sun, Z.; Deal, B.; Pallathucheril, V.G. The land-use evolution and impact assessment model: A comprehensive urban planning support system. *URISA J.* **2009**, *21*, 55–62.

37. Deal, B.; Schunk, D. Spatial dynamic modeling and urban land use transformation: A simulation approach to assessing the costs of urban sprawl. *Ecol. Econ.* **2004**, *51*, 79–95. [CrossRef]

38. Forester, J. *The Argumentative Turn in Policy Analysis and Planning*; Duke University Press: Durham, NC, USA, 1993.

39. Stein, S.M.; Harper, T.L. Creativity and Innovation Divergence and Convergence in Pragmatic Dialogical Planning. *J. Plan. Educ. Res.* **2012**, *32*, 5–17. [CrossRef]

40. Fischler, R. Communicative planning theory: A Foucauldian assessment. *J. Plan. Educ. Res.* **2000**, *19*, 358–368. [CrossRef]

41. Geertman, S. Participatory planning and GIS: A PSS to bridge the gap. *Environ. Plan. B Plan. Des.* **2002**, *21*, 21–35. [CrossRef]

42. Deal, B.; Pallathucheril, V. Developing and Using Scenarios. In *Engaging the Future: Forecasts, Scenarios, Plans, and Projects*; Hopkins, L.D., Zapata, M.A., Eds.; Lincoln Institute for Land Policy: Cambridge, MA, USA, 2007; pp. 222–242.

43. Deal, B.; Jenicek, E.; Goran, W.; Myers, N.; Fittipaldi, J. Strategic Sustainability Assessment. *Geol. Soc. Am. Spec. Pap.* **2011**, *482*, 41–57.

44. Deal, B.; Pallathucheril, G.; Pan, H. Planning Support Systems in the Rear View Mirror. In Proceedings of the 49th Annual Conference of the Association of Collegiate Schools of Planning, Philadelphia, PA, USA, 30 October–2 November 2014.

45. McHenry County Regional Planning Commission McHenry County 2030 Comprehensive Plan. 2010. Available online: https://www.co.mchenry.il.us/county-government/departments-j-z/planning-development/2030-beyond/project-introduction (accessed on 13 December 2016).

46. Coughlin, R.E.; Pease, J.R.; Steiner, F.; Papazian, L.; Pressley, J.A.; Sussman, A.; Leach, J.C. The status of state and local LESA programs. *J. Soil Water Conserv.* **1994**, *49*, 6–13.

47. McHenry County Farm Bureau Update on the Regional Planning Commission's 2030 Plan, by Jim McNutt. 2008. Available online: http://www.mchenrycfb.org/drupal/news/update-regional-planning-commission%E2%80%99s-2030-plan-jim-mcnutt (accessed on 13 December 2016).

48. Senbel, M.; Church, S.P. Design empowerment: The limits of accessible visualization media in neighborhood densification. *J. Plan. Educ. Res.* **2011**, *31*, 423–437. [CrossRef]

4

Exploring a Novel Agricultural Subsidy Model with Sustainable Development: A Chinese Agribusiness in Liaoning Province

Li Cui [1], Kuo-Jui Wu [1,*] and Ming-Lang Tseng [2]

[1]   School of Business, Dalian University of Technology, Panjin 124221, China; cuili@dlut.edu.cn
[2]   Department of Business Administration, Lunghwa University of Science and Technology, Taoyuan 33306, Taiwan; tsengminglang@gmail.com
*   Correspondence: wukuojui@dlut.edu.cn

Academic Editors: Alessio Cavicchi and Cristina Santini

Abstract: To improve the incomes of farmers in China, the Chinese government is paying increased attention to the reform of its agricultural subsidy policy. However, the effectiveness of the subsidy remains insufficient and thus fails to encourage farmers to cultivate their land and develop sustainability. Thus, there is a need for a novel model that will improve the effectiveness and efficiency of subsidies. The proposed novel agricultural subsidy model comprises four major actors: farmers, specialized farmers' cooperatives, agribusiness and government. Furthermore, the subsidy in this novel model would no longer go directly to farmers but to the agribusiness. To develop the model, the empirical data for this study are obtained from a Chinese agribusiness in Liaoning Province that was selected as a benchmark. With this novel model, farmers receive triple rebates: the price received when the rice is initially sold; a share of the profits of the specialized farmers' cooperatives; and a share of the profits of the agribusiness. Accordingly, exploring the optimal subsidy rate for agribusinesses is the critical task of this study, and the results demonstrate that agribusinesses must use the government subsidy policy as the basis for a dynamic subsidy model that ensures the income of farmers and encourages sustainable development.

Keywords: novel agricultural subsidy model; sustainable development; Chinese agribusiness; Liaoning

---

1. Introduction

The Chinese government's focus on industrialization policy has seriously impacted agriculture. By failing to acknowledge environmental issues, the promotion of agricultural technology leads to the destruction of land and sustainable resources. Moreover, due to rapid urbanization, increasing numbers of farmers are giving up farming to work in the city, leaving large expanses of land to become barren. In fact, the records indicate that there are almost a hundred million acres of fallow fields in southern China. In the northern part of China, approximately 40% of the land is infertile. This situation restricts the development of sustainability. Although government policies have been established to promote agricultural development and increase farmers' income, the results have fallen short of expectations [1]. Beginning in 2004, the Chinese government implemented the three-item subsidies policy (which includes a crop seed subsidy, subsidies to cultivating farmers and general subsidies for agricultural production supplies), which is herein referred to as the traditional subsidy model [2,3]. According to this policy, farmers must sign a contract with the government, under which they receive an annual government subsidy of ¥90 (equal to 15 USD), regardless of whether they cultivate their land. However, farmers who cultivate land that they do not own are not entitled to this subsidy. Hence, this model contains inequities and does not motivate land cultivation.

To promote sustainable development, interventions have been introduced in the agricultural sector by certain global governments that are focused on promoting their respective agricultural sectors [4]. Based on the experiences of the United States, agricultural subsidies are considered a transformative process. Between 1998 and 2004, U.S. farmers received $17 billion annually on average, which is more than the grants for federal temporary assistance for needy families, which averaged $13.6 billion, and the federal aid to postsecondary students, which averaged $16.1 billion [5]. In the recently proposed Thirteen Five Project (2016–2020) (the Thirteen Five Project is a Chinese government project that proposes to improve people's lives by providing stability in a transitional economy. This project includes developments in innovation, coordination, sustainability, opening up and sharing), the Chinese government significantly modified its agricultural subsidy policy to implement the three-item subsidies policy, signaling the support of large-scale producers and operators. Agribusinesses have attempted to adopt the new national policy and thus to seek sustainable development. The novel agricultural subsidy model proposed in this study requires the large-scale development of competitive modern agriculture, which means that agribusinesses must possess sufficient size and capital [6]. In addition, environmental pressures highly influence agribusinesses to engage in reactive internal practices as they develop agricultural sustainability and creativity [4,7].

The key to successfully developing sustainable agriculture is to convince farmers to surrender their direct subsidies from the government and cooperate with agribusinesses. For the farmers, the choice depends on their income, i.e., if the agribusiness can offer a better income than the government subsidy, farmers are willing to collaborate. Many studies have discussed the effects of agricultural subsidies. Chen et al. [8] examined the effects of a government subsidy (including area and price subsidies) on the total cultivated area of crops based on the assumption that a subsidy has both stimulating and inhibiting effects. Wang and Yang [9] examined the relationship between agricultural subsidies and moderate-scale land operations. Furthermore, Zhao et al. [10] argued that the government must distribute a portion of the subsidy to consumers. However, these studies focused on the government and did not consider agribusinesses as the subsidizer subject. When considering agribusinesses as subsidizers, a dilemma is generated whereby the agribusiness must provide an attractive subsidy rate for farmers while simultaneously ensuring sufficient profits to maintain its business operations. It is critical that agribusinesses find an optimal subsidy rate based on the triple rebates (the triple rebate includes three types of revenue to farmers under the proposed model. First, farmers can obtain the initial sale price from specialized farmers' cooperatives; second, farmers can obtain a percentage of the profits from the sales of the specialized farmers' cooperatives; and third, agribusinesses provide a percentage of their total sales to subsidize farmers) to farmers in this model and that they create a dynamic system to comply with government subsidy policies.

Therefore, the objective of this study is to develop a novel agricultural subsidy model to assist agribusinesses as they explore the optimal subsidy rate. This subsidy rate must allow for an increase in the income of farmers and promote the sustainable development of agriculture. To provide a precise guideline for agribusinesses, this study compares the traditional subsidy model (farmers receive the subsidies directly) with the novel model (the subsidies go to the agribusinesses instead of the farmers). The empirical data are obtained from a Chinese agribusiness in the city of Panjin, Liaoning Province that is used as a benchmark. The findings demonstrate that an agribusiness subsidy must use the government subsidy as a base and then build a dynamic subsidy model to respond to adjustments in the government subsidy policy. The remaining part of this study consists of five sections. A comprehensive literature review is provided in the next section. Notations and the proposed novel agricultural subsidy model are presented in Section 3. Section 4 presents the background of the case and the empirical results, and Section 5 discusses the implications. The conclusion, research limitations and suggestions for future studies are discussed in the final section.

2. Literature Review

This section provides a comprehensive review of the literature on agricultural subsidies and sustainable development.

2.1. Agricultural Subsidies

As a result of food shortages and climate change, agriculture is receiving global attention for its role in ensuring the survival of humanity. Governments provide subsidies to encourage farmers to increase productivity and improve their cultivation techniques [11]. However, it is challenging to find a balance between a reasonable agricultural subsidy and farmers' income; indeed, it is a problem analogous to the double-edged sword. Hence, starting in the twentieth century, the European Union focused on exporting grain and agricultural products to reduce the pressure for subsidies. Moreover, several relevant polices underwent significant reforms before being included on the European Union agenda for 2000. This agenda emphasized that the developed policy must ensure that the productivity of the food supply is maintained; the value of agricultural products and competition among them are enhanced; the rural environment is improved; the service industry and agricultural businesses are supported; employment in the new and growing rural economic sector is expanded; reasonable income for farmers is protected; intervention prices from governments are reduced; and compensation is increased from 54 to 63 euros per ton of grain [12].

Regarding the relationship between subsidies and farm performance, Banse et al. [13] revealed that the level of producer subsidies has a negative correlation with the cost of domestic resources in Hungary. Because subsidization ensures a certain return to farmers, they may reduce their cultivation efforts. In the U.S., the aim of agricultural subsidies is to protect farmers from the risks of the industry by ensuring a minimum level of economic growth and stability. Thus, U.S. agricultural subsidies include direct and countercyclical payments [14]. Many other countries have applied the target zone policy to stabilize the market and protect farmers. Chen et al. [15] adopted this concept to investigate the relationship between product purchasing and price subsidy strategies to assess the effect of target zones with different operating strategies. In addition, Turkey separated its subsidy policy into two types of instruments: output-based instruments in the form of deficiency payments for specific crops to increase farm revenue and input support instruments that include subsidies for fuel oil, fertilizer and soil analysis [16].

The Chinese government has long been concerned with its agricultural sector [17]. For example, several practices have been implemented by the government to reduce or eliminate the tax burden on farmers [18]. After joining the World Trade Organization in 2001, China began searching for the optimal process to subsidize farmers who might directly suffer from foreign competition [19]. The Ministry of Finance realized that impacts on the agricultural sector have changed dramatically in terms of direction and quantity, as has the nature of payments to farmers. To address these issues, the government established several policies intended to increase the benefits to the farmer, reversing its centuries-old practice of taxing agriculture [20,21]. In recent years, the government has raised subsidies for the agricultural sector several times to reduce negative impacts on farmers [22]. Accordingly, the records indicate that the largest portion of the subsidies are paid directly to the farmers. There are four types of agricultural subsidies in China: subsidies for grain, inputs, quality seeds and agricultural machinery. The subsidies in grain and inputs account for more than 70% of total subsidies [23].

An increasing number of researchers are focusing on the topic of agricultural subsidies. Previous studies have found significant results from the establishment of new policies and have provided new direction to governments. Gale et al. [21] found that subsidies are spread thinly over a substantial agricultural population and thus have had only a minor impact on rural incomes. Huang et al. [24] and Huang et al. [23] administered a household-level survey to explore the influence of China's subsidy program on household behavior and found that although agricultural subsidies per farm are low, the subsidies per unit of cultivated area and the total budget amount are high and that all producers received subsidies. Liu [25] conducted an empirical study of a direct food subsidy policy in Shandong Province to analyze the efficiency of policy implementation in different regions. Yi et al. [26] argued that, in general, the grain subsidy policy contributes to improvements in farm households' grain planting areas in liquidity-constrained households. In addition, Ito [27] applied a stochastic frontier output distance function to investigate the rationality of Chinese farmers' crop selection and found strong evidence that the Chinese policy of grain self-sufficiency exemplifies the technical and allocative efficiencies of agricultural production.

Grain production in China is confronting tremendous challenges, including increases in demand, resource constraints and rural labor shortages. These challenges create barriers to the reform and development of the rural economy, for example, by slowing the growth of farmers' income, increasing the need for migrant rural laborers and increasing the outflow of rural resources [28]. Although increased subsidies have a positive and direct effect on farm households by increasing the income of farmers, it may also result in agricultural intensification and thus challenge agricultural sustainability and create or enhance food supply problems [29,30]. In addition, because the use of agricultural land is affected by government policies, its use is supervised to develop regional sustainability [31]. Therefore, there is an essential need to seek sustainable development in the agricultural sector to overcome these challenges. The Chinese government is aggressively reforming the agricultural subsidy policy to promote sustainable development, which entails the simultaneous consideration of environmental issues, land governance, food security and productivity enhancement.

2.2. Sustainable Development

Sustainable development is receiving increased attention in the Chinese agricultural sector. Far-reaching programs have been launched by businesses, governments, social reformers and environmental activists, all of whom have their own interpretation of sustainability and their own ideas about how it should be developed [32]. There has been much debate about the sustainable development of agriculture in the context of policy development due to the diverse characters and characteristics of the agricultural sector, which make it difficult to achieve a consensus in advance of implementation [33,34]. Moreover, previous studies have doubted the feasibility of sustainable development in the Chinese agricultural sector. For example, Brown [35] raised the question, "Who will feed China?" Xu et al. [36] emphasized that the Chinese grain supply might be further undermined by constraints on land and water resources and in the long term by environmental degradation. For these reasons, the Chinese government must consider these concerns as it searches for a way to develop sustainable agriculture.

Sustainable agriculture differs from the traditional approach [37] in that it not only offers sufficient food for an increasing world population while simultaneously preventing harm or risk to the environment but also ensures economic returns for the farmers [38]. Because farmers play an important role in agriculture, enhancing their income is a first step in the development of sustainable agriculture. Thus, the Chinese government proposed a series of policy reforms to the agricultural subsidy in the Thirteen Five Project. Prior studies that investigated the relationship between enhancing the farmers' income and the optimal subsidy policy have offered significant evidence in support of policy reforms. Pan et al. [31] used simulations in cropping patterns, including acreage, cropping locations and management-related environmental impacts, under various policy scenarios for Quzhou County, China. Although the studied subsidy policy ensures farmers' income, it does not encourage water conservation in sustainable crop production and thus may lead to the abandonment of land due to water shortages. Qian et al. [39] indicated that agricultural subsidy policies contribute to increases in the market price of grain as well as to increases in farmers' income.

Overall, the government's agricultural policy is shifting from one of enhancing traditional agricultural productivity to a focus on developing sustainable agriculture. The Thirteen Five Project emphasizes that agribusinesses must advance the development of sustainability to increase farmers' income. However, the majority of current studies only analyze the strengths and weaknesses from the perspective of government policy development; few studies have adopted the agribusiness perspective of policy implementation. If agribusinesses could offer subsidies when they collect grain from farmers, the motivation for farmers to cooperate with them would be strengthened. With such cooperation, agribusinesses would not only receive subsidies and support from the government due to their enhanced performance but also stand with the farmers in the search for benefits and well-being. Together, these factors play an important role in the successful development of sustainable agriculture.

3. A Novel Agricultural Subsidy Model

To explore the relationship between government subsidies and farmer income, the appropriate mathematical model must be applied. Thus, the relevant notations, income function and proposed analytical procedure are discussed in this section.

3.1. Notations

This study develops a mathematical model to support optimal decision making for agribusinesses. Before formulating the mathematical model, the relevant notations are provided:

P_m: the market price when the farmer sells rice directly to the market

c_m: the unit cost of cultivation without any control by agribusiness

G: the government subsidy per unit of rice

q: the annual productivity of rice

P_h: the acquisition price when the farmer joins the specialized farmers' cooperative; normally, $P_h > P_m$

c_n: the cultivation cost under the novel agricultural model

P_r: the acquisition price based on the price of rice purchased by the agribusiness from the specialized farmers' cooperative (a specialized farmers' cooperative is a cooperative economic organization developed to resolve conflicts between the farmers' small production and the large market.)

P_s: the selling price of rice from the agribusinesses to customers

ω_h: the percentage of the net profit that specialized farmers' cooperatives return to farmers, $\omega_h \in (0,1)$

ω_r: the percentage of net profit that the agribusinesses return to farmers, $\omega_r \in (0,1)$

π_m: the profit of farmers under the traditional subsidy model

π_h: the profit of specialized farmers' cooperatives

π_r: the profit of the agribusiness

π_n: the profit of farmers under the novel agricultural model

3.2. Farmers' Income Function

In the traditional model, the government subsidy is given directly to the farmer. Accordingly, the profit function of the farmer under the traditional model is presented below:

$$\pi_m = (P_m - c_m + G)q. \tag{1}$$

In the novel agricultural model, the government gives the subsidy to the agribusiness. Farmers' income is increased by agribusiness sales. The farmers' profit function under this novel model is as follows:

$$\pi_n = (P_h - c_n)q + \omega_h \pi_h + \omega_r \pi_r. \tag{2}$$

The profits of specialized farmers' cooperatives are defined as follows:

$$\pi_h = (P_r - P_h)q - \omega_h \pi_h. \tag{3}$$

The equation π_h is rewritten as follows:

$$\pi_h = \frac{(P_r - P_h)q}{1 + \omega_h}. \tag{4}$$

The profit of the agribusiness is obtained using the following equation:

$$\pi_r = (P_s - P_r + G)q - \omega_r \pi_r. \tag{5}$$

The equation for π_r is as follows:

$$\pi_r = \frac{(P_s - P_r + G)q}{1 + \omega_r}. \tag{6}$$

Subsequently, adopting the profits of specialized farmers' cooperatives and of the agribusiness associated with the farmers' profit function, the profit of the farmers under the novel agricultural subsidy model is obtained using the following equation:

$$\pi_n = (P_h - c_n)q + \frac{\omega_h(P_r - P_h)q}{1 + \omega_h} + \frac{\omega_r(P_s - P_r + G)q}{1 + \omega_r}, \tag{7}$$

where $P_m < P_h < P_r < P_s$.

3.3. The Proposed Analytical Procedure for Selecting the Optimal Subsidy Rate

Agribusinesses can help to decide how to subsidize farmers' income under different subsidy models. The proposed analytical procedure includes five steps:

(1) Identify the major participants of the different models: the major participants in the traditional subsidy model are farmers and the government, and the distribution of subsidies depends on the area of land under contract with the government. Thus, farmers receive compensation directly from the government. In the novel agricultural subsidy model, in addition to the farmers and the government, the agribusiness and specialized farmers' cooperatives are also participants. In this novel model, the farmers no longer sign a contract with the government; instead, the specialized farmers' cooperatives sign the contract. The cooperatives must associate with the agribusiness to develop agriculture and promote sales. The government gives the subsidies to the agribusiness to maximize performance.

(2) Confirm the agricultural product price in each node: agribusiness research facilitates the acquisition of the market price of a product. For example, the market price P_m, direct purchase price P_h, and indirect purchase price P_r of rice can be obtained from the rice subsidy. A detailed discussion about the collection of data on the price of rice is provided in Section 4.2.

(3) Gather annual productivity and government subsidy data: the annual productivity for rice is obtained from internal agribusiness information. The government subsidy is stated in the relevant policy document.

(4) Identify the farmers' income function: the farmers' income function under different models is obtained through the different subsidy models and survey responses.

(5) Determine the optimal subsidy rate for the agribusiness: the appropriate subsidy for farmers is determined based on farmers' income and requires a comparison between the traditional model and the novel model to explore the effects of the government subsidy, the agribusiness subsidy and the farmers' income. The optimal subsidiary rate for the agribusiness is selected by examining these influences. This ensures that the rate offered provides the optimal income for farmers.

4. Empirical Study

This section discusses the background of the empirical case, the novel agricultural subsidy model, the data collection methodology and empirical results. The case background introduces the Liaoning agribusiness. Data gathering involves a unique cooperative model between farmers, specialized farmers' cooperatives and the agribusiness. Finally, the empirical results provide evidence allowing the agribusiness to select the optimal subsidy rate based on quantitative analysis.

4.1. Case Background

The Liaoning Jin She Yu Nong Supply and Marketing Group (LJSYN) (The LJSYN can also be considered a system of "Gong Xiao She" under the planned economy. This system has decreased due to the market economy transition but still possesses certain infrastructure and government resources in the north of China. For this reason, the LJSYN can obtain government support and launch this novel model in Panjin City.) is an innovative and aggressive agribusiness in Liaoning Province, China (as shown in Figure 1). They adopted a new agricultural subsidy model to enhance the income of farmers and promote sustainable development throughout the agricultural supply chain, particularly focusing on rural areas. The LJSYN, located in Panjing City, was established in 2010 with 11.9 hundred million RMB in capital. The group has ten independent subsidiary companies, seven controlling specialized farmers' cooperatives, five industry associations, four distribution centers, seven agricultural bases and three hundred and thirty-one supermarkets. The principle of the LJSYN is to enhance the quality of life for farmers and to promote its supply and marketing business. In addition, the LJSYN is devoted to the integration of resources along the supply chain to create a unified structure and form an innovative agricultural eco-system.

Figure 1. Liaoning Jin She Yu Nong Supply and Marketing Group.

Recently, the LJSYN completed its business function by building an innovative agricultural production system, a direct sales system with production information and a distribution system. Specialized farmers' cooperatives play an important role in monitoring and guiding farmers, providing a basis for sustainable development. Cooperatives not only mean that farmers no longer cultivate alone, but they also create a known representative brand of local agricultural products, as presented in Figure 2. Although the LJSYN adheres to the rules of the market economy to strengthen the circulation of agricultural products, it continues to have difficulty finding an optimal subsidy rate for enhancing farmers' income. However, the proposed novel agricultural subsidy model of the LJSYN encourages adherence to China's policy reforms and to sustainable development.

Figure 2. Famous rice brand in Liaoning—King of Fong Jing (Yan Fong-47).

4.2. Data Collection

This study adopts the LJSYN as a benchmark agribusiness and uses Yan Fong-47 rice to examine the income of farmers under the novel subsidy model. Yan Fong-47 is a middle–early maturing cultivated variety of round-grained rice. It is characterized by high productivity and high quality and has been certified at the second level according to national quality standards. The cultivating period of this rice is 157.2 days, and the cultivating area is distributed throughout southern Liaoning, Beijing and Tianjin. Figure 3 displays the agricultural subsidy model of Yan Fong-47 rice in Liaoning Province.

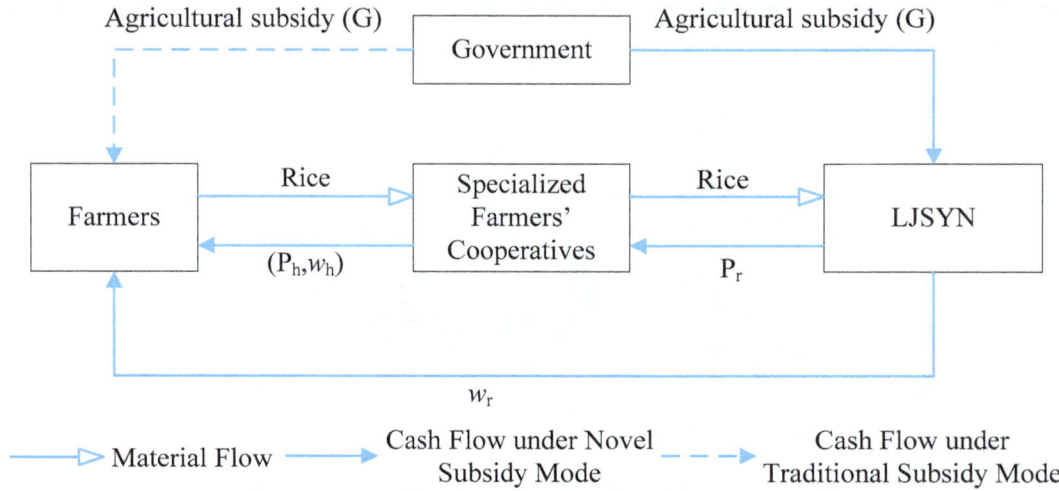

Figure 3. The novel subsidy model of the LJSYN.

In Figure 3, the dotted line indicates that the government directly gives the subsidies to the farmers under the traditional model. Conversely, the solid line represents the novel subsidy model, under which the government gives the subsidy to the LJSYN rather than directly to the farmers. Farmers must sell the rice to the specialized farmers' cooperatives in accordance with signed contracts. The cooperatives use price P_h to purchase the rice. They will then return a portion of the profit ω_h to farmers at the end of the year. The LJSYN purchases the rice at price P_r from the cooperatives, and at the end of year, a percentage of the profit ω_r of the LJSYN is distributed to the farmers. This novel model results in triple rebates for the farmers, which greatly enhances their income and motivates them to be productive.

The data regarding Yan Fong-47 are as follows (2015 data):

(1)   If farmers directly sell the rice in the market, the price is ¥1.62 per half-kilogram.

(2)   Annual production is 1300 half-kilograms.

(3)	The cost to farmers to cultivate the rice without any control by the LJSYN is ¥700 per acre; if each acre can generate 1300 half-kilograms, the cost of cultivation is ¥0.54 per half-kilogram.

(4)	After farmers join the specialized farmers' cooperatives, the cooperatives require ¥0.05 to acquire the rice from the farmers, in addition to the purchase price of ¥1.67 per half-kilogram.

(5)	The farmers' cultivation cost is ¥300 per acre and ¥0.23 per half-kilogram under the novel model.

(6)	The LJSYN's selling price for rice is ¥3 per half-kilogram.

(7)	The LJSYN acquires the rice from the specialized farmers' cooperatives. The acquisition price is higher than the market price of ¥0.05. Thus, *the market price of rice = current market price × rice milling yield = U'3 × 0.7 = U'2.1*. Accordingly, the acquisition price is ¥2.15 per half-kilogram.

(8)	The specialized farmers' cooperatives offer a 25% net profit to farmers.

4.3. Empirical Results

By incorporating the above data into Equations (1) and (7) and creating the diagram using MATLAB R2012a (MathWorks, Natick, MA, USA, 2012), the relationships between the subsidy rate of the LJSYN, the government subsidy and farmers' income under the traditional model become apparent. Within this model, the government distributes the subsidies directly to the farmers, and the agribusiness has no power to intervene or enhance farmers' income through subsidies. The interactions between the agribusiness and farmers are limited. The agribusiness possesses substantial bargaining power when collecting the rice from the farmers, but under the traditional model, the profits of the agribusiness are not shared with the farmers. In other words, the farmers' only income is the government subsidy. Accordingly, under the traditional model, there exists a highly positive relationship between the government subsidy per half-kilogram and the income of the farmer, as shown in Figure 4.

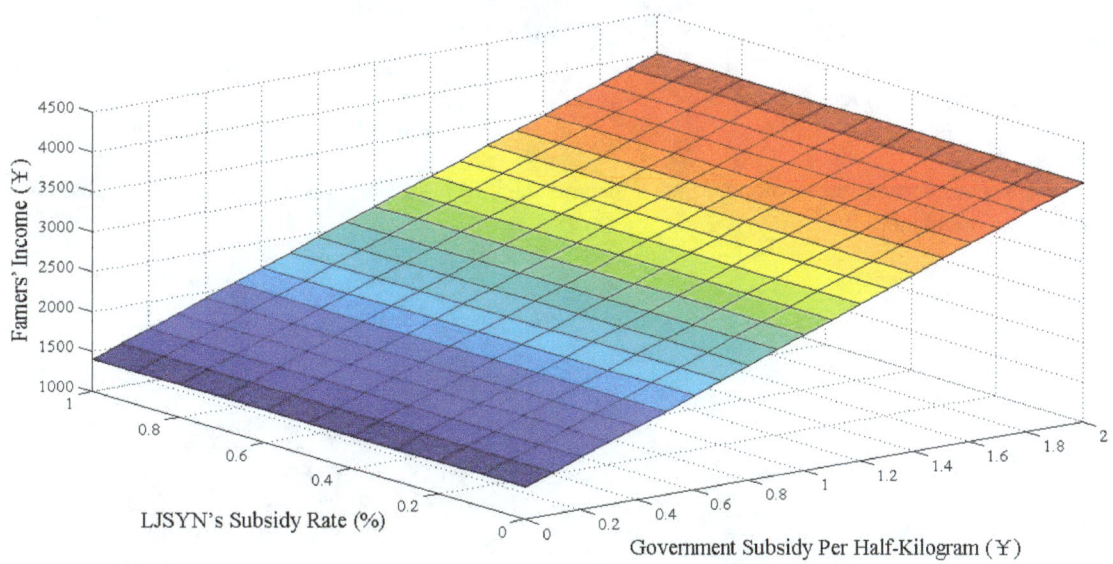

Figure 4. The effect diagram of farmers' income, the LJSYN subsidy rate and government subsidies under the traditional model.

Figure 5 presents the effects of farmers' income, the LJSYN subsidy rate and government subsidies in the novel model. It further indicates that if farmers want to increase their income, they must rely on the subsidy rate of the LJSYN rather than on the government subsidy. Simultaneously, the agribusiness must establish a subsidy rate to reimburse farmers based on the government subsidy. Under the novel model, farmers' income includes the government subsidy and a share of the agribusiness profits. Furthermore, farmers' income is correlated with the LJSYN subsidy rate as the government subsidy increases. That is, when the government subsidy is maintained at a steady level, the farmers' income is directly proportionate to the agribusiness subsidy rate. When the government subsidy rate increases to a higher level, the farmers' income increases aggressively as the agribusiness subsidy rate also increases. If the government subsidy rate does not increase, the farmers' income is slow to increase.

To clarify the major difference between the traditional model and the novel model, additional comparative analysis is conducted. Figure 6 combines Figures 4 and 5 and represents the integration of the two models. Based on this integration, it is evident that the LJSYN subsidy rate must adjust in response to government subsidy changes to ensure that the farmers' income increases accordingly. In Figure 6, to the left side of the intersection line, the novel model exhibits better performance than the traditional model in enhancing farmers' income. On the right side of the intersection line, the farmers' income in the novel model is less than that in the traditional model. This suggests that agribusiness must establish a break-even point to simultaneously maintain profits and increase farmers' income.

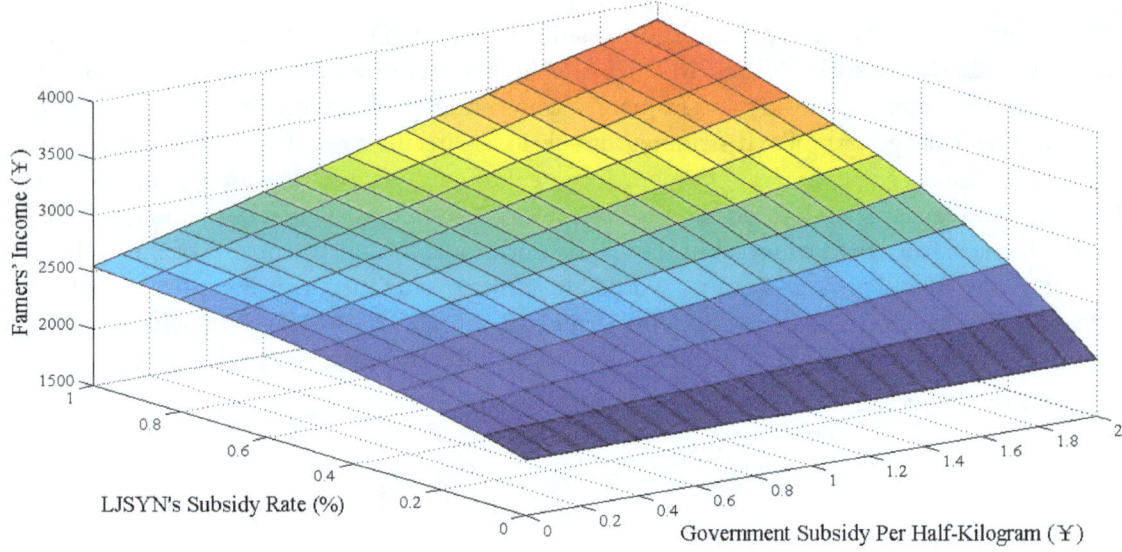

Figure 5. The effect diagram among farmers' income, the LJSYN subsidy rate and government subsidies under the novel model.

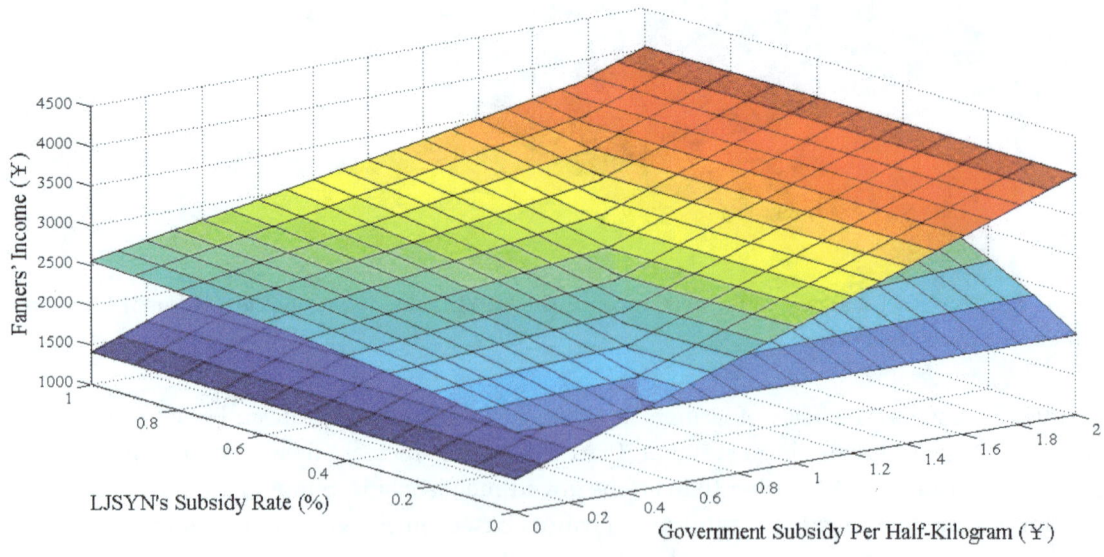

Figure 6. Comparison analysis between the traditional and novel models.

To confirm the optimal subsidy rate of the LJSYN, this study adopts ω_r = 0.1, 0.3, 0.5, 0.7 and 0.9 and compares these values with those of the traditional government subsidy to identify a better alternative for increasing farmers' income. Figure 7 demonstrates that the LJSYN must adjust their subsidy rate consistently with increases in the government subsidy. For example, when the government subsidy is ¥0.58 per half-kilogram, the subsidy rate of the LJSYN cannot be more than 10% below that rate. In other words, the LJSYN must use 10% of its net profit as a rebate for farmers to ensure that farmers' income is better than it was under the traditional subsidy model. When the government subsidy reaches ¥0.82 per half-kilogram, the LJSYN's optimal subsidy rate should be adjusted to 30%. These findings confirm that the novel model provides farmers with better benefits than does the traditional government subsidy. Specifically, the novel model generates triple rebates to farmers.

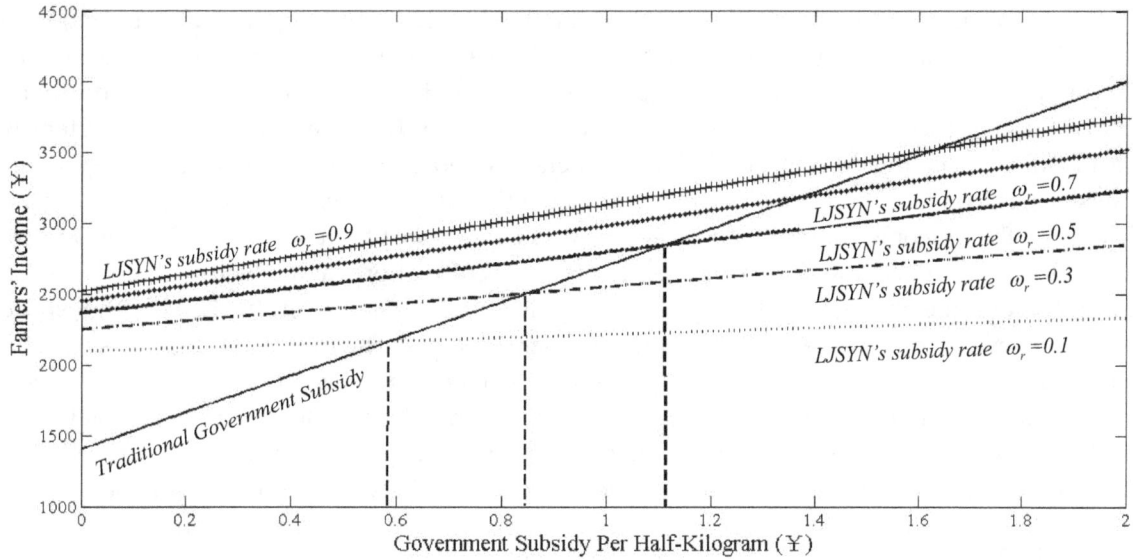

Figure 7. Comparison of the traditional government subsidy with different LJSYN subsidy rates.

5. Implications

Based on the analytical results, the LJSYN can serve as a benchmark agribusiness. Different types of subsidies generate diverse effects for farmers and impact the profits of the agribusiness in different ways [4,17,36]. The proposed novel model presents a type of subsidy intended to improve farmers' income and to enable agribusinesses to launch a sustainable development program. In this model, specialized farmers' cooperatives play a mediating role by monitoring and educating farmers in the cultivation of rice. For example, cooperatives provide non-toxic pesticides at lower costs to their member farmers to improve rice productivity, rice milling yield, and rice quality in the face of land exhaustion and water pollution. Thus, a unique ecological cultivation approach is required in the city of Panjin in Liaoning. Farmers simultaneously cultivate rice and feed river crabs on the same cropland, which means that both rice and river crabs can become popular products nationwide. As a result, the agribusiness can implement higher prices when selling these products in the market, and customers are willing to pay these higher prices to purchase green products. Consequently, the agribusiness can then provide better rebates to the farmers and improve their standard of living. This type of positive cooperative loop generates profits for the agribusiness, cooperatives and farmers; once the loop is sufficiently mature and self-sustainable, government subsidies may no longer be needed.

For agribusinesses to improve performance and enhance sales, farmers must be motivated to be a part of that change. This motivation requires a novel subsidy model that offers a better rebate than the direct government subsidy in order to strengthen farmers' initiative. Although previous studies have discussed the strengths and weaknesses of government subsidies [37–39], there is a lack of research from the agribusiness perspective. Thus, this is the first study to propose a novel subsidy model to replace the traditional government subsidy model and enhance farmers' income. The novel subsidy model transfers the government subsidy from the farmers to the agribusiness to maximize performance and effects. This is also the first study to model the agricultural subsidy based on China's agricultural subsidy policy. A comparison of the traditional model with the novel model reveals that the two models affect farmers' income differently. In the traditional model, the government issues the subsidy directly to the farmers and thus there is a high positive correlation between farmers' income and the government subsidy at the per half-kilogram level. In the novel model, increases in farmers' income depend on their receipt of the agribusiness subsidy. As the amount of the government subsidy increases, the agribusiness must respond to this increase by adjusting farmers' income. Accordingly, agribusinesses that adopt the novel model must be familiar with the government subsidy policy to promote a dynamic strategy that motivates farmers.

The government must recognize and understand the difference between the traditional model and the novel model because the government subsidy could be eliminated when the novel model matures. In the initial phase of the novel model, the government can assist the agribusiness in the performance of its functions and the implementation of the rebate system, because only a mature system will attract loyal farmers. In addition, specialized farmers' cooperatives must share information with the agribusinesses to enable these businesses to understand the difficulties encountered by farmers during cultivation. To overcome this gap and achieve information transparency, the LJSYN is establishing its own electronic information platform [40] to record all cultivation data, which will then be shared among all group members. This study provides significant insight into agribusiness with respect to sustainable development while also considering farmer subsidies. The role of agribusinesses can be to create value for the customer by maximizing sales and then returning this profit to the farmer. This will encourage farmers to improve their cultivation skills, which will promote sustainable development. In addition, the government should consider this proposed novel model to improve the current subsidy policy based on environmental, social and economic considerations (the triple bottom line).

6. Conclusions

The Chinese government has always been concerned about its agricultural sector and people's livelihoods [41,42]. In the traditional subsidy policy, farmers only needed to sign a contract with the government. Thus, regardless of whether they cultivate the land, farmers can still receive a per acreage subsidy from the government. However, farmers who cultivate rice on land that they do not own cannot receive this subsidy. Although the traditional subsidy model attempts to solve the issue of farmers' income, it lacks the ability to encourage farmers to aggressively cultivate rice. With the external environment rapidly changing, this traditional model is unable to satisfy the needs of farmers and provide a fair subsidy. The new subsidy model aims to support large-scale producers and operators, including large farmers, family farms, specialized farmers' cooperatives, and agricultural social service organizations. This model embodies the "whoever can produce more rice will receive priority subsidy support" concept. In response to the national policy, agribusiness must launch this new model to motivate farmers to adopt sustainable cultivation and to improve farmers' standard of living.

This study assists agribusiness by exploring the optimal subsidy rate to stimulate farmers' initiative, increase their income and encourage sustainable development. The proposed novel model allows for triple rebates. First, the specialized farmers' cooperatives purchase the rice from farmers and pay a one-time price to the farmers; second, a percentage of the cooperatives' profits is rebated to farmers at the end of the year; and third, a percentage of the annual profits of the agribusiness is rebated to the farmers. The purpose of this novel model is to increase the income of farmers. As the government distributes the subsidy to the agribusiness, it confirms the performance of the agribusiness, which forms a positive loop that plays an important role in developing sustainable agriculture. Government subsidies should support this loop until its maturation, at which point the government can gradually terminate its support until the agribusiness can sustain the entire loop independently.

The findings of this study not only indicate that the novel model has different effects on farmers' income than the traditional subsidy model but also contribute to the development of sustainability by agribusinesses under the triple bottom line considerations. The economic consideration is that farmers should be able to obtain a higher income than that obtained under the traditional subsidy model; the environmental consideration involves the transfer of knowledge regarding sustainable cultivation from specialized farmers' cooperatives to farmers; and finally, farmers become members of the agribusiness, which allows them to afford good-quality necessities for living. With respect to the traditional subsidy model, there is a positive correlation between farmers' income and government subsidies. With respect to the novel model, farmers' income relies primarily on the subsidy from the agribusiness. Hence, the agribusiness must establish a dynamic subsidy model that permits immediate adjustments to the subsidy rate as government policy changes. When farmers engage in sustainable development, they can obtain the necessary support from specialized farmers' cooperatives and the agribusiness to improve cultivation techniques and reduce negative environmental impacts; the traditional subsidy model is unable to accomplish these benefits. Based on the significant findings of this study, the government must consider this novel subsidy model to support the development of agribusiness, and subsequently, agribusiness can use government resources to create a nationwide brand that will enhance farmer well-being.

There are several limitations regarding the content of this study, and several future studies are proposed to address these limitations. Although farmers are considered a low-income group in the Chinese context, certain farmers have their own opinions about [43,44] and reactions to government subsidy policy. This study does not examine that issue but rather explores a novel model for enhancing farmers' income. Accordingly, future studies can examine the differences between farmers when determining the appropriate subsidy type to meet their needs. In the proposed novel model, this study only considers triple rebates. Once additional organizations join the current loop, the subsidy model must be further clarified. Because of these multiple rebates, future studies must develop an optimal model that considers all roles, including those of the agribusiness, specialized farmers' cooperatives, farmers and the government, to maximize subsidy effectiveness. In addition, this study adopts Yan Fong-47 rice as an example to develop the decision-making model for agribusinesses and to explore the subsidy rate; future studies should include all agricultural products to obtain more comprehensive findings. Because subsidy models vary across countries, this case focuses on China, in particular on the institution of specialized farmers' cooperatives. Thus, future studies can apply the concept proposed in this study to other countries for purposes of comparison.

Acknowledgments: This study was supported by the National Social Science Funds Projects (15BGL023 and 13&ZD147), the Dalian University of Technology Fundamental Research Fund (DUT16RC(3)038) and (DUT16RC(4)72) and the National Natural Science Foundation (71671022). In addition, the authors appreciate the support from Gong Jinsheng and Gao Wei of Liaoning Jin She Yu Nong Supply and Marketing Group.

Author Contributions: Li Cui discovered this novel agricultural subsidy model from LJSYN and developed the mathematic model. Kuo-Jui Wu was in charge of collecting the data and integrated all analytical results for writing this study. Ming-Lang Tseng played an important role for structuring the concept of this study with logical and systematic thinking.

Conflicts of Interest: The authors declare no conflict of interest.

References

1. Li, J.Q. The present situation and prospect China's agriculture. *China Agric. Inform.* **2013**, *17*, 142.
2. Zhao, S.K. The farmers' political: Confusion and thoughts. *China Dev. Observ.* **2009**, *9*, 7–13.
3. Zhu, M.D.; Li, X.Y.; Cheng, G.Q. Comprehensive income support impact analysis of China's corn total factor productivity: Based on the provincial panel data of DEA—Tobit two-stage method. *Chin. Rural Econ.* **2015**, *11*, 4–14. (In Chinese)

4.   Anderson, K.; Rausser, G.; Swinnen, J. Political economy of public policies: Insights from distortions to agricultural and food markets. *J. Econ. Lit.* **2013**, *51*, 423–477. [CrossRef]

5.   Kirwan, B. The incidence of U.S. agricultural subsidies on farmland rental rates. *J. Political Econ.* **2009**, *117*, 138–164. [CrossRef]

6.   Chen, C. The agricultural subsidy policy impact on the farmers' land circulation behavior mechanism. *Econ. Res. Guide* **2014**, *22*, 36–37.

7.   Zhu, Q.; Geng, Y.; Sarkis, J. Shifting Chinese organizational responses to evolving greening pressures. *Ecol. Econ.* **2016**, *121*, 65–74. [CrossRef]

8.   Chen, Y.H.; Wan, J.Y.; Wang, C. Agricultural subsidy with capacity constraints and demand elasticity. *Agric. Econ. (Czech Repub.)* **2015**, *61*, 39–49. [CrossRef]

9.   Wang, Y.; Yang, J. Investigation and analysis of Land moderate scale operation under the perspective of agricultural subsidy—A Case of Manas in Xinjiang. *J. Shanxi Agric. Sci.* **2016**, *44*, 541–544. (In Chinese)

10. Zhao, S.; Zhu, Q.; Cui, L. A decision-making model for remanufacturers: Considering both consumers' environmental preference and the government subsidy policy. *Resour. Conserv. Recycl.* **2016**, in press. [CrossRef]

11. Kroupová, Z.; Malý, M. Analysis of agriculture subsidy policy tools-application of production function. *Politická Ekon.* **2010**, *6*, 774–794. [CrossRef]

12. Mao, Z.Y.; Gao, P. National food security and grain subsidies. Jiangxi. *Soc. Sci.* **2004**, *11*, 240–246. (In Chinese)

13. Banse, M.; Gorton, M.; Hartel, J.; Hughes, G.; Köckler, J.; Möllman, T.; Münch, W. The evolution of competitiveness in Hungarian agriculture: From transition to accession. *MOCT-MOST* **1999**, *9*, 307–318. [CrossRef]

14. Franck, C.; Grandi, S.M.; Eisenberg, M.J. Agricultural subsidies and the American obesity epidemic. *Am. J. Prev. Med.* **2013**, *45*, 327–333. [CrossRef] [PubMed]

15. Chen, L.J.; Hu, S.W.; Wang, V.; Wen, J.; Ye, C. The effects of purchasing and price subsidy policies for agricultural products under target zones. *Econ. Model.* **2014**, *43*, 439–447. [CrossRef]

16. Demirdöğen, A.; Olhan, E.; Chavas, J.P. Food vs. fiber: An analysis of agricultural support policy in Turkey. *Food Policy* **2016**, *61*, 1–8. [CrossRef]

17. Sicular, T. Agricultural planning and pricing in the post-Mao period. *China Q.* **1988**, *116*, 671–705. [CrossRef]

18. Bernstein, T.P.; Lü, X. *Taxation without Representation in Contemporary Rural China*; Cambridge University Press: Cambridge, UK, 2003.

19. Liu, Z.W.; OuYang, H.H.; Zhang, Z.X. Discussion on the grain subsidy way reforms. *Prob. Agric. Econ.* **2003**, *5*, 4–9. (In Chinese)

20. Tao, R.; Lin, J.Y.; Liu, M.; Zhang, Q. Rural taxation and government regulation in China. *Agric. Econ.* **2004**, *31*, 161–168. [CrossRef]

21. Gale, H.F.; Lohmar, B.; Tuan, F.C. *China's New Farm Subsidies*; USDA-ERS WRS-05-01; U.S. Department of Agriculture (USDA): Washington, DC, USA, 2005.

22. Chen, L. The effect of China's RMB exchange rate movement on its agricultural export: A case study of export to Japan. *China Agric. Econ. Rev.* **2011**, *3*, 26–41. [CrossRef]

23. Huang, J.; Wang, X.; Rozelle, S. The subsidization of farming households in China's agriculture. *Food Policy* **2013**, *41*, 124–132. [CrossRef]

24. Huang, J.; Wang, X.; Zhi, H.; Huang, Z.; Rozelle, S. Subsidies and distortions in China's agriculture: Evidence from producer-level data. *Aust. J. Agric. Res. Econ.* **2011**, *55*, 53–71. [CrossRef]

25. Liu, Y. Empirical analysis of China's direct Food subsidy policy based on DEA model: A Case study of direct Food subsidy policy in Shandong Province. *Asian Agric. Res.* **2014**, *6*, 23–28. (In Chinese)

26. Yi, F.; Sun, D.; Zhou, Y. Grain subsidy, liquidity constraints and food security—Impact of the grain subsidy program on the grain-sown areas in China. *Food Policy* **2015**, *50*, 114–124. [CrossRef]

27. Ito, J. Diversification of agricultural production in China: Economic rationality of Crop choice under the producer subsidy program. *Jpn. J. Rural Econ.* **2015**, *17*, 1–17. [CrossRef]

28. Chen, X. Review of China's agricultural and Rural Development: Policy changes and current issues. *China Agric. Econ. Rev.* **2009**, *1*, 121–135. [CrossRef]

29. Trade, O. *Agricultural Policy Monitoring and Evaluation 2011: OECD Countries and Emerging Economies*; Organisation for Economic Co-Operation and Development (OECD) Publishing: Paris, France, 2011.

30. Trade, O. *Agricultural Policy Monitoring and Evaluation 2013: OECD Countries and Emerging Economies*; Organisation for Economic Co-Operation and Development (OECD): Paris, France, 2013.

31. Pan, Y.; Yu, Z.; Holst, J.; Doluschitz, R. Integrated assessment of cropping patterns under different policy scenarios in Quzhou County, North China Plain. *Land Use Policy* **2014**, *40*, 131–139. [CrossRef]

32. Giddings, B.; Hopwood, B.; O'brien, G. Environment, economy and society: Fitting them together into sustainable development. *Sustain. Dev.* **2002**, *10*, 187–196. [CrossRef]

33. Lin, B.B. Resilience in agriculture through crop diversification: Adaptive management for environmental change. *BioScience* **2011**, *61*, 183–193. [CrossRef]

34. Davis, A.S.; Hill, J.D.; Chase, C.A.; Johanns, A.M.; Liebman, M. Increasing cropping system diversity balances productivity, profitability and environmental health. *PLoS ONE* **2012**, *7*, 1–8. [CrossRef] [PubMed]

35. Brown, L.R. Who will feed China. *World Watch* **1994**, *7*, 10–19.

36. Xu, Z.; Zhang, W.; Li, M. China's grain production: A decade of consecutive growth or stagnation? *Mon. Rev.* **2014**, *66*, 25–37. (In Chinese) [CrossRef]

37. Singh, J.S.; Pandey, V.C.; Singh, D.P. Efficient soil microorganisms: A new dimension for sustainable agriculture and environmental development. *Agric. Ecosyst. Environ.* **2011**, *140*, 339–353. [CrossRef]

38. Wezel, A.; Casagrande, M.; Celette, F.; Vian, J.; Ferrer, A.; Peigné, J. Agroecological practices for sustainable agriculture. A review. *Agron. Sustain. Dev.* **2014**, *34*, 1–20. [CrossRef]

39. Qian, J.; Ito, S.; Zhao, Z.; Mu, Y.; Hou, L. Impact of agricultural subsidy policies on grain prices in China. *J. Fac. Agric. Kyushu Univ.* **2015**, *60*, 273–279.

40. Zhu, Q.; Li, H.; Zhao, S.; Lun, V. Redesign of service modes for remanufactured products and its financial benefits. *Int. J. Prod. Econ.* **2016**, *171*, 231–240. [CrossRef]

41. Bermouna, S.; Li, J. China's agricultural project finance and support policies. *Eur. Food Feed Law Rev.* **2014**, *9*, 171–178.

42. Qin, T.; Gu, X.; Tian, Z.; Deng, J. Comparison of agriculture and Forestry fiscal subsidy policies in China. *J. Sustain. For.* **2015**, *34*, 683–697. [CrossRef]

43. Gold, C.S.; Kiggundu, A.; Abera, A.M.K.; Karamura, D. Diversity, distribution and farmer preference of Musa cultivars in Uganda. *Exp. Agric.* **2002**, *38*, 39–50. [CrossRef]

44. Anglaaere, L.C.N.; Cobbina, J.; Sinclair, F.L.; McDonald, M.A. The effect of land use systems on tree diversity: Farmer preference and species composition of cocoa-based agroecosystems in Ghana. *Agrofor. Syst.* **2011**, *81*, 249–265. [CrossRef]

5

Moving Low-Carbon Transportation in Xinjiang: Evidence from STIRPAT and Rigid Regression Models

Jiefang Dong [1,2], Chun Deng [2,3,*], Rongrong Li [4] and Jieyu Huang [2]

[1] State Key Laboratory of Desert and Oasis Ecology, Xinjiang Institute of Ecology and Geography, Chinese Academy of Sciences, Urumqi 830011, China; dongjiefang-2005@163.com
[2] Department of Economics and Management, Yuncheng University, Yuncheng 044000, China; nanbei1028@sina.com
[3] School of Economics & Management, Northwest University, Xi'an 710127, China
[4] School of Economic & Management, China University of Petroleum (Huadong), Qingdao 266580, China; lirr@upc.edu.cn
* Correspondence: dengchun-2005@163.com

Academic Editor: Giuseppe Ioppolo

Abstract: With the rapid economic development of the Xinjiang Uygur Autonomous Region, the area's transport sector has witnessed significant growth, which in turn has led to a large increase in carbon dioxide emissions. As such, calculating of the carbon footprint of Xinjiang's transportation sector and probing the driving factors of carbon dioxide emissions are of great significance to the region's energy conservation and environmental protection. This paper provides an account of the growth in the carbon emissions of Xinjiang's transportation sector during the period from 1989 to 2012. We also analyze the transportation sector's trends and historical evolution. Combined with the STIRPAT (Stochastic Impacts by Regression on Population, Affluence and Technology) model and ridge regression, this study further quantitatively analyzes the factors that influence the carbon emissions of Xinjiang's transportation sector. The results indicate the following: (1) the total carbon emissions and per capita carbon emissions of Xinjiang's transportation sector both continued to rise rapidly during this period; their average annual growth rates were 10.8% and 9.1%, respectively; (2) the carbon emissions of the transportation sector come mainly from the consumption of diesel and gasoline, which accounted for an average of 36.2% and 2.6% of carbon emissions, respectively; in addition, the overall carbon emission intensity of the transportation sector showed an "S"-pattern trend within the study period; (3) population density plays a dominant role in increasing carbon dioxide emissions. Population is then followed by per capita GDP and, finally, energy intensity. Cargo turnover has a more significant potential impact on and role in emission reduction than do private vehicles. This is because road freight is the primary form of transportation used across Xinjiang, and this form of transportation has low energy efficiency. These findings have important implications for future efforts to reduce the growth of transportation-based carbon dioxide emissions in Xinjiang and for any effort to construct low-carbon and sustainable environments.

Keywords: transportation; carbon emissions; STIRPAT model; ridge regression model; Xinjiang

1. Introduction

The CO_2 emissions generated by human activities constitute one of the most significant contributory factors to global warming. As pointed out by the IPCC in its fifth assessment report, of the total global greenhouse gas emissions in 2014, urban transportation accounted for 13.1%. This made urban transportation the third highest emission sector, behind only energy supply and industrial production [1]. In 2010, the petroleum consumption of the transportation sector accounted for 38.25% of China's total petroleum consumption. This substantial consumption of petroleum resulted in the continuing increase of CO_2 emissions [2,3]. Located in the northwest border area of China, Xinjiang Uygur Autonomous Region is China's largest provincial-level administrative region, and major energy supply base. Moreover, this region is home to ethnic minorities, such as Uighur and Kazaks. In addition, Xinjiang is also an important channel for economic exchanges between China and Central Asia. Since the implementation of the Western Development Strategy in 2001 and the Jumping Development Strategy in 2010 [4], Xinjiang's transportation sector has witnessed rapid and sustained development. The total output value of the transportation sector increased from USD 305.8 million in 1990 to USD 748.5 million in 2012. This represented an average annual growth rate of 14.9%. During the same period, the energy consumption of the transportation sector also experienced a rapid rise, from 10.71 million tons of standard coal to an amazing 85.54 million tons of standard coal. This translates to an average annual growth rate of 9.5% [5]. Accompanied by the substantial consumption of energy, CO_2 emissions also inevitably increased at a rapid rate. In 2013, the Chinese government put forward the One Belt, One Road Initiative [6]. By virtue of its unique geographic location, Xinjiang will undoubtedly see a rapid development of its transportation sector after the implementation of this initiative [7]. In the short term, the development of the transportation sector will inevitably give rise to even more carbon emissions. Consequently, accurate monitoring and accounting of the transportation sector's carbon emissions and a quantitative analysis of those factors that influence carbon emissions will provide important policy implications for the green and low-carbon development of transportation in Xinjiang.

With the continuous advance of economic globalization, the energy consumption of the transportation sector has received growing attention. As a key component of sustainable development, reducing the level of energy use in the transportation sector would both tackle energy security and address climate change concerns [8–19]. Researchers have analyzed the carbon emissions of the transportation sector from various perspectives. Several studies have made creditable attempts to accurately calculate transportation-related carbon emissions and build models of the influencing factors [20–37]. Chandran et al. [25] introduced a co-integration analysis and Granger causality analysis to study the influence of energy-related CO_2 emissions in the transportation sector on five Association of Southeast Asian Nations (ASEAN) countries. The results indicated that reducing the energy consumption of the transportation sector would undoubtedly reduce carbon emissions in the short term. However, in the long run, the most fundamental way to reduce carbon emissions is to improve the transportation sector's efficiency in terms of energy utilization and to optimize energy structures. Saboori et al. [22] adopted the "fully modified ordinary least square method" (FMOLS) and generalized impulse response to explore the relationships between energy consumption in the road transport sector, CO_2 emissions and the economic growth in Organization for Economic Co-operation and Development (OECD) countries. The results indicated the existence of a positive, significant, long-run and bi-directional relationship between CO_2 emissions and economic growth, road sector energy consumption and economic growth and CO_2 emissions and road sector energy consumption in all OECD countries. Moreover, in most cases, any effort on carbon emissions caused by changes in the road transport sector's energy consumption lasts longer than effects brought about due to economic growth. In addition, many scholars have also studied the CO_2 reduction potential in the transport sector at the national level [2,33,38–50]. For instance, Xu et al. [46,51] introduced the vector auto-regression model and the dynamic non-parametric additive regression model as a means to analyze the factors that influenced the CO_2 emissions of China's transportation sector. This study concluded that improving energy efficiency will reduce CO_2 emissions, but increasing the total number of private vehicles and promoting the progress of urbanization will significantly increase CO_2 emissions. Ratanavaraha et al. [29] considered five independent variables, namely (1) the size of the population, (2) gross domestic product (GDP) and the number of (3) small, (4) medium and (5) large-sized registered vehicles, and employed four different measurement techniques (log-linear regression, path analysis, time series and curve estimation) to forecast the carbon emissions coming from Thailand's transportation sector. The researchers claimed that the primary means of reducing carbon emissions will be to improve the energy efficiency of motor vehicles and to transform the current highway freight-based mode of transportation. Shahbaz et al. [52] applied combined co-integration tests and Autoregressive Distributed Lag (ARDL) bound tests to investigate the causal relationships between transportation-related energy consumption, CO_2 emissions, fuel prices and transport sector added value in Tunisia. The test results indicated that the energy consumption and added value output of the transportation sector promotes CO_2 emissions, but increases in fuel prices reduce the level of CO_2 emissions.

In terms of the content of prior research, all of the studies mentioned above focus on the macro-level (specifically, the international or national level), but research on a local level is rare. Taking China as an example, many studies have been conducted at a national level, but few have addressed the provincial level [33,44–46,51]. Given that the carbon emissions of the transportation sector are restrained by many region-specific factors (such as topography and geomorphology, energy endowments and regional energy policies, which differ significantly from region to region), it is necessary to carry out a microscopic analysis. Furthermore, with regard to the research methods, two approaches are widely used at present. They are the index decomposition method [6–8,11,20,23,53] and the econometric method [2,4,6,24,28,54,55]. Due to the constraints of the Kaya identity, the factors that are considered in the index decomposition method are limited. There are also defects in the econometric method. In particular, the econometric method usually explores the relationships among variables from the perspectives of co-integration and causality, but neglects the multicollinearity problems, which prevail in macroeconomic data.

Compared with previous studies, this paper fills the above-mentioned gaps in the following three ways: Firstly, our research empirically studies the carbon emissions of the transport sector at the local level, while at the same time taking into consideration each local area's significant uniqueness. Xinjiang is China's largest provincial-level administrative region, with an area of approximately 1.66 million km^2. The region accounts for one-sixth of China's total land area. Remarkably, transportation in Xinjiang is mainly based on road freight. Moreover, the One Road, One Belt Initiative [6] aims to promote communication and cooperation between the countries along the Silk Road. The initiative also aims to promote the construction of transportation infrastructure in Xinjiang, which will in turn result in a rapid rise in transportation-related carbon emissions. Secondly, in terms of methodology, the STIRPAT (Stochastic Impacts by Regression on Population, Affluence and Technology) model and the ridge regression model were combined in this paper to thoroughly analyze the influencing factors of carbon emissions. Our approach effectively overcomes the multicollinearity problem inherent in macroeconomic variables and, thus, guarantees the objectivity and reliability of our estimated results. Finally, the accounting of the carbon emissions of the transportation sector in this study is both comprehensive and accurate. Specially, our study respectively calculates the carbon emissions from nine types of energy, namely coal, coke, crude oil, gasoline, kerosene, diesel, fuel oil, natural gas and electricity. Thus, the results of our study better reflect Xinjiang's conditions and are more scientific.

2. Methodology and Data

2.1. Accounting of Carbon Emissions

Currently, two methods are primarily used for the accounting of the carbon emissions of the transportation sector, namely the "bottom-up distance-based" method and the "top-down fuel-based" method [56]. The former method calculates the total carbon emissions according to the vehicle mileage of the various means of transportation, as well as the energy consumption per unit of mileage and the carbon emission coefficients of the various types of energies in the region studied. The latter method calculates the total carbon emissions by multiplying the energy consumption of the transportation sector by the carbon emission coefficients of various energy types. Considering the difficulty of fully and reliably collecting the mileage data of different vehicle models, as required by the "bottom-up distance-based" method, this paper adopted the "top-down fuel-based" method to calculate the CO_2 emissions of Xinjiang's transportation sector. The calculation formula is as follows:

$$C = \sum_{i=0}^{n} C_i = \sum_{i=0}^{n} E_i \times LCV_i \times PCC_i \times O_i \times 44/12 \tag{1}$$

where C represents the total CO_2 emissions of the energy consumption of the transportation sector; C_i denotes the CO_2 emissions based on fuel type i; E_i is the consumption of fuel type i; LCV_i and PCC_i represent the low calorific value and the potential carbon content of fuel type i, respectively; O_i represents the oxidation rate of fuel type i; $44/12$ is the coefficient of conversion from C to CO_2. See the CO_2 emissions factors of the various energy types in Table 1. Based on the baseline emissions factor of the power grid in northwest China and the related literature [57,58], the CO_2 emissions factor of electricity in Xinjiang was determined as being 1.0174 tCO_2/MWh.

Table 1. CO_2 emission factors of various energy types.

Fuel Type	Coal	Coke	Crude Oil	Gasoline	Kerosene	Diesel Oil	Fuel Oil	Natural Gas
Low calorific value ($TJ/10^3$ t or $TJ/10^4$ m^3) [59]	20.908	28.435	41.816	43.070	43.070	42.652	41.816	38.93
Potential carbon content (kg C/GJ) [60]	26.37	29.5	20.1	18.9	19.6	20.2	21.1	15.3
Oxidation rate [60]	0.98	0.93	0.98	0.98	0.98	0.98	0.98	0.99

2.2. STIRPAT (Stochastic Impacts by Regression on Population, Affluence and Technology) Model

The STIRPAT model was proposed by Dietz and Rosa [61]. It was extended on the basis of the IPAT (Impact = Population × Affluence × Technology) model, which was put forward by Ehrlich and Holden [62]. The STIRPAT model has overcome the limitations of IAPT's hypothesis that "various factors influence the environment by the same proportion" [54]. The STIRPAT model can also better reflect the non-monotonic or non-proportional functional relationships between the factors that influence the natural environment [54,55,63]. The STIRPAT model is as follows:

$$I_t = a P_t^b A_t^c T_t^d e_t \tag{2}$$

where I represents the environmental influence; P represents the population size; A represents the wealth level, generally measured by per capita GDP; T is the technical index, usually measured by the effect on the environment per output; a, b, c and d represent the model coefficients to be estimated; e_t represents the random error term. The subscript t denotes the time, which usually is the corresponding year. The STIRPAT model combines economic activities and environmental influence and is widely applied as a means to analyze factors influencing the environment. In order to eliminate the heteroscedasticity, which could possibly exist in the model, as well as to facilitate the testing of hypotheses, all of the factors take a logarithmic form. Because e_t is the random error term, we do not need to distinguish between e_t and Le_t. Then, we rewrote Equation (2) as follows:

$$LnI_t = Lna + bLnP_t + cLnA_t + dLnT_t + e_t \tag{3}$$

where P, A and T are the same as in Equation (2). In order to probe the influencing factors of the transport sector's CO_2 emissions, we use the total transport-related CO_2 emissions to represent the environmental influence. Equation (3) can then be rewritten as follows:

$$LnCO_{2t} = a + \beta_1 LnP_t + \beta_2 LnA_t + \beta_3 LnT_t + e_t \tag{4}$$

where CO_2 represents the total CO_2 emissions of the transportation industry (10^4 t), and this implies environmental impact; P represents the population size (10^4 persons); A represents the economic development level, which is expressed in this paper by per capita GDP (10^4 yuan/person, converted by 1990 as the constant price level); T represents the energy intensity, that is the ratio of total energy consumption to the added value output of the transportation sector (tce/10^4 yuan). Herein, total energy consumption refers to the sum of the main nine types of energy, which had been converted to tons of standard coal equivalent (tce), respectively.

To further analyze the driving forces of the transport sector's CO_2 emissions and considering the specific situation in Xinjiang, we expand Equation (4) by incorporating CT (Cargo Turnover) and PC (Private Vehicle Population) into the model. There are two main reasons for incorporating these two variables. On the one hand, Xinjiang is a vast territory, with great distances between cities. Moreover, the main mode of transport in Xinjiang is road freight, which relies chiefly on heavy lorries. The extensive use of heavy lorries means higher energy consumption and higher carbon emissions. Therefore, cargo turnover is an important factor affecting the carbon emissions from the transport sector. On the other hand, due to the increase in residents' incomes in recent years, the demand for private vehicles continues to rise. The rapid growth in private vehicle ownership has resulted in the corresponding and continued increase in energy consumption and, in turn, energy-related CO_2 emissions. Thus, cargo turnover and private vehicles were incorporated into the estimated model.

Based on the STIRPAT model and the above analysis, the econometric model of the transport sector's CO_2 emissions is established as follows:

$$LnCO_{2t} = a + \beta_1 LnP_t + \beta_2 LnA_t + \beta_3 LnT_t + \beta_4 LnCT_t + \beta_5 LnPC_t + e_t \tag{5}$$

where CO_2, P, A and T are the same as in Equation (4). CT denotes cargo turnover (100 million ton-km), and PC represents private vehicle population (by unit); β_1, β_2, β_3, β_4 and β_5, respectively, represent the elasticity coefficients of the various variables corresponding to CO_2 emissions.

2.3. Multicollinearity Diagnostics and Ridge Regression

Multicollinearity is a phenomenon in which two or more predictor variables in a multiple regression model are highly correlated. This means that one variable can be linearly predicted from the others with a substantial degree of accuracy [64,65]. In this situation, the coefficient estimates of the multiple regression may change erratically in response to small changes in either the model or the data. Multicollinearity affects calculations regarding individual predictors [66]. That is, a multiple regression model with correlated predictors may not give valid results pertaining to any individual predictor, or about which predictors are redundant with respect to others. To determine whether or not there was multicollinearity existing between independent variables, a multivariate linear regression analysis using least squares was conducted. If the Variance Inflation Factor (VIF) of independent variables was greater than the maximum tolerance of 10, this indicates the existence of multicollinearity between explanatory variables [66–68]. A multiple regression model can be expressed as follows:

$$Y = X\beta + \varepsilon \tag{6}$$

where Y is an $n \times 1$ observation vector; $X = [x_1, x_2, ..., x_n]^T$ is an $n \times q$ full rank matrix; $\beta = [\beta_1, \beta_2, ..., \beta_q]^T$ is a $q \times 1$ parameter vector to be estimated. By using the least square method, the estimated value of β can be obtained from Equation (7). The mean square error of $\hat{\beta}$ is calculated by Equation (8); wherein λ_i is q the characteristic root of the non-negative symmetric matrix $X^T X$.

$$\hat{\beta} = \left(X^T X\right) X^T Y \tag{7}$$

$$\hat{\beta}_{MSK} = E\left|\|\hat{\beta} - \beta\|^2\right| = E\left|(\hat{\beta} - \beta)^T.(\hat{\beta} - \beta)\right| = \sigma^2 \sum_{i=1}^{q} \frac{1}{\lambda_i} \tag{8}$$

When multicollinearity existed between independent variables, the matrix $X^T X$ is singular, and some of the matrix's characteristic roots are close to zero. Under these conditions, the value of $\hat{\beta}_{MSK}$ will be especially large, which indicates a larger deviation between the estimated values and observed values. Thus, the ordinary least squares (OLS) method loses its stability and reliability. Ridge estimation (RE) is an alternative method to the OLS method and can be used when a collinearity problem exists in a linear regression model [64–67].

To address the aforementioned problem, the ridge regression method substitutes $X^T X$ for $X^T X + kI$ to ensure the characteristic roots of matrix $X^T X + kI$ are far from zero. Then, the value of $\hat{\beta}_{MSK}$ will be significantly reduced. Finally, the estimated value of β can be solved by Equation (9).

$$\hat{\beta}(k) = \left(X^T X + kI\right)^{-1} X^T Y \tag{9}$$

Herein, I is an identity matrix, k is a ridge parameter and $\hat{\beta}(k)$ is the ridge estimated value for β. In this paper, we determined the ridge parameters by means of the ridge trace method.

2.4. Data Sources and Description

The data used in this paper include annual observations of the CO_2 emissions, population size, per capita GDP, energy intensity, cargo turnover and private vehicle population in Xinjiang during the period from 1990 to 2014. In order to eliminate the effect of price changes, per capita GDP is calculated at a constant price (1990 = 100). All data used in this paper are obtained from 50 Years of Glories of Xinjiang [69], the Xinjiang Statistical Yearbook (XSY) (1989 to 2014) [5] and the China Energy Statistical Yearbook (1990 to 2014) [70]. Data on the level of energy consumption of the various types of energy used by the transport sector were derived from the table of "Energy Consumption by Sector and Major Energy Consumption" provided by the Xinjiang Statistical Yearbook [5]. Total energy consumption data came from the China Energy Statistical Yearbook [70]. The data relating to GDP, population size, total number of private vehicles, cargo turnover and transportation sector's added value output came from the 50 Years of Glories of Xinjiang and the Xinjiang Statistical Yearbook [5]. In order to eliminate the effects of inflation, GDP is again calculated at a constant price (1990 = 100). Cargo turnover represents total freight ton-kilometers, which included four categories, namely railways, highways, civil aviation and petroleum and gas pipelines. It is calculated as being transport mileage multiplied by freight volume. In view of the classification standards of the National Bureau of Statistics, there are two types of private cars, namely passenger cars and freight cars. Moreover, passenger cars are divided into four sub-categories: large, medium, small and micro cars. Freight car categories include heavy, medium, light and miniature. Private car ownership in this paper is calculated as the total amount of the above-mentioned types of motor vehicles.

3. Results and Discussion

3.1. Features of Carbon Emissions from the Transport Sector

3.1.1. Macro-Level: Total Energy-Related Carbon Emissions

The estimated results of the total carbon emissions and per capita carbon emissions of Xinjiang's transportation sector during the period of from 1989 to 2012 are shown in Figure 1. As shown, on the whole, the two indicators display a gradually rising trend. Based on the changes in trend, the study period could be divided into two phases: 1989 to 2000 and 2001 to 2012. In the first phase, both the total carbon emissions and per capita carbon emissions presented a slowly rising trend with their average annual growth rates of 6.8% and 4.5%, respectively. In the second phase, the total carbon emissions increased from 5.35 million tons to 16.53 million tons. This represented a total growth of

310% over the duration of the period and an average annual growth rate of 10.8%. During the same phase, the per capita carbon emissions increased from 0.29 t to 0.74 t, or a total growth of 260% and an average annual growth rate of 9.1%. The enormous differences between the two phases in terms of growth rate can be explained by the implementation by the central government of the Western Development Strategy in 2001 [4]. The implementation of this strategy has promoted the construction of transportation infrastructure and thus facilitated the development of the logistics sector. With the rapid development of the logistics sector, the carbon emissions caused by energy consumption have also increased.

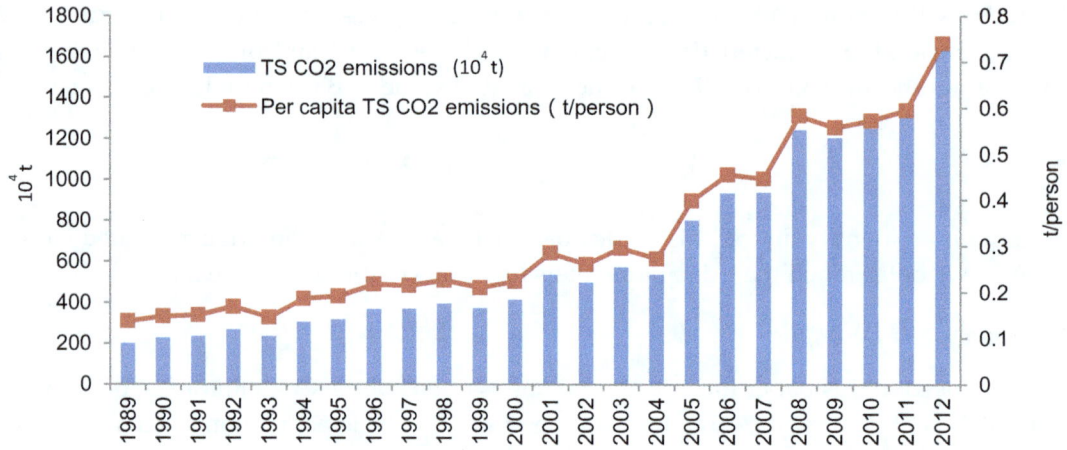

Figure 1. Changes of the total CO_2 emissions and per capita CO_2 emissions of Xinjiang's transportation sector (1989 to 2012).

3.1.2. Micro-Level: Carbon Emissions Structure and Intensity

The structure of carbon emissions can be analyzed from different aspects. Due to the fact that energy mix has an important influence on carbon emissions, energy mix thus became the most commonly-used means to illustrate the changes of the carbon emissions structure. Figure 2 shows the CO_2 emissions from the five major energy types during the period from 1989 to 2012. Three interesting results can be drawn from this figure. Firstly, as a whole, the level of carbon emissions from all fuel types continued to increase during the study period, especially since 2004. According to XSY [5], the length of road transport lines has increased markedly since 2004, which in turn indicates that the construction of traffic facilities can lead to a rapid growth in energy consumption (for transportation). Secondly, diesel oil, rather than gasoline, turns out to be the biggest emitter of carbon emissions. In the period from 1989 to 2012, the highest carbon emissions were generated by the consumption of diesel oil, with an average value of 2.60 million tons per year and a proportion as high as 36.2% of all emissions. Thirdly, the use of cleaner energies such as electricity experienced steady growth. However, clean energy still only accounts for a very small slice of the total energy "pie", with an average annual value of 0.42 million tons, representing only 7.1% of usage among all fuel types. However, it should be noted that, since 2012, almost half a million cars and buses and more than 100,000 private cars in Xinjiang have been using Liquid Natural Gas (LNG) for eco-friendliness and higher efficiency purposes. The use of LNG is potentially a fundamental way to reduce CO_2 emissions in Xinjiang. At present, there are three types of railway locomotives in Xinjiang, which are steam locomotives, diesel locomotives and electric locomotives. Steam locomotives use coal as the driving energy. For example, in Sandaoling coal mine, which is located in Hami Prefecture, the steam locomotive still bears the task of coal transportation. However, it should be noted that the coal consumption in the transport sector is gradually decreasing. As for kerosene, it is mainly used in air transport. Because there is great distance from one city to another in Xinjiang, travel by air is preferred by more and more people. Therefore, the consumption of kerosene showed a rising trend in the research period.

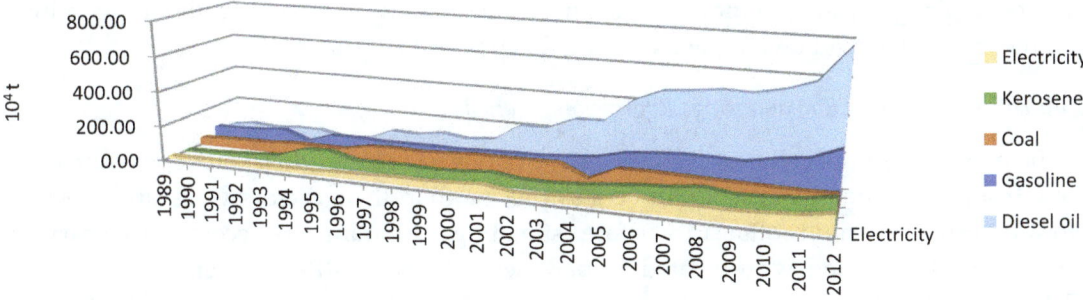

Figure 2. Carbon emissions from five main energy types from 1989 to 2012.

In addition to the energy mix, we further analyzed the characteristics of traffic carbon emissions in Xinjiang from the perspective of carbon emission intensity. Carbon emission intensity is defined as the amount of carbon dioxide emissions per unit of GDP growth. Herein, for the purpose of this study, both carbon dioxide emissions and GDP are limited to the transport sector. The calculation results of carbon emission intensity are shown in Table 2.

Table 2. The carbon emission intensity in Xinjiang's transport sector.

Year	Total Carbon Emissions (10^4 t)	Value Added Output (10^4 Yuan)	Emission Intensity (t/10^4 Yuan)	Year	Total Carbon Emissions (10^4 t)	Value Added Output (10^4 Yuan)	Emission Intensity (t/10^4 Yuan)
1989	199.26	12.90	15.45	2001	535.29	148.38	3.61
1990	225.94	14.63	15.44	2002	496.71	168.58	2.95
1991	233.82	22.63	10.33	2003	572.34	159.43	3.59
1992	266.79	28.32	9.42	2004	596.08	186.70	3.87
1993	234.29	32.58	7.19	2005	800.98	149.61	5.35
1994	304.22	44.40	6.85	2006	934.27	165.60	5.64
1995	317.89	59.63	5.33	2007	936.43	177.28	5.28
1996	366.69	73.74	4.97	2008	1243.52	191.84	6.48
1997	368.48	85.70	4.30	2009	1204.18	209.10	5.76
1998	394.47	106.87	3.69	2010	1249.61	222.47	5.62
1999	372.73	129.60	2.88	2011	1313.75	256.72	5.12
2000	412.93	148.63	2.78	2012	1653.05	357.90	4.62

As can be seen from Table 2, the changes in trends were relatively complicated. According to the characteristics of numerical value change, the study period was divided into three phases for the convenience of analysis. In the first phase (1989 to 2000), the intensity of carbon emissions steadily declined, from 15.45 t/10^4 yuan in 1989 to 2.78 t/10^4 yuan in 2000 with an average annual rate of decline of 16.9%. This phase corresponded to China's "Eighth Five-year Plan" and "Ninth Five-year Plan" [71], in which the government's energy policies focused on the control of energy consumption and improvements in efficiency. These plans and policies were the primary cause of the year-by-year decline in the intensity of carbon emissions [72,73]. In the second phase (2001 to 2008), the intensity of carbon emissions rose, although the degree of increase fluctuated. To be specific, the intensity of carbon emissions reached 6.48 t/10^4 yuan in 2008. The implementation of the Western Development Strategy [4] and China's accession to the World Trade Organization (WTO) may be the main causes for this rise. According to XSY [5], during this period, the level of cargo turnover increased sharply, with an annual growth rate of 12.76%. Moreover, the average share of highway use for freight traffic reached as high as 81.0%. A high proportion of highway transport means a corresponding increase in energy consumption. During the third phase (2009 to 2012), the intensity of carbon emissions experienced a declining trend. Along with the economic development of Xinjiang, the government attached greater importance to efficient energy utilization. Meanwhile, the cooperation between China and Central Asia in the field of energy (especially natural gas) accelerated the optimization of the energy consumption structure of Xinjiang's transportation sector. This optimization, in turn, directly and continuously, reduced the intensity of carbon emissions during this period [3,74,75].

3.2. Multicollinearity Detection and Ridge Regression Analysis

In the presence of multicollinearity, the estimate of one variable's impact on dependent variable Y while controlling for the others tends to be less precise than if predictors were uncorrelated [76]. Therefore, detecting any multicollinearity before estimating the parameters becomes necessary. First of all, by using the OLS method, we estimated the parameters in the STIRPAT model. Then, in accordance with the VIF, we can determine whether multicollinearity exists between the variables [77]. Finally, one typical remedy for multicollinearity will be adopted in this paper. The results of OLS regression and the VIF values of each variable are listed in Table 3.

Table 3. The OLS regression results of transport's carbon emission in Xinjiang.

Variables	Parameters	Standard Error	t Statistics	p-Value	Variance Inflation Factor (VIF)
Constant	−17.152	6.546	−2.620	0.017 **	—
lnP	2.411	0.853	2.837	0.011 **	56.437
lnA	0.498	0.237	2.101	0.050 *	34.681
lnT	0.293	0.065	4.490	0.000 ***	4.649
lnCT	0.105	0.269	0.391	0.700	155.264
lnPC	0.095	0.159	0.601	0.556	226.711

Note: ***, ** and * denote significant level at 1%, 5% and 10%, respectively.

As can be seen from Table 3, the VIF values of population size, per capita GDP, cargo turnover and private vehicles were far greater than 10. This finding indicates the existence of multicollinearity between the explanatory variables. Therefore, the OLS method was not suitable for making an unbiased estimation. In order to obtain more accurate results, a ridge regression estimation was used to re-estimate the model in Equation (5). Ridge regression estimation involves an improved algorithm of least squares and can address the previous inability to inversely solve the matrix for the coefficient vector by using least squares [68]. By adding a non-negative factor K to the element on the main diagonal of a standardized matrix of independent variables, the ridge regression algorithm was able to significantly improve the stability of estimation [4]. Since the ridge regression is a biased estimate, to retain as much information as possible, the value of K should not be overly large. The ridge parameter K fell within the range of (0, 1), and the step size of 0.005 was adopted for the purpose of valuation. When $k = 0.02$, the coefficient of determination R^2 was 0.987, and the regression coefficient of each explanatory variable was stabilized. The ridge regression estimation results are listed in Table 4.

Table 4. Ridge regression estimation results of the carbon emissions of Xinjiang's transportation sector.

Variables	Parameters	Standard Error	Standardized Coefficients	t Statistics	p-Value
Constant	−12.504	2.207	0.000	−5.665	0.000 ***
P	1.777	0.344	0.358	5.159	0.000 ***
A	0.416	0.127	0.236	3.284	0.004 ***
T	0.261	0.038	0.197	6.831	0.000 ***
CT	0.224	0.055	0.237	4.061	0.001 ***
PC	0.110	0.024	0.238	4.558	0.000 ***
Adjusted R^2 = 0.987		F statistics = 360.31		Significance (F statistics) = 0.000 ***	

Note: ***, ** and * denote significant level at 1%, 5% and 10%, respectively.

As shown in Table 4, the general coefficient of the model's determination R^2 was 0.987, with a relatively high degree of fit. Every explanatory variable passed the t-test significantly. Therefore, the regression coefficients were valid. See the specific ridge regression equation below:

$$LnCO_2 = -12.504 + 1.777LnP + 0.416LnA + 0.261LnT + 0.224LnCT + 0.110LnPC \qquad (10)$$

As shown by the results of ridge regression, population size is the most important driver of the carbon emission increases of Xinjiang's transportation sector. Specifically, every 1% growth in population size would cause the transportation sector's carbon emissions to increase by approximately 1.78%. To the best of our knowledge, there are three main reasons for this phenomenon. Firstly, since the implementation of the family planning policy in China, the natural growth rate of the population has gradually declined, year on year [78]. However, due to more liberal childbearing policies for minorities, the population of Xinjiang has grown more rapidly than in other regions. In the study period, the natural growth rate of Xinjiang's population was approximately 12.48%. This rate is 4.2% higher than the national average, which was only 8.28%. The growth in population inevitably drives the increase of transportation-related energy consumption, which correspondingly elevates the level of CO_2 emissions. Secondly, the rate of population flow (which included the migration from rural areas to the city both within and outside Xinjiang) is growing faster in recent years. According to XSY [5], Xinjiang is currently experiencing a process of rapid urbanization. The urbanization level of Xinjiang increased from 33.8% in 1989 to 44.0% in 2012. It is recognized that urban form features affect the distance people travel each day, as well as their choice of transportation mode and ultimately the level of CO_2 emissions. Thirdly, due to the long distances between the various prefectures and cities in Xinjiang, compared to inland provinces, Xinjiang experiences higher levels of energy consumption and CO_2 emissions as a direct result of population flow.

In addition, the increase in per capita income constitutes another important factor influencing the carbon emissions of Xinjiang's transportation sector. The elasticity coefficient of transportation-related carbon emissions for per capita GDP is 0.42%. This finding indicates that, with the continuous rise in social and economic development levels, transportation-related carbon emissions are correspondingly gradually increasing. These increases in per capita income and carbon emissions can be interpreted from two aspects. On the one hand, higher incomes cause more people to buy private cars. The increased number of vehicles on the road naturally leads to increased energy consumption, which in turn results in higher CO_2 emissions. On the other hand, the increased per capita income encourages more people to travel. The increased number of people's trips will also increase energy consumption, which then boosts the level of CO_2 emissions coming from the transport sector.

As indicated by the energy intensity coefficient, for every 1% increase in energy consumption per unit GDP, carbon emissions increase by 0.26%. In other words, energy intensity and carbon emissions are positively correlated. That is, a reduction in energy intensity will effectively decrease the amount of carbon emissions. In the study period of 1989 to 2012, the energy intensity of Xinjiang's transportation department continuously declined. In turn, these declines contributed to the reduction of carbon emissions. However, in fact, the total carbon emissions in Xinjiang continue to grow rather than decline. This growth might be explained by the fact that the inhibiting effect of energy intensity cannot offset the driving forces, namely the size of population, per capita GDP and cargo turnover. The coefficient also highlights the positive effects on low carbon traffic of reducing energy intensity.

For every 1% increase of both cargo turnover and the total number of private vehicles, their transportation-related CO_2 emissions correspondingly increase by 0.22% and 0.11%, respectively. As shown by a comparison between the two coefficients, cargo turnover exerts a more significant influence on transportation-related CO_2 emissions in Xinjiang. On the one hand, the vastness of Xinjiang (in terms of territory and the long distances between its prefectures and cities) constitutes a basic reality for this province. In addition, with the gradual improvements being made to the transportation network and the rapid development of the logistics sector, cargo turnover has necessarily increased at a fast pace. On the other hand, freight transport vehicles in Xinjiang mainly consume diesel, which is a fuel that generates more carbon emissions than vehicles that use other types of energy. These two aspects combined suggest that cargo turnover more significantly drives the increase of transportation-related CO_2 emissions in Xinjiang than do private vehicles.

4. Conclusions and Policy Suggestions

Based on the Guidelines for National Greenhouse Gas Inventories [56] and the baseline emission factor of the regional power grid in northwest China, this paper calculated the total carbon emissions of Xinjiang's transportation sector during the period from 1989 to 2012. On the basis of the results, by applying a STIRPAT model and rigid regression method, an in-depth econometric analysis was conducted, in order to clarify the influencing factors of transportation-related carbon emissions. The results of our study indicate that, during the study period, the total carbon emissions and per capita carbon emissions of Xinjiang's transportation sector both exhibited an upward trend. We found that the total carbon emissions increased from 1.99 million tons in 1989 to 16.53 million tons in 2012, representing an average annual growth rate of 30%. Per capita carbon emissions during the same period increased from 0.14 t to 0.74 t, representing an average annual growth rate of 7.6%.

Our analysis of the structure of carbon emissions revealed that diesel consumption accounted for both the highest amount and largest proportion of carbon emissions. This fact is explained by the dominant position of large trucks in Xinjiang's transportation system. Although the absolute quantities of clean energies, such as natural gas and electricity, are constantly on the rise, they still account for an extremely low proportion of total energy use. Given the geographical uniqueness of Xinjiang, the dominant position of diesel in the energy consumption structure of Xinjiang's transportation sector will not change substantially, at least in the short term.

As shown by the results of ridge regression, every 1% increase in population size, per capita GDP, energy intensity, cargo turnover and total number of private vehicles has resulted in increases of transportation-related carbon emissions of 1.78%, 0.42%, 0.26%, 0.22% and 0.11%, respectively. This finding clearly indicates that the expansion of Xinjiang's population exerted the most significant influence on the area's transportation-related carbon emissions. Given that there are many minorities living in Xinjiang and China has implemented more liberal childbearing policies for minorities, the population of Xinjiang has grown at a rate much higher than the national average. Moreover, massive domestic migration is another major reason for the increase in Xinjiang's population. While the expansion of the population and the rise of per capita income levels both contributed to the increase in carbon emissions, we found that energy intensity did not play its anticipated inhibitory role for transportation-related carbon emissions during the study period. In addition, cargo turnover more significantly promoted the increase of carbon emissions in Xinjiang than did the total number of private vehicles. This finding can be explained by Xinjiang's highway, freight-based mode of transportation and the structure of diesel consumption-based energy utilization in the region.

Based on the conclusions reached and the actual situation in Xinjiang, this paper puts forward the following suggestions:

(1)   More attention should be placed on the promotion of clean and renewable energy in the transport sector. Diesel and gasoline are still the main energies used in the most recent period (especially diesel). Reducing the consumption of diesel is of great significance to creating low carbon transportation. Therefore, with Xinjiang's unique geographical advantages and driven by the One Belt, One Road Initiative [6], cooperation with Central Asia in the energy field should be reinforced, thus increasing the consumption of natural gas, which emits less carbon.

(2)   Rigid regression results show that population size is one of the key factors driving Xinjiang's traffic carbon emissions. Therefore, the natural population growth rate should be appropriately controlled. In addition, the flow of the population should be guided reasonably and effectively. Reasonable and orderly migration could effectively reduce the population's moving distance and thereby reduce transport sector carbon emissions. Moreover, raising people's awareness of low carbon travel could also be an important way to achieve low carbon transport.

(3)   The intensity of scientific and technological input into the energy utilization field should be strengthened, in order to improve the utilization efficiency of traditional energies. For instance, improving the utilization efficiency of diesel could effectively reduce the carbon emissions caused by the transport of bulk cargo in highway freight vehicles.

(4)   Efforts should be made to realize supply side reform, promote high-speed railway construction, increase railway network density and effectively reduce the proportion of high carbon-emission highway freight vehicles in Xinjiang. The government should increase investment in public transportation facilities and non-motorized transportation facilities as one means to reduce the excessive use of private vehicles.

(5)   Preferential policies should be implemented and promoted to encourage the use of hybrid energy motor vehicles. Specifically, appropriate financial subsidies should be given to buyers of hybrid motor vehicles in terms of purchase tax, fuel tax and use tax. Efforts should also be made to encourage people to purchase low-carbon and environmentally-friendly vehicles.

Despite the contributions presented by this paper, there are also some limitations that would warrant further discussion. Firstly, due to the constraints of the STIRPAT model, the factors that may affect CO_2 emissions were selected based on regional features and reference to relevant literature, rather than statistical testing methods. Thus, there may be some influencing factors that were ignored; for instance, urbanization level, trade openness, transportation infrastructure investment, and so on. These factors may also play an important role in increasing the transport sector's CO_2 emissions. Secondly, even though the rigid regression model in this paper is reasonable, to some extent, the results obtained from this method are not unbiased. Therefore, further studies are needed to identify to what extent each factor plays its role in increasing carbon emissions. In other words, other econometric models, for example the nonparametric additive regression model and the vector autoregression model, may also be applicable for the analysis of driving factors of CO_2 emissions in Xinjiang's transport sector. For the above-mentioned limitations, further in-depth research should be conducted. Specifically, considering more influencing factors and seeking more suitable methods are the two key points of further study. Comprehensive consideration for influencing factors and a better model to estimate the coefficients of dependent variables could make sure that the results are more accurate and practical.

Acknowledgments: This paper is supported by the Initial Founding of Scientific Research for the Introduction of Talents of China University of Petroleum (Huadong) (05Y16060020) and the Ring-Fenced Funding of research on the endogenous development path of Yuncheng under the regional cooperation of Golden Triangle in the Yellow River (2014. No. 4).

Author Contributions: Chun Deng and Jiefang Dong conceived of, designed and performed the experiments. Jiefang Dong, Rongrong Li and Jieyu Huang analyzed the data and wrote the paper. All authors have read and approved the final manuscript.

Conflicts of Interest: The authors declare no conflict of interest.

References

1.   Edenhofer, O.; Pichs-Madruga, R.; Sokona, Y.; Farahani, E.; Kadner, S.; Seyboth, K.; Adler, A.; Baum, I.; Brunner, S.; Eickemeier, P.; et al. (Eds.) *Climate Change 2014: Mitigation of Climate Change. Working Group III Contribution to the Fifth Assessment Report of the Intergovernmental Panel on Climate Change, 2014*; Cambridge University Press: Cambridge, UK; New York, NY, USA, 2014; Available online: http://www.ipcc.ch/report/ar5/wg3/ (accessed on 25 May 2016).

2.   Lin, B.; Xie, C. Reduction potential of CO_2 emissions in China's transport industry. *Renew. Sustain. Energy Rev.* **2014**, *33*, 689–700. [CrossRef]

3.   Ju, J.; Wang, Q.; Liang, L.; Chen, X. International Carbon Trading: A Game Changer for Climate Change? *Environ. Sci. Technol.* **2014**, *48*, 14069. [CrossRef] [PubMed]

4.   Huo, J.; Yang, D.; Zhang, W.; Wang, F.; Wang, G.; Fu, Q. Analysis of influencing factors of CO_2 emissions in Xinjiang under the context of different policies. *Environ. Sci. Policy* **2015**, *45*, 20–29. [CrossRef]

5.   Statistics Bureau of Xinjiang Uygur Autonomous Region. *Xinjiang Statistical Yearbook (1990–2014)*; China Statistics Press: Beijing, China, 2014.

6.   Tsao, R. One Belt One Road. *Chinese American Forum* **2015**, *31*, 11.

7.   Wang, C.; Zhang, X.; Wang, F.; Lei, J.; Zhang, L. Decomposition of energy-related carbon emissions in Xinjiang and relative mitigation policy recommendations. *Front. Earth Sci.* **2014**, *9*, 65–76. [CrossRef]

8.   Voigt, S.; Cian, E.D.; Schymura, M.; Verdolini, E. Energy intensity developments in 40 major economies: Structural change or technology improvement? *Energy Econ.* **2014**, *41*, 47–62. [CrossRef]

9.   Wang, X.; Zhang, C. The impacts of global oil price shocks on China's fundamental industries. *Energy Policy* **2014**, *68*, 394–402. [CrossRef]

10. Wang, Q. China has the capacity to lead in carbon trading. *Nature* **2013**, *7432*, 273. [CrossRef] [PubMed]

11. Wang, Q.; Li, R. Drivers for energy consumption: A comparative analysis of China and India. *Renew. Sustain. Energy Rev.* **2016**, *62*, 954–962. [CrossRef]

12. Wang, Q.; Li, R.; Liao, H. Toward Decoupling: Growing GDP without Growing Carbon Emissions. *Environ. Sci. Technol.* **2016**, *50*, 11435–11436. [CrossRef] [PubMed]

13. Wang, Q.; Chen, Y. Energy saving and emission reduction revolutionizing China's environmental protection. *Renew. Sustain. Energy Rev.* **2010**, *14*, 535–539. [CrossRef]

14. Wang, Q.; Li, R. Sino-Venezuelan oil-for-loan deal—The Chinese strategic gamble? *Renew. Sustain. Energy Rev.* **2016**, *64*, 817–822. [CrossRef]

15. Wang, Q.; Li, R. Impact of cheaper oil on economic system and climate change: A SWOT analysis. *Renew. Sustain. Energy Rev.* **2016**, *54*, 925–931. [CrossRef]

16. Dong, J.-F.; Wang, Q.; Deng, C.; Wang, X.-M.; Zhang, X.-L. How to Move China toward a Green-Energy Economy: From a Sector Perspective. *Sustainability* **2016**, *8*, 337. [CrossRef]

17. Wang, Q.; Li, R. Cheaper Oil: A turning point in Paris climate talk? *Renew. Sustain. Energy Rev.* **2015**, *52*, 1186–1192. [CrossRef]

18. Wang, Q. Cheaper Oil—Challenge and Opportunity for Climate Change. *Environ. Sci. Technol.* **2015**, *49*, 1997–1998. [CrossRef] [PubMed]

19. Wang, Q.; Chen, X.; Xu, Y.C. Pollution protests: Green issues are catching on in China. *Nature* **2012**, *489*, 502. [CrossRef] [PubMed]

20. Scholl, L.; Schipper, L.; Kiang, N. CO_2 emissions from passenger transport: A comparison of international trends from 1973 to 1992. *Energy Policy* **1996**, *24*, 17–30. [CrossRef]

21. Greening, L.A.; Ting, M.; Davis, W.B. Decomposition of aggregate carbon intensity for freight: Trends from 10 OECD countries for the period 1971–1993. *Energy Econ.* **1999**, *21*, 331–361. [CrossRef]

22. Saboori, B.; Sapri, M.; Baba, M. Economic growth, energy consumption and CO_2 emissions in OECD (Organization for Economic Co-operation and Development)'s transport sector: A fully modified bi-directional relationship approach. *Energy* **2014**, *66*, 150–161. [CrossRef]

23. Timilsina, G.R.; Shrestha, A. Transport sector CO_2 emissions growth in Asia: Underlying factors and policy options. *Energy Policy* **2009**, *37*, 4523–4539. [CrossRef]

24. Timilsina, G.R.; Shrestha, A. Factors affecting transport sector CO_2 emissions growth in Latin American and Caribbean countries: An LMDI decomposition analysis. *Int. J. Energy Res.* **2009**, *66*, 396–414. [CrossRef]

25. Chandran, V.G.R.; Tang, C.F. The impacts of transport energy consumption, foreign direct investment and income on CO_2 emissions in ASEAN-5 economies. *Renew. Sustain. Energy Rev.* **2013**, *24*, 445–453. [CrossRef]

26. Mazzarino, M. The economics of the greenhouse effect: Evaluating the climate change impact due to the transport sector in Italy. *Energy Policy* **2000**, *28*, 957–966. [CrossRef]

27. Lakshmanan, T.R.; Han, X. Factors underlying transportation CO_2 emissions in the U.S.A.: A decomposition analysis. *Transport. Res. D Transp. Environ.* **1997**, *2*, 1–15. [CrossRef]

28. McKinnon, A.C.; Piecyk, M.I. Measurement of CO_2 emissions from road freight transport: A review of UK experience. *Energy Policy* **2009**, *37*, 3733–3742. [CrossRef]

29. Ratanavaraha, V.; Jomnonkwao, S. Trends in Thailand CO_2 emissions in the transportation sector and Policy Mitigation. *Transp. Policy* **2015**, *41*, 136–146. [CrossRef]

30. Stelling, P. Policy instruments for reducing CO_2-emissions from the Swedish freight transport sector. *Res. Transp. Bus. Manag.* **2014**, *12*, 47–54. [CrossRef]

31. Johansson, B. Will restrictions on CO_2 emissions require reductions in transport demand? *Energy Policy* **2009**, *37*, 3212–3220. [CrossRef]

32. Lu, I.J.; Lewis, C.; Lin, S.J. The forecast of motor vehicle, energy demand and CO_2 emission from Taiwan's road transportation sector. *Energy Policy* **2009**, *37*, 2952–2961. [CrossRef]

33. Yin, X.; Chen, W.; Eom, J.; Clarke, L.E.; Kim, S.H.; Patel, P.L.; Yu, S. China's transportation energy consumption and CO_2 emissions from a global perspective. *Energy Policy* **2015**, *82*, 233–248. [CrossRef]

34. Yang, W.; Li, T.; Cao, X. Examining the impacts of socio-economic factors, urban form and transportation development on CO_2 emissions from transportation in China: A panel data analysis of China's provinces. *Habitat Int.* **2015**, *49*, 212–220. [CrossRef]

35. Dai, Y.; Gao, H.O. Energy consumption in China's logistics industry: A decomposition analysis using the LMDI approach. *Transport. Res. D Transp. Environ.* **2016**, *46*, 69–80. [CrossRef]

36. Wang, Q. Effective policies for renewable energy—The example of China's wind power—lessons for China's photovoltaic power. *Renew. Sustain. Energy Rev.* **2010**, *14*, 702–712. [CrossRef]

37. Wang, Q.; Li, R. Journey to burning half of global coal: Trajectory and drivers of China's coal use. *Renew. Sustain. Energy Rev.* **2016**, *58*, 341–346. [CrossRef]

38. Zhang, C.; Nian, J. Panel estimation for transport sector CO_2 emissions and its affecting factors: A regional analysis in China. *Energy Policy* **2013**, *63*, 918–926. [CrossRef]

39. Liu, J. Energy saving potential and carbon emissions prediction for the transportation sector in China. *Res. Sci.* **2011**, *33*, 640–646. (In Chinese)

40. Wang, W.; Zhang, M.; Zhou, M. Using LMDI method to analyze transport sector CO_2 emissions in China. *Energy* **2011**, *36*, 5909–5915. [CrossRef]

41. Cai, B.; Yang, W.; Cao, D.; Liu, L.; Zhou, Y.; Zhang, Z. Estimates of China's national and regional transport sector CO_2 emissions in 2007. *Energy Policy* **2012**, *41*, 474–483. [CrossRef]

42. Mao, X.; Yang, S.; Liu, Q.; Tu, J.; Jaccard, M. Achieving CO_2 emission reduction and the co-benefits of local air pollution abatement in the transportation sector of China. *Environ. Sci. Policy* **2012**, *21*, 1–13. [CrossRef]

43. Wei, Q.; Zhao, S.; Xiao, W. A quantitative qnalysis of carbon emissions reduction ability of transportation structure optimization in China. *J. Transp. Eng. Inf. Technol.* **2013**, *13*, 10–17.

44. Guo, B.; Geng, Y.; Franke, B.; Hao, H.; Liu, Y.; Chiu, A. Uncovering China's transport CO_2 emission patterns at the regional level. *Energy Policy* **2014**, *74*, 134–146. [CrossRef]

45. Liu, Z.; Li, L.; Zhang, Y. Investigating the CO_2 emission differences among China's transport sectors and their influencing factors. *Nat. Hazards* **2015**, *77*, 1323–1343. [CrossRef]

46. Xu, B.; Lin, B. Carbon dioxide emissions reduction in China's transport sector: A dynamic VAR (vector autoregression) approach. *Energy* **2015**, *83*, 486–495. [CrossRef]

47. Ma, J.; Heppenstall, A.; Harland, K.; Mitchell, G. Synthesising carbon emission for mega-cities: A static spatial microsimulation of transport CO_2 from urban travel in Beijing. *Comput. Environ. Urban* **2014**, *45*, 78–88. [CrossRef]

48. Liu, X.; Ma, S.; Tian, J.; Jia, N.; Li, G. A system dynamics approach to scenario analysis for urban passenger transport energy consumption and CO_2 emissions: A case study of Beijing. *Energy Policy* **2015**, *85*, 253–270. [CrossRef]

49. Wang, C.; Wang, F.; Wang, Q.; Yang, D.; Li, L.; Zhang, X. Preparing for Myanmar's environment-friendly reform. *Environ. Sci. Policy* **2013**, *25*, 229–233. [CrossRef]

50. Wang, Q.; Chen, Y. Barriers and opportunities of using the clean development mechanism to advance renewable energy development in China. *Renew. Sustain. Energy Rev.* **2010**, *14*, 1989–1998. [CrossRef]

51. Xu, B.; Lin, B. Factors affecting carbon dioxide (CO_2) emissions in China's transport sector: A dynamic nonparametric additive regression model. *J. Clean. Prod.* **2015**, *101*, 311–322. [CrossRef]

52. Shahbaz, M.; Tiwari, A.K.; Nasir, M. The effects of financial development, economic growth, coal consumption and trade openness on CO_2 emissions in South Africa. *Energy Policy* **2013**, *61*, 1452–1459. [CrossRef]

53. Wang, Q.; Chen, X.; Jha, A.N.; Rogers, H. Natural gas from shale formation—The evolution, evidences and challenges of shale gas revolution in United States. *Renew. Sustain. Energy Rev.* **2014**, *30*, 1–28. [CrossRef]

54. Shahbaz, M.; Loganathan, N.; Muzaffar, A.T.; Ahmed, K.; Ali Jabran, M. How urbanization affects CO_2 emissions in Malaysia? The application of STIRPAT model. *Renew. Sustain. Energy Rev.* **2016**, *57*, 83–93. [CrossRef]

55. Tan, X.; Dong, L.; Chen, D.; Gu, B.; Zeng, Y. China's regional CO_2 emissions reduction potential: A study of Chongqing city. *Appl. Energy* **2016**, *162*, 1345–1354. [CrossRef]

56.   IPCC. International Panel on Climate Change (IPCC)'s Task Force on National Greenhouse Gas Inventories (TFI). IPCC Guidelines for National Greenhouse Gas Inventories. 2006. Available online: http://www.ipcc-nggip.iges.or.jp/public/2006gl/pdf/2_Volume2/V2_3_Ch3_Mobile_Combustion.pdf (accessed on 15 June 2016).

57.   Wang, C.; Xie, H. Analysis on dynamic characteristics and influencing factors of carbon emissions from electricity in China. *China Pop. Res. Environ.* **2015**, *25*, 21–27. (In Chinese)

58.   Li, X.; Wang, H.; Chen, Z.; Liu, Q.; Yu, X. Inter-provincial discrepancy and spatiotemporal characteristics of carbon dioxide emission intensity from power energy consumption in China. *J. Arid Land Res. Environ.* **2015**, *29*, 43–47. (In Chinese)

59.   National Bureau of Statistics, China. *China Statistical Yearbook (1990–2013)*; China Statistics Press: Beijing, China, 2014.

60.   Song, R.; Yang, S.; Sun, M. *GHG Protocol Tool for Energy Consumption in China (Version 2.1)*; World Resources Institute (WRI): Washington, DC, USA, 2013.

61.   Dietz, T.; Rosa, E.A. Effects of population and affluence on CO_2 emissions. *Proc. Natl. Acad. Sci. USA* **1997**, *94*, 175–179. [CrossRef] [PubMed]

62.   Ehrlich, P.R.; Holdren, J.P. Impact of population growth. *Science* **1971**, *171*, 1212–1217. [CrossRef] [PubMed]

63.   York, R.; Rosa, E.A.; Dietz, T. STIRPAT, IPAT and ImPACT: Analytic tools for unpacking the driving forces of environmental impacts. *Ecol. Econ.* **2003**, *46*, 351–365. [CrossRef]

64.   García, C.B.; García, J.; Martín, M.M.L.; Salmerón, R. Collinearity: Revisiting the variance inflation factor in ridge regression. *J. Appl. Stat.* **2015**, *42*, 648–661. [CrossRef]

65.   Alkhamisi, M.A.; Macneill, I.B. Recent results in ridge regression methods. *Metron* **2015**, *73*, 359–376. [CrossRef]

66.   García, J.; Salmerón, R.; García, C.; Martín, M.D.M.L. Standardization of Variables and Collinearity Diagnostic in Ridge Regression. *Int. Stat. Rev.* **2015**, *84*, 245–266. [CrossRef]

67.   Hoerl, A.E.; Kennard, R.W. Ridge regression: Biased estimation for nonorthogonal problems. *Technometrics* **2000**, *42*, 80–86. [CrossRef]

68.   Inman, J.R. Resistivity Inversion with Ridge Regression. *Geophysics* **2012**, *40*, 798–817. [CrossRef]

69.   Statistics Bureau of Xinjiang Province. *50 Years of Glories of Xinjiang (1949–1999)*; Xinjiang People's Press: Urumuqi, China, 2000.

70.   The Institute of Contemporary China Studies. The History of the People's Republic of China. Available online: http://www.hprc.org.cn/wxzl/wxysl/wnjj/ (accessed on 15 June 2016).

71.   Departmeng of Energy Statistics, National Bureau of Statistics. *China Energy Statistical Yearbook (1990–2014)*; China Statistics Press: Beijing, China, 2014.

72.   Wang, Z.; Yang, L. Delinking indicators on regional industry development and carbon emissions: Beijing–Tianjin–Hebei economic band case. *Ecol. Indic.* **2015**, *48*, 41–48. [CrossRef]

73.   Wang, Q.; Chen, X. Energy policies for managing China's carbon emission. *Renew. Sustain. Energy Rev.* **2015**, *50*, 470–479. [CrossRef]

74.   Wang, Q.; Li, R. Natural gas from shale formation: A research profile. *Renew. Sustain. Energy Rev.* **2016**, *57*, 1–6. [CrossRef]

75.   Wang, Q. China should aim for a total cap on emissions. *Nature* **2014**, *512*, 115. [CrossRef] [PubMed]

76.   Kock, N.; Lynn, G. Lateral Collinearity and Misleading Results in Variance-Based SEM: An Illustration and Recommendations. *J. Assoc. Inf. Syst.* **2012**, *13*, 546–580.

77.   O'Brien, R.M. A Caution Regarding Rules of Thumb for Variance Inflation Factors. *Qual. Quant.* **2007**, *41*, 673–690. [CrossRef]

78.   Nie, J. China's one-child policy, a policy without a future. Pitfalls of the "common good" argument and the authoritarian model. *Camb. Q. Healthc. Ethics* **2014**, *23*, 272–287. [CrossRef] [PubMed]

6

Application of Emergy Analysis to the Sustainability Evaluation of Municipal Wastewater Treatment Plants

Shuai Shao [1,2,3,*], Hailin Mu [1,*], Fenglin Yang [3], Yun Zhang [3] and Jinhua Li [3]

[1] Key Laboratory of Ocean Energy Utilization and Energy Conservation of Ministry of Education, School of Energy and Power Engineering, Dalian University of Technology, Linggong Road 2, Dalian 116024, Liaoning, China

[2] School of Innovation and Entrepreneurship, Dalian University of Technology, Linggong Road 2, Dalian 116024, Liaoning, China

[3] School of Environmental Science and Technology, Dalian University of Technology, Linggong Road 2, Dalian 116024, Liaoning, China; yangfl@dlut.edu.cn (F.Y.); zhangyun@dlut.edu.cn (Y.Z.); lijinhua@mail.dlut.edu.cn (J.L.)

* Correspondence: shaoshuai24015@mail.dlut.edu.cn (S.S.); hailinmu@dlut.edu.cn (H.M.)

Academic Editor: Vincenzo Torretta

Abstract: Municipal wastewater treatment plants consume much energy and manpower, are expensive to run, and generate sludge and treated wastewater whilst removing pollutants through specific treatment regimes. The sustainable development of the wastewater treatment industry is therefore challenging, and a comprehensive evaluation method is needed for assessing the sustainability of different wastewater treatment processes, for identifying the improvement potential of treatment plants, and for directing policymakers, management measures and development strategies. This study established improved evaluation indicators based on Emergy Analysis that place total wastewater, resources, energy, economic input and emission of pollutants on the same scale compared to the traditional indicators. The sustainability of four wastewater treatment plants and their associated Anaerobic-Anoxic-Oxic (A2O), Constant Waterlevel Sequencing Batch Reactor (CWSBR), Cyclic Activated Sludge Technology (CAST) and Biological Aerated Filter (BAF) treatment processes were assessed in a city in northeast China. Results show that the CWSBR process was the most sustainable wastewater treatment process according to its largest calculated value of Improved Emergy Sustainable Index (2.53×10^0), followed by BAF (1.60×10^0), A2O (9.78×10^{-1}) and CAST (5.77×10^{-1}). Emergy Analysis provided improved indicators that are suitable for comparing different wastewater treatment processes.

Keywords: emergy analysis; wastewater treatment process; improvement indicator; sustainability evaluation

1. Introduction

Municipal wastewater treatment plants utilize specific technologies to reduce or eliminate pollutants in wastewater resulting from human activities and other sources, and therefore play a critical role in reducing water pollution. In 2015, discharge of wastewater (including industrial wastewater, municipal sewage, and pollution control facility wastewater) pollutants in China alone resulted in chemical oxygen demand (COD) emissions of 22.235 million tons and ammonia nitrogen (NH_3-N) emissions of 2.299 million tons [1]. Furthermore, urbanization processes are rapidly accelerating in China, and water consumption and wastewater treatment demand are subsequently increasing. By the end of 2015, the treatment capacity of national municipal wastewater treatment plants reached 140 million cubic meters per day, cumulative wastewater disposal reached 41.03 billion cubic meters, and the national municipal wastewater treatment rate reached 91.97% [2]. Wastewater treatment facilities are therefore an important part of municipal construction in China.

Wastewater treatment plants consume a lot of energy, chemicals, manpower, and are associated with various environmental issues. Furthermore, they generate sludge and treated wastewater during the process of collecting, treating, and discharging municipal wastewater to acceptable permit standards. Municipal wastewater treatment plants must simultaneously consider environmental, energy and economic impacts according to municipal structures and developmental planning constraints [3], and these considerations influence the choice of treatment processes and screen wastewater treatment technologies employed. Sustainable wastewater treatment processes aiming to address the problems above include hydrodynamic cavitational technique [4], molecular imprinting [5] and microbial fuel cells [6]. The future wastewater treatment industry in China must address: (1) how to build new wastewater treatment plants with appropriate treatment processes; and (2) how to maximize the potential of existing wastewater treatment plants by implementing energy-saving and emission-reduction technologies to meet the needs of sustainable development. To achieve these aims, a systematic evaluation method is needed to determine the improvement potential of wastewater treatment plants, to evaluate the sustainability of different wastewater treatment processes, and to guide policymakers, management measures and development strategies.

The sustainable development of an industrial system (the industrial entirety with specific functions which is composed of several interactional and interdependent components) needs to be technologically feasible, environmentally friendly and economically attractive. For wastewater treatment plants, technological and environmental feasibility is dependent on adopting advanced and stable treatment processes that are appropriate for different wastewater treatments and that generate treated water meeting national discharge standards. However, a system that is technologically and environmentally feasible but not economically competitive is likely to be unsustainable.

Currently, researchers apply various methods to evaluate the resource use, energy and environmental performance of industrial systems, including Ecological Footprint Analysis [7–9], Substance Flow Analysis [10–13], Energy Flow Analysis [14–16], Exergy Analysis [17–19], Life Cycle Assessment [20–22], Emergy Analysis [23–26] and other methods. For a method to be suitable for evaluating the sustainable development of wastewater treatment processes, it needs to be possessing clear indicator system, quantifiable and able to simultaneously consider the social, economic and environmental impacts.

Emergy Analysis is a system analysis method based on energy analysis which has been widely applied to countries and regions [27–29], ecological systems [30–32], and industrial systems [33–35].

The concepts of Emergy Theory and the Emergy Analysis method were proposed by the American ecologist H.T. Odum in the 1980s. Emergy is defined as the total amount of solar energy directly or indirectly input into any given resource, product or service, in units of Solar Emjoules (sej) [36]. Different types of energy differ quantitatively and qualitatively, and they can be related using a specific conversion relationship termed solar transformity, which refers to the amount of solar energy required to form each unit of product or service. The solar emergy of a particular energy can be calculated by multiplying the basic amount of energy by its corresponding solar transformity using the expression $M = T \times B$ [37], where M represents emergy in sej, T represents solar transformity, and B refers to the basic amount of energy. Emergy is an important object function which is used to study self-organization processes in ecological systems, ecosystem function, depicting and simulating ecosystem behavior, predicting evolutionary trends, and evaluating ecosystem sustainability [38].

Emergy Analysis introduces solar transformity into the calculation based on Emergy Theory. In order to assess the function and status of energy in a system, it is quantitatively analyzed by converting different types and levels of energy into solar emergy.

Emergy Analysis can provide a common scale for measuring and comparing different substances, energy types, environmental impacts and economic indicators. The method can therefore be used to evaluate the sustainability of wastewater treatment processes, and for determining the sustainable development potential of wastewater treatment systems (the entirety reasonably formed by several unit treating procedures). To date, most Emergy Analysis of wastewater treatment processes has focused on different aspects of a single process [39], and comparison of the sustainability of different treatment processes has received less attention.

The present study established improved emergy evaluation indicators (Improved Emergy Yield Ratio, Improved Environment Load Ratio, Improved Emergy Sustainable Index) for wastewater treatment systems that place total wastewater, resources, energy, economic input and emission of pollutants on the same scale compared to the traditional indicators (Emergy Yield Ratio, Environment Load Ratio, and Emergy Sustainable Index). The coastal city in northeast China studied in the present work has more than 20 large- and medium-sized wastewater treatment plants, at which the main treatment processes are Anaerobic-Anoxic-Oxic (A2O), Constant Waterlevel Sequencing Batch Reactor (CWSBR), Cyclic Activated Sludge Technology (CAST) and Biological Aerated Filter (BAF). This study focused on four wastewater treatment plants running the treatment processes described below (Cases 1–4). This method, based on emergy flow analysis per ton of wastewater treated, provides technical guidance for policymakers to compare wastewater treatment technologies for new plants, and for upgrading existing plants.

2. Materials and Methods

2.1. Methods

2.1.1. Emergy Evaluation Indicators

Emergy evaluation indicators are established based on the outputs of Emergy Analysis, and they can unify different types of ecological flows at the emergy scale, as and quantitatively evaluate the structure and function of a system. Emergy indicators therefore offer a useful basis for evaluating the sustainability of a system.

Traditional Emergy Evaluation Indicators

Traditional emergy indicators [37] that are suitable for evaluating ecological-economic systems are shown in Table 1, where R refers to input emergy from renewable resources in the natural environment, N refers to input emergy from non-renewable resources, and F represents the purchased input emergy from the socioeconomic system.

Table 1. Traditional emergy evaluation indicators and calculation formulas.

Indicators	Formula	Implications
Emergy Yield Ratio (EYR)	$(N + R + F)/F$	Emergy efficiency and economic competitiveness of the system
Environment Load Ratio (ELR)	$(F + N)/R$	Environmental loading exerted by the system
Emergy Sustainable Index (ESI)	EYR/ELR	Sustainability of the system

The emergy yield ratio (EYR) of the system reflects the feedback emergy from the economic system, and is a measure of how much the emergy yield contributes to the economic system. EYR is used to judge the emergy efficiency and economic competitiveness of a system based on purchased inputs. ELR (environment load ratio) is the ratio of non-renewable input emergy (including purchased emergy and emergy from non-renewable resources) to renewable resources input emergy, and this parameter denotes the pressure economic activities place on the environment. Systems should avoid a high ELR status for a long period of time in order to avoid causing irreversible functional degradation to the environment. The emergy sustainability index (ESI) is the ratio of EYR to ELR, and reflects both ecological and economic benefits in terms of the overall sustainability of the system.

Various studies have employed emergy theory and analysis based on the indicators described above [40,41], whereas others have developed improved indicators based on traditional emergy evaluation indicators [42,43]. Industrial systems have their own characteristics compared with ecological-economic systems, therefore, it is necessary to explicit the connotation, mechanism and application situations of "emergy" in order to establish the improved and specific emergy evaluation indicators suitable for different industrial systems.

Improved Emergy Evaluation Indicators

Wastewater is the main input into the wastewater treatment system, together with auxiliary inputs from natural and social resources, while the main outputs are drainage (treated water) and sludge. It should be noted that although the discharge concentration of chemicals in the drainage must meet national standards, COD, NH_3-N and other pollutants present in the treated water may still impact the environment, and should therefore be taken into account in the emergy indicators.

Figure 1 shows the wastewater treatment system emergy flow diagram, in which R refers to renewable natural resources emergy, N denotes non-renewable natural resources emergy, F_R refers to purchased renewable resources emergy, F_N denotes purchased non-renewable resources emergy, F_S refers to capital emergy, and W_{in}, W_{out} and S represent the emergy of wastewater input, drainage and sludge, respectively. The total input emergy of renewable natural resources, non-renewable natural resources, purchased renewable resources, purchased non-renewable resources and capital combine to support the treatment of wastewater (input) to drainage (output), while simultaneously meeting the discharge standards and generation of sludge.

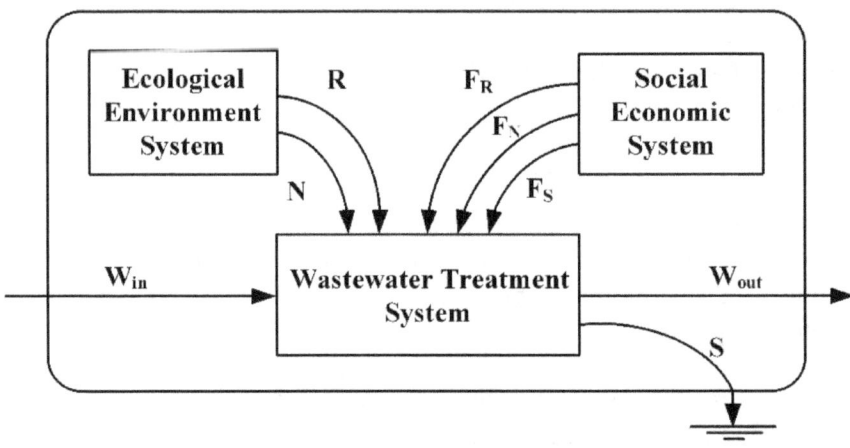

Figure 1. Emergy flow diagram of wastewater treatment systems.

(1) Improved Emergy Yield Ratio (IEYR) denotes the ratio of the net emergy input to the economic feedback emergy in the wastewater treatment system, where "input" is defined as renewable resources, non-renewable resources, capital and other types of input emergy invested in order to obtain target products or meet economic interests. Wastewater input emergy also reflects total input emergy in terms of unique raw materials used in the wastewater treatment system, whereas "net emergy" is embodied within the sludge emergy and the emergy of pollutants in drainage, both of which have a negative influence on the economics of the system and should therefore be deducted from the total input emergy. Economic feedback emergy, in addition to the emergy of purchased renewable resources, purchased non-renewable resources and capital, as well as the money invested by the economic system for sludge disposal should also be considered. The capital emergy for sludge disposal is indicated by P, the cost of treating a unit of sludge, which equals 270 Yuan/ton [44].

When the net emergy input of the wastewater treatment system is larger and the total economic feedback emergy is smaller, the Improved Emergy Yield Ratio (IEYR) is greater. The expression is as follows:

$$IEYR = (W_{in} + R + N + F_R + F_N + F_S - W_{out} - S)/(F_R + F_N + F_S + P) \qquad (1)$$

IEYR can reflect the relationship between the emergy efficiency and the market competitiveness of different wastewater treatment processes. The larger the value, the more net emergy yield can be obtained by investing per unit of economic feedback emergy. In this case the competitiveness of the system is greater, and the economic benefit of the process is higher. Conversely, a smaller value indicates a lower emergy yield efficiency and lower competitiveness.

(2) Improved Environment Load Ratio (IELR) reflects the impact on the environment and the resource dependence of the main operation. The dependence on resources can be divided into positive benefit and negative benefit, where positive benefit refers to the dependence on renewable resources, and negative benefit reflects the dependence on non-renewable resources and capital. The impact on the environment caused by the output of drainage and sludge both fall into the negative benefit category, since industrial discharge and pollutants have a negative impact on the environment.

IELR equals the ratio of the sum of non-renewable natural resources emergy, purchased non-renewable resources emergy, capital emergy, sludge emergy and the emergy of pollutants in drainage over the sum of renewable natural resources emergy and purchased renewable resources emergy. This parameter describes the effect that the wastewater treatment system has on the local environmental ecosystem. The expression is as follows:

$$IELR = (N + F_N + F_S + W_{out} + S)/(R + F_R) \qquad (2)$$

IELR represents the negative impact on the environment caused by non-renewable resources consumption, economic burden and waste discharge. The larger the value of this parameter, the pressure on the surrounding environment is greater. Conversely, a smaller value indicates less pressure on the surrounding environment, in cases where the system has a more eco-friendly operational mode, and the local surroundings have adequate time and space to dilute the environmental impact and can recycle resources.

(3) Improved Emergy Sustainable Index (IESI) denotes the ratio of the Improved Emergy Yield Ratio (IEYR) and the Improved Environment Load Ratio (IELR). This composite indicator reflects the emergy yield efficiency under a certain environmental load, and can simultaneously evaluate the sustainability of different wastewater treatment processes. The expression is as follows:

$$IESI = IEYR/IELR \qquad (3)$$

IESI reflects the sustainability of the wastewater treatment system, and a larger value indicates a higher level of sustainability.

The formula comparison of the traditional and improved emergy evaluation indicators is shown in Table 2.

Table 2. Formula comparison of traditional and improved emergy evaluation indicators.

Indicators	Formula	Indicators	Formula
EYR	$(N + R + F)/F$	IEYR	$(W_{in} + R + N + F_R + F_N + F_S - W_{out} - S)/(F_R + F_N + F_S + P)$
ELR	$(F + N)/R$	IELR	$(N + F_N + F_S + W_{out} + S)/(R + F_R)$
ESI	EYR/ELR	IESI	IEYR/IELR

2.1.2. Environmental Impact Emergy of Pollutants in Drainage

Drainage outputs may contain different types of pollutants such as COD, BOD_5, NH_3-N and TP. Furthermore, even the input wastewater treated by the wastewater treatment system may cause pollution to the environment when the concentration of drainage pollutants is greater than the environmental background concentration. Water in the natural environment decreases the concentration of pollutants through autopurificaton and dilution, and this reduces damage to the environment. Therefore, the impact of discharge pollutants on the environment can be calculated in terms of the emergy associated with dilution of pollutants to their environmental background concentration [45]. The expression for calculating the water requirement for diluting pollutants is as follows:

$$M_{w,i} = d \times (W_i/c_i) - M_w \tag{4}$$

In this expression, $M_{w,i}$ refers to the water requirement for diluting pollutants in units of g, d denotes the density of water in units of 1×10^6 g/m^3, W_i refers to the discharge quantity of pollutant i in units of g, c_i denotes the environmental background concentration of pollutant i in units of g/m^3, and M_w refers to the total discharge quantity of the wastewater treatment system in units of g.

Table 3. Environmental safety concentrations for selected pollutants.

Item	COD	BOD_5	NH_3-N	TP
Concentration (mg/L) [46]	15	3	0.15	0.02

In our calculations, we adopted the environmental safety concentrations of different types of pollutants stipulated by regulations in China (Table 3) rather than the environmental background concentrations due to the difficulty of accessing actual environmental background concentration data. The expression used to calculate the emergy of the water requirement for diluting pollutants is as follows:

$$ECEW_i = M_{w,i} \times 4.92J/g \times 4.48 \times 10^4 sej/J \tag{5}$$

In this expression, $ECEW_i$ refers to the emergy consumption of diluting the emission of water pollutant i in units of sej, the Gibbs free energy of water is 4.92 J/g, and the emergy transformity of surface water is 4.48×10^4 sej/J. Since different types of pollutants are diluted by water simultaneously, we selected the maximal emergy value of a certain pollutant ($ECEW_{max}$) as the environmental impact emergy of all pollutants in drainage.

2.2. Materials

2.2.1. Description of the Study Objects

Current municipal wastewater treatment processes adopted in China include A/O, A2O, Oxidation Ditch (OD), SBR, and Biofilms and their variants [47]. Emergy Analysis was performed for all four cases studied in present work using actual (collected) production data, natural conditions and economic indicators. The basic information for each case is described as follows:

Case 1: The Anaerobic-Anoxic-Oxic (A2O) process. This process involved a complete denitrification and dephosphorization regime based on the biological environment and alternating anaerobic, anoxic and aerobic reactions. The appropriate reaction conditions for denitrification and dephosphorization are required in the anaerobic and anoxic steps, respectively, and reaction conditions must also be conducive for treating COD and BOD in the aerobic steps. The wastewater treatment capacity of Case 1 was 30,000 m^3/day, the occupied area was 4.8 hectares, and the sludge yield per day was 14,280 kg.

Case 2: The Constant Waterlevel Sequencing Batch Reactor (CWSBR) process. This process overcomes the disadvantages of intermittent inflow, intermittent drainage and waterlevel sequencing changes by retaining the advantages of the traditional SBR process and combining with constant waterlevel sequencing and continuous operation throughout the entire wastewater treatment process. The wastewater treatment capacity of Case 2 was 30,000 m^3/day, the occupied area was 3.5 hectares, and the sludge yield per day was 9540 kg.

Case 3: The Cyclic Activated Sludge Technology (CAST) process. This process is renowned for its sludge sedimentation and dehydration performance. Wastewater is treated repeatedly by aeration, precipitation and skimming, and sludge inverse and discharge systems are incorporated at the end of the wastewater treatment system. The wastewater treatment capacity of Case 3 was 80,000 m^3/day, the occupied area was 2.5 hectares, and the sludge yield per day was 21,280 kg.

Case 4: The Biological Aerated Filter (BAF) process. This process utilizes a denitrifying biological filter and applies a new microbial adhesion wastewater treatment technology to accomplish biological treatment and solid-liquid separation simultaneously. The equipment required for this treatment is compact, biochemical reactions and filtration occur in the same unit, the occupied area is reduced due to not requiring a secondary sedimentation tank, and the filter needs aeration in order to biologically oxidize organic compounds and ammonia nitrogen. The wastewater treatment capacity of Case 4 was 120,000 m^3/day, the occupied area was 3.5 hectares, and the sludge yield per day was 58,440 kg.

The main treatment procedures of wastewater and sludge of the four cases are shown in Table 4.

Table 4. Main treatment procedures of wastewater and sludge of the four cases.

Case No.	Wastewater Treatment			Sludge	
	Preliminary	Secondary	Tertiary	Treatment	Disposal
1	Coarse and Fine Screens, Grit Chamber	A2O Process, Secondary Sedimentation Tank	Advanced Treatment	Thickening, Dewatering	Outward Transport
2	Coarse and Fine Screens, Aerated Grit Chamber	CWSBR Process	Disinfected	Dewatering	Outward Transport
3	Fine Screen, Vortex Grit Chamber	CAST Process	Advanced Treatment	Stabilization, Thickening, Dewatering	Outward Transport
4	Coarse, Intermediate and Fine Screens, Vortex Grit Chamber, Sedimentation Tank	BAF Process	Ultraviolet, Disinfected	Dewatering	Outward Transport

2.2.2. Process Flow Diagram

Process flow diagrams for the four wastewater treatment Processes are shown in Figures 2–5.

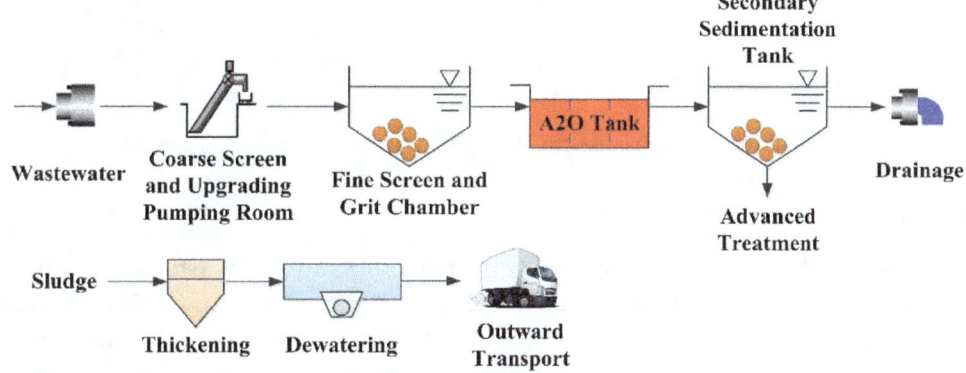

Figure 2. Flow diagram for the Anaerobic-Anoxic-Oxic (A2O) process.

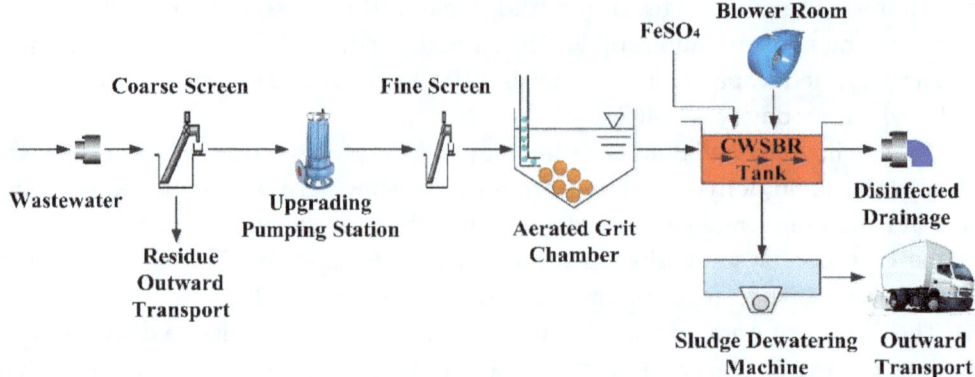

Figure 3. Flow diagram for the Constant Waterlevel Sequencing Batch Reactor (CWSBR) process.

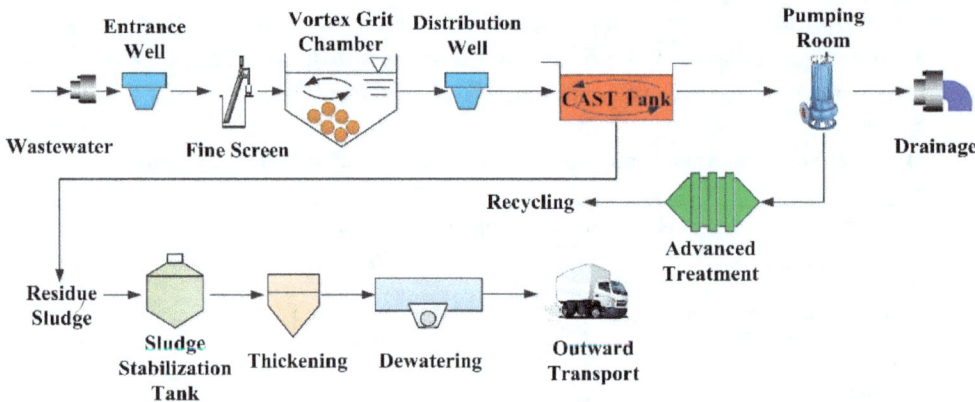

Figure 4. Flow diagram for the Cyclic Activated Sludge Technology (CAST) process.

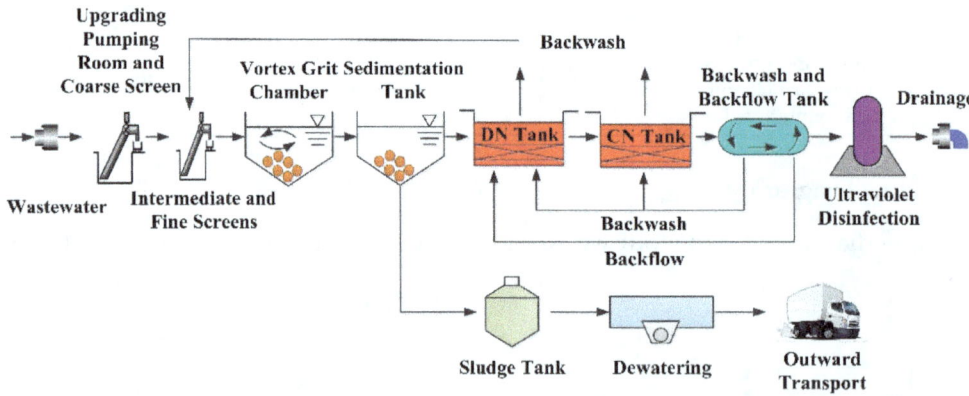

Figure 5. Flow diagram for the Biological Aerated Filter (BAF) process.

2.2.3. Basic Natural, Economic and Technical Data

The data used in this study were derived from actual operational wastewater treatment plants. All four plants studied were situated in the same city and were close geographically, therefore the same set of natural conditions were used for all calculations and analysis. Specifically, the annual average precipitation was 658.6 mm, the average wind speed was 4.4 m/s, the relative humidity was 68%, the average sunshine duration was 2650 h, and the annual average evaporation was 1615 mm.

The main economic and technical indicators of the four cases are shown in Table 5.

Table 5. Main economic and technical indicators of the four cases.

Case No.	Working Time (Days/Year)	Manpower (Persons)	Investment (Ten Thousand Yuan/Year)	Cost (Ten Thousand Yuan/Year)
1	360	35	531.93	1343.42
2	360	20	220.55	584.14
3	360	30	469.07	1310.00
4	360	44	535.35	1456.97

The design discharge standards of the four cases all met the national wastewater discharge primary standards (GB18918-2002) [48]. Case 1 met the primary B standards while Cases 3 and 4 met the primary A standards, and Case 2 broadly met the discharge standards for NH_3-N based on primary B standards. The specific input and output water quality indicators of the four cases were derived from actual operation data meeting the design discharge standards, as shown in Table 6.

Table 6. Input and output water quality indicators based on actual operation of the four cases.

Case No.	Input/Output	BOD_5 (mg/L)	COD (mg/L)	NH_3-N (mg/L)	TP (mg/L)
1	Input	180	360	35	5
	Output	10	50	5(8)	0.5
2	Input	200	400	30	3
	Output	10	50	5(8)	0.5
3	Input	150	350	26	3
	Output	20	60	15	1
4	Input	180	400	25	3.5
	Output	20	60	8	1

2.2.4. Synthesized Emergy System Diagram

The synthesized emergy system diagram of the wastewater treatment system was drawn based on the Energy System Language Legend [37] introduced by Odum. It includes technological processes and data collected on substances, energy, currency and other information shown in Figure 6. It should be noted that "Wastewater Treatment System" was replaced by each of the four treatment processes described above during analysis.

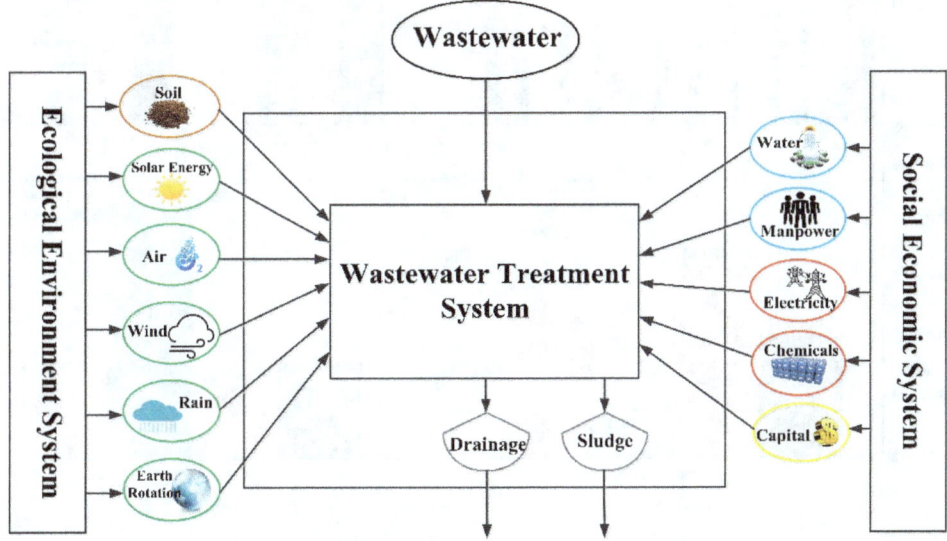

Figure 6. Diagram of the synthesized emergy system for wastewater treatments.

3. Results and Discussions

3.1. Results

3.1.1. Results of Emergy Analysis

The ECEW values of drainage pollutants (Table 7) were calculated according to Equations (4) and (5), and Table 6.

Table 7. ECEW values of drainage pollutants.

Case No.	$ECEW_{COD}$ (sej/Year)	$ECEW_{BOD5}$ (sej/Year)	$ECEW_{NH_3-N}$ (sej/Year)	$ECEW_{TP}$ (sej/Year)	$ECEW_{max}$ (sej/Year)
1	5.55×10^{18}	5.55×10^{18}	1.25×10^{20}	5.71×10^{19}	1.25×10^{20}
2	5.55×10^{18}	5.55×10^{18}	1.25×10^{20}	5.71×10^{19}	1.25×10^{20}
3	1.90×10^{19}	3.60×10^{19}	6.28×10^{20}	3.11×10^{20}	6.28×10^{20}
4	2.86×10^{19}	5.40×10^{19}	4.98×10^{20}	4.67×10^{20}	4.98×10^{20}

Tables S1−S4 list the main resources and energy inputs and outputs of each wastewater treatment system, and the emergy of disposing wastewater in units of per ton, for the different treatments. Different units of ecological flows were converted to common solar emergy units based on the corresponding transformities of different kinds of resources and energy. Transformities used in this study are as described by H.T. Odum and other emergy researchers [37,42,49–52]. Transformities of substances with similar properties were applied in instances where actual transformities values were unknown. The emergy inventory of each wastewater treatment system is discussed in the Supplementary Materials.

3.1.2. Emergy Flows Analysis

Elaborated data of emergy flows for each case are shown in Figure 7.

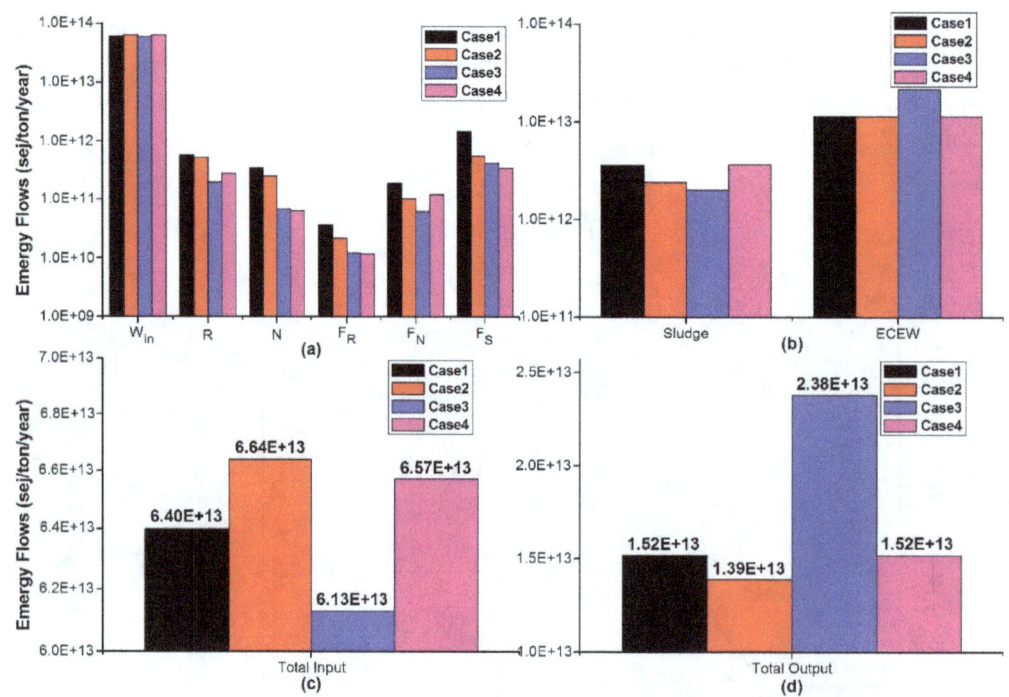

Figure 7. Elaborated data of emergy flows: (**a**) Input emergy flows for each case; (**b**) Output emergy flows for each case; (**c**) The total input emergy flow; and (**d**) The total output emergy flow.

In terms of input emergy flow, the total input emergy of Case 2 was the highest of the four cases, with a value of 6.64×10^{13} sej/ton/year, followed by Case 4 and Case 1, with values of 6.57×10^{13} sej/ton/year and 6.40×10^{13} sej/ton/year, respectively, while Case 3 had the smallest total input emergy value of 6.13×10^{13} sej/ton/year. The total input emergy of wastewater treatment systems is composed of W_{in}, R, N, F_R, F_N and F_S. W_{in} was highest for Cases 2 and 4 (6.49×10^{13} sej/ton/year), and lowest for Case 3 (6.05×10^{13} sej/ton/year). R and F_N emergy inputs were largest for Case 1 (5.79×10^{11} sej/ton/year and 1.91×10^{11} sej/ton/year, respectively) and smallest for Case 3 (1.98×10^{11} sej/ton/year and 6.25×10^{10} sej/ton/year, respectively). Similarly, N, F_R and F_S were all largest in Case 1 (3.48×10^{11} sej/ton/year, 3.69×10^{10} sej/ton/year and 1.46×10^{12} sej/ton/year, respectively) but were smallest in Case 4 (6.34×10^{10} sej/ton/year, 1.18×10^{10} sej/ton/year and 3.48×10^{11} sej/ton/year, respectively).

In terms of output emergy flow, the total output emergy of Case 3 was largest, with the value of 2.38×10^{13} sej/ton/year, followed by Cases 1 and 4 (both 1.52×10^{13} sej/ton/year), and smallest for Case 2 (1.39×10^{13} sej/ton/year). The total output emergy of all four cases is composed of the sludge emergy and the ECEW of drainage. Regarding the sludge emergy output, this was highest in Case 4 (3.69×10^{12} sej/ton/year), closely followed by Case 1 (3.62×10^{12} sej/ton/year), and was lowest for Case 3 (2.02×10^{12} sej/ton/year). Meanwhile, ECEW was highest for Case 3 (2.18×10^{13} sej/ton/year) and comparable (1.15×10^{13} sej/ton/year) for the other three cases.

3.1.3. Emergy Indicators Analysis

Wastewater input emergy accounts for over 90% of the total input emergy, due to the nature of wastewater treatment systems. Therefore, the wastewater input emergy was excluded during emergy flow proportion analysis in order to more accurately compare the relative proportion of other types of input emergy flows. The relative proportions of emergy flows for each case are listed in Table 8.

Table 8. Emergy flow contributions for the four cases (%).

Item	Case 1	Case 2	Case 3	Case 4
R	22.16	35.70	25.85	34.20
N	13.30	17.31	8.87	7.65
F_R	1.41	1.48	1.62	1.42
F_N	7.30	7.14	8.17	14.72
—Electricity Emergy	7.30	5.93	8.04	11.80
—Chemicals Emergy	—	1.21	0.13	2.92
F_S	55.83	38.37	55.49	42.01

The F_S emergy flow accounted for the largest proportion of input emergy flow in all four cases (Case 2 = 38.37%; Case 4 = 42.01%; Case 3 = 55.49%; Case 1 = 55.83%). The R emergy flow was the second highest input emergy flow contributor in all four cases, and this was significantly higher for Case 2 (35.70%) and Case 4 (34.20%) than for Case 3 (25.85%) and Case 1 (22.16%). The proportion of N emergy flow for Cases 1−3 (13.30%, 17.31% and 8.87%) was larger than the contribution made by F_N emergy flow (7.30%, 7.14% and 8.17%), whereas the N emergy flow of Case 4 (14.72%) was higher than the F_N emergy flow (7.65%). Within the F_N emergy flow parameter, the electricity emergy flow made the largest contribution in Case 4 (11.80%) and the smallest contribution in Case 2 (5.93%), and this parameter also had the largest chemicals emergy flow proportion (Case 4 = 2.92%), followed by Case 2 (1.21%), while Cases 1 and 3 had a smaller chemicals emergy flow proportion. Finally, the contribution of F_R emergy to input emergy flow was minimal in all four cases (Case 3 = 1.62%; Case 2 = 1.48%; Case 4 = 1.42%; Case 1 = 1.41%).

It can be seen from the input emergy flows (not including wastewater emergy input) that in all four cases, the F_S emergy input made the largest contribution, followed by the R emergy input and N and F_N emergy inputs, while the F_R emergy input had minimal impact on the total input emergy.

The improved emergy evaluation indicators and the order of sustainability of each case are listed in Table 9. IEYR values ranged between 2.72×10^1 and 8.53×10^1, and were largest in Case 4, followed by Cases 2, 3 and 1. The emergy efficiency and economic competitiveness of the four treatment processes are therefore ordered Case 4 > Case 2 > Case 3 > Case 1. IELR values ranged between 2.73×10^1 and 1.16×10^2, and were ordered Case 2 < Case 1 < Case 4 < Case 3, suggesting Case 2 placed the least pressure on the surrounding environment, while Case 3 had the largest impact. IESI values ranged between 5.77×10^{-1} and 2.53×10^0 and were ordered Case 2 > Case 4 > Case 1 > Case 3. The CWSBR process used in Case 2 was therefore the most sustainable, while the CAST process used in Case 3 was the least sustainable process employed in the four cases. The overall order of sustainability for the four wastewater treatment processes was CWSBR > BAF > A2O > CAST.

Table 9. Improved emergy evaluation indicators for the four cases.

Case No.	Process	Improved Emergy Indicator			Sustainability Order
		IEYR	IELR	IESI	
1	A2O	2.72×10^1	2.79×10^1	9.78×10^{-1}	3
2	CWSBR	6.90×10^1	2.73×10^1	2.53×10^0	1
3	CAST	6.69×10^1	1.16×10^2	5.77×10^{-1}	4
4	BAF	8.53×10^1	5.34×10^1	1.60×10^0	2

3.2. Discussions

This study attempted to establish improved emergy evaluation indicators for different wastewater treatment processes based on existing emergy indicators and the results of four case studies.

Case 1 employed the A2O process and was ranked third based on IESI value. Even though the scale of this treatment scale was not particularly large, the contribution from the capital emergy F_S (55.83%) was the largest of the four cases. regarding emergy flow, the electricity emergy and capital emergy indicators of Case 1 (1.91×10^{11} sej/ton/year and 1.46×10^{12} sej/ton/year, respectively) were the largest of the four processes, mainly because the A2O process involves increased investment and energy consumption due to the addition of sludge recirculation and internal reflux systems. However, the A2O process can be improved by energy-saving and consumption-reducing processes by adjusting methods and controlling aeration [53]. One improvement is reducing the aeration length by transforming the aerobic zone into the anoxic zone, which reduces energy consumption by decreasing aeration. Additionally, strict control of DO quantity in the aeration zone can minimize waste caused by excessive aeration.

The CWSBR process employed in Case 2 had the largest IESI value and hence the highest sustainability of the four processes. Compared with the other three processes, CWSBR relies more on renewable natural resources input emergy R (35.70%), and renewable resources emergy R and non-renewable resources emergy N inputs are well balanced in this process. The contributions from both electricity emergy (5.93%) and capital emergy (38.37%) are minimal in all four cases. Regarding emergy flow, the total waste emergy flow output was the smallest for the CWSBR process (1.39×10^{13} sej/ton/year, and sludge emergy output and electricity emergy input were also relatively small (2.41×10^{12} sej/ton/year and 8.68×10^{10} sej/ton/year, respectively). This can be explained by the characteristics of the CWSBR process. Firstly, this process improves on the conventional SBR process by switching form a variable waterlevel to a constant waterlevel mode of operation with the help of the reciprocating motion of a hydro sail. This modification allows the CWSBR process to periodically accomplish the entire process of filling, stirring, aeration, precipitation and decanting with continuous input and output water in a single pond. Secondly, this process reduces the hydraulic loss and energy consumption used during constant waterlevel decanting. The CWSBR process therefore covers a smaller area and involves higher system integration, stronger impulsion load resistance, and lower sludge retention, and the number of ponds can be increased or decreased according to water quality and quantity, which avoids unnecessary energy consumption.

The CAST process employed in Case 3 improves nitrogen and phosphorus removal by increasing sludge recirculation compared with the conventional SBR process. This process had the smallest IESI value and hence the lowest sustainability of the four processes. Regarding emergy flow, both electricity emergy (6.16×10^{10} sej/ton/year) and capital emergy (4.25×10^{11} sej/ton/year) inputs were lower than Cases 1 and 2, because the CAST process has no primary sedimentation or secondary sedimentation tanks, and involves a higher degree of automation and thus less capital investment. Meanwhile, the ECEW of drainage (2.18×10^{13} sej/ton/year) for the CAST process was markedly higher than for the other three processes. Regarding the relative emergy proportions, the CAST process relied more on capital emergy (8.04%) and electricity emergy (55.49%) than the other three processes, but less on renewable natural resources emergy R (25.85%). To improve the CAST process, a blower with better adjustment should be utilized, or combinations of blowers could be used to facilitate easier control, which would likely save energy and reduce consumption. Additionally, stricter control of the water discharge quality could decrease the emission of NH_3-N and reduce the discharge of other pollutants in drainage.

The sustainability of the BAF process employed in Case 4 was ranked second in terms of IESI value. Capital emergy F_S (3.48×10^{11} sej/ton/year) input was the lowest of the four processes, but electricity emergy was the highest of the four processes (9.78×10^{10} sej/ton/year), accounting for 11.80%. This is mainly because the size of the filler particles used in the BAF process are generally smaller, and the filter therefore reaches the designed head loss in a shorter time period when the amount of suspended solids (SS) in the input water is high, which inevitably leads to frequent backwash and higher energy consumption. The purchased chemicals emergy of Case 4 was 2.42×10^{10} sej/ton/year, accounting for 2.92% of the total emergy input, and both values are larger than the other three processes. Meanwhile, the sludge emergy output was a relatively high 3.69×10^{12} sej/ton/year, because the BAF process must remove the chemicals added to remove SS in the primary sedimentation tank in order to prevent suspended particles from clogging the filter during operation. This increases the cost of chemicals and sludge produced by adding chemicals and precipitation. Improvements could be made by incorporating technologies and methods that enhance the performance of filler particles. Additionally, selection of suitable pretreatment technologies to match the BAF process could provide further improvements that would allow better control of backwash, reduce energy consumption and decrease the addition of $FeCl_3$. In addition, since sludge quantity affects the dosage of chemicals needed, particularly during sludge bulking, the stability of the biochemical system should receive special attention in order to avoid or minimize sludge bulking.

4. Conclusions

This study established improved emergy evaluation indicators for wastewater treatment systems, and evaluated the sustainability of four wastewater treatment plants and their typical treatment processes (A2O, CWSBR, CAST, and BAF) in a city in northeast China. Improvement measures were subsequently suggested that may overcome existing problems in the wastewater treatment processes based on the results of Emergy Analysis. The conclusions are as follows:

(1) This study successfully established improved emergy evaluation indicators (IEYR, IELR, and IESI) introducing W_{in}, F_R, F_N, F_S, W_{out}, S and P into the formulas of traditional emergy indicators (EYR, ELR, and ESI), as the traditional indicators were not suitable for evaluating wastewater treatment systems. The novel indicators can be used to comprehensively compare the sustainability of different wastewater treatment systems and processes based on wastewater emergy as a resource input and the emergy output of sludge and ECEW in drainage.

(2) In terms of input emergy flow, the total input emergy of the CWSBR process was the largest of the four, while the CAST process was the smallest, and the relative contributions of R and F_N were largest in the A2O process, and smallest in the CAST process. N, F_R and F_S were largest in the A2O

process and smaller in the BAF process. In terms of output emergy flow, the total output emergy was largest in the CAST process and smallest in the CWSBR process. Of the contributors to this parameter, sludge emergy output was highest in the BAF process and lowest in the CAST process, but the ECEW indicator was largest in the CAST process.

(3) Analysis of emergy flow revealed that F_S made the biggest contribution to input emergy flows (not including wastewater emergy input) for all wastewater treatment plants and processes studied, followed by R, N and F_N, while F_R accounted for the lowest proportion of the total input emergy.

(4) Evaluation of the emergy efficiency provided information on environmental load and sustainability for the four wastewater treatment processes based on the results of the improved emergy evaluation indicators. IEYR values were ordered Case 4 > Case 2 > Case 3 > Case 1. Emergy efficiency and economic competitiveness were therefore highest for the BAF process employed in Case 4. IELR values were ordered Case 2 < Case 1 < Case 4 < Case 3. The CWSBR process employed in Case 2 therefore placed the least pressure on the surrounding environment. IESI values were ordered Case 2 > Case 4 > Case 1 > Case 3. The CWSBR process employed in Case 2 was therefore the most sustainable, while the CAST process employed in Case 3 was the least sustainable, and the overall order of sustainability for the four wastewater treatment processes was CWSBR > BAF > A2O > CAST.

Therefore, under comparable effluent conditions, the CWSBR process appears to be the most appropriate choice based on emergy evaluation using the improved indicators.

Acknowledgments: This study was supported by the Major Science and Technology Program for Water Pollution Control and Treatment (2012ZX07202-001). The authors wish to acknowledge the anonymous reviewers for their suggestions that have greatly improved our study.

Author Contributions: S.S. designed the study, analyzed the data and wrote the manuscript; H.M. and F.Y. provided good advice throughout the paper; Y.Z. helped revise the manuscript; and J.L. did the statistics work.

Conflicts of Interest: The authors declare no conflict of interest.

References

1. Ministry of Environmental Protection of the People's Republic of China. 2015 Report on the State of Chinese Environment. Available online: http://www.mep.gov.cn/hjzl/zghjzkgb/lnzghjzkgb/201606/P020160602333160471955.pdf (accessed on 20 May 2016).
2. China Water Web. Big Data Analysis of Wastewater Treatment in 2015. Available online: http://www.h2o-china.com/news/241538.html (accessed on 13 June 2016).
3. Huang, J.; Cha, A.P. Study on environmental impact and Countermeasures of engineering construction of wastewater treatment plants. *Res. Econ. EnvProtect.* **2016**, *4*, 104.
4. Sivakumar, M.; Pandit, A.B. Wastewater treatment: A novel energy efficient hydrodynamic cavitational technique. *Ultrason. Sonochem.* **2002**, *9*, 123–131. [CrossRef]
5. Razali, M.; Kim, J.F.; Attfield, M.; Budd, P.M.; Drioli, E.; Lee, Y.M.; Szekely, G. Sustainable wastewater treatment and recycling in membrane manufacturing. *Green Chem.* **2015**, *17*, 5196–5205. [CrossRef]
6. Wen, W.L.; Han, Q.Y.; Zhen, H. Towards sustainable wastewater treatment by using microbial fuel cells-centered technologies. *Energy Environ. Sci.* **2014**, *7*, 911–924.
7. Cerutti, A.K.; Bagliani, M.; Beccaro, G.L.; Bounous, G. Application of ecological footprint analysis on nectarine production: Methodological issues and results from a case study in Italy. *J. Clean. Prod.* **2010**, *18*, 771–776. [CrossRef]
8. Herva, M.; Hernando, R.; Carrasco, E.F.; Roca, E. Development of a methodology to assess the footprint of wastes. *J. Hazard. Mater.* **2010**, *180*, 264–273. [CrossRef] [PubMed]
9. He, C.; Wu, J.; Liu, W. Calculation method of cement ecological footprint. *Acta Ecol. Sin.* **2009**, *29*, 3549–3558.
10. Yellishetty, M.; Mudd, G.M. Substance flow analysis of steel and long term sustainability of iron ore resources in Australia, Brazil, China and India. *J. Clean. Prod.* **2014**, *84*, 400–410. [CrossRef]

11. Arena, U.; Gregorio, F.D. A waste management planning based on substance flow analysis. *Resour. Conserv. Recy.* **2013**, *85*, 54–66. [CrossRef]

12. Gurauskiene, I.; Stasiskiene, Z. Application of material flow analysis to estimate the efficiency of e-waste management systems: The case of Lithuania. *Waste Manag. Res.* **2011**, *29*, 763–777. [CrossRef] [PubMed]

13. Torretta, V.; Ragazzi, M.; Trulli, E.; Feo, G.D.; Urbini, G.; Raboni, M.; Rada, E.C. Assessment of biological kinetics in a conventional municipal WWTP by Means of the oxygen uptake rate method. *Sustainability* **2014**, *6*, 1833–1847. [CrossRef]

14. Mohamed, Z.; Wang, X. A deterministic and statistical energy analysis of tyre cavity resonance noise. *Mech. Syst. Signal Process.* **2016**, *70*, 947–957. [CrossRef]

15. Yau, Y.H.; Lim, K.S. Energy analysis of green office buildings in the tropics-Photovoltaic system. *Energ. Build.* **2016**, *126*, 177–193. [CrossRef]

16. Peng, T.; Xu, X. An interoperable energy consumption analysis for CNC machining. *J. Clean. Prod.* **2016**, *140*, 1828–1841. [CrossRef]

17. Hou, D.; Shao, S.; Zhang, Y.; Liu, S.L.; Chen, Y.; Zhang, S.S. Exergy analysis of a thermal power plant using a modeling approach. *Clean Technol. Environ. Policy* **2012**, *14*, 805–813. [CrossRef]

18. Sui, X.W.; Zhang, Y.; Shao, S.; Zhang, S.S. Exergetic life cycle assessment of cement production process with wasteheat power generation. *Energy. Convers. Manag.* **2014**, *88*, 684–692. [CrossRef]

19. Ameri, M.; Ahmadi, P.; Khanmohammadi, S. Exergy analysis of a 420 MW combined cycle power plant. *Int. J. Energy. Res.* **2008**, *32*, 175–183. [CrossRef]

20. Li, J.H.; Zhang, Y.; Shao, S.; Zhang, S.S. Comparative life cycle assessment of conventional and new fused magnesia production. *J. Clean. Prod.* **2015**, *91*, 170–179. [CrossRef]

21. Zhang, Y.; Liang, K.M.; Li, J.H.; Zhao, C.C.; Qu, D.L. LCA as a decision support tool for evaluating cleaner production schemes in iron making industry. *Environ. Prog. Sustain. Energy* **2016**, *35*, 195–203. [CrossRef]

22. Righi, S.; Oliviero, L.; Pedrini, M.; Buscaroli, A.; Casa, C.D. Life Cycle Assessment of management systems for sewage sludge and food waste: centralized and decentralized approaches. *J. Clean. Prod.* **2013**, *44*, 8–17. [CrossRef]

23. Li, C.F.; Cao, Y.Y. Visualization analysis of hot topics and frontiers on international emergy research. *Ecol. Environ. Sci.* **2014**, *23*, 1084–1092.

24. Brown, M.T.; Herendeen, R.A. Embodied energy analysis and emergy analysis: A comparative view. *Ecol. Econ.* **1996**, *19*, 219–235. [CrossRef]

25. Campbell, D.E.; Ohrt, A. *Environmental Accounting Using Emergy: Evaluation of Minnesota*; Environmental Protection Agency, Office of Research and Development, National Health and Environmental Effects Research Laboratory, Atlantic Ecology Division: Narragansett, RI, USA, 2009.

26. Brown, M.T.; Ulgiati, S. Emergy-based indices and ratios to evaluate sustainability: monitoring economics and technology toward environmentally sound innovation. *Ecol. Eng.* **1997**, *9*, 51–69. [CrossRef]

27. Lomas, P.L.; Alvarez, S.; Rodríguez, M.; Montes, C. Environmental accounting as a management tool in the Mediterranean context: The Spanish economy during the last 20 years. *J. Environ. Manag.* **2008**, *88*, 326–347. [CrossRef] [PubMed]

28. Sweeney, S.; Cohen, M.J.; King, D.; Brown, M.T. Creation of a global emergy database for standardized national emergy synthesis. In Proceedings of the 4th Biennial Emergy Research Conference, Gainesville, FL, USA, 19–21 January 2006.

29. Gao, L.H.; Gao, Q. The Emergy Analysis and sustainability evaluation of ecological economic system of coastal areas in China. *Environ. Pollut. Control* **2012**, *34*, 86–93.

30. Zhu, Y.L.; Zhou, J.; Li, S.; Liu, Y. On eco-efficiency of the agricultural eco-economic system in Hunan based the emergy theory. *J. Hunan Univ. Sci. Technol.* **2011**, *14*, 86–89.

31. Dong, X.B.; Gao, W.S.; Yan, M.C. Emergy analysis of agroecosystem productivity of Typical Valley in Loess Hilly-gully Region of the Loess Plateau: A case study in Zhifanggou Valley of Ansai County. *Acta Geogr. Sin.* **2004**, *59*, 223–229.

32. Xu, M.C. Study on sustainability of agricultural ecosystem in Liaocheng city based on emergy analysis. *Tianjin Agri. Sci.* **2015**, *21*, 29–33, 43.

33. Brown, M.T.; Ulgiati, S. Emergy evaluations and environmental loading of electricity production systems. *J. Clean. Prod.* **2002**, *10*, 321–334. [CrossRef]

34. Marchettini, N.; Ridolfi, R.; Rustici, M. An environmental analysis for comparing waste management options and strategies. *Waste Manag.* **2007**, *27*, 562–571. [CrossRef] [PubMed]

35. Li, J.J. Emergy Account and Analysis in China's Industrial System. Master's Thesis, Nanjing University of Finance and Economics, Nanjing, China, September 2011.

36. Odum, H.T. *Environmental Accounting: Emergy and Environmental Decision Making*; John Wiley & Sons: New York, NY, USA, 1996.

37. Lan, S.F.; Qin, P.; Lu, H.F. *Emergy Synthesis of Ecological Economic Systems*; Chemical Industry Press: Beijing, China, 2002.

38. Fu, X.; Wu, G.; Liu, Y. Analytical theories of exergy and emergy for ecological research. *Acta Ecol. Sin.* **2004**, *24*, 2621–2626.

39. Vassallo, P.; Paoli, C.; Fabiano, M. Emergy required for the complete treatment of municipal wastewater. *Ecol. Eng.* **2009**, *35*, 687–694. [CrossRef]

40. Zhang, X.X.; Ma, F. Emergy evaluation of different straw reuse technologies in northeast China. *Sustainability* **2015**, *7*, 11360–11377. [CrossRef]

41. Liu, C.; Shi, X.Y.; Qu, L.L.; Li, B.Y. Comparative analysis for the urban metabolic differences of two types of cities in the resource-dependent region based on emergy theory. *Sustainability* **2016**, *8*. [CrossRef]

42. Zhang, X.H.; Jiang, W.J.; Wu, J.; Zhang, T.H. Improved emergy indices for analyzing sewage treatment ecosystems. *Resour. Sci.* **2009**, *31*, 250–256.

43. Li, M.; Zhang, X.H.; Li, Y.W.; Xiao, H.; Qi, H.; Deng, S.H. Emergy and economic evaluations of two sewage treatment systems. *Acta Ecol. Sin.* **2012**, *32*, 6936–6945.

44. China Environment Web. How Much Does It Cost to Dispose a Ton of Sludge? Available online: http://www.cenews.com.cn/qy/qygc/201509/t20150929_797767.html (accessed on 29 September 2015).

45. Li, M.; Zhang, X.H.; Li, Y.W.; Zhang, H.; Zhao, M.; Deng, S.H. Environmental impacts of sewage treatment system based on emergy analysis. *Chin. J. Appl. Ecol.* **2013**, *24*, 488–496.

46. Ministry of Environmental Protection of the People's Republic of China. Environmental Quality Standards for Surface Water (GB3838-2002). Available online: http://kjs.mep.gov.cn/hjbhbz/bzwb/shjbh/shjzlbz/200206/W020061027509896672057.pdf (accessed on 1 June 2002).

47. Liu, H.; Luo, L.X.; Lin, M.; Li, X.N.; Zhao, Z.Y. The present situation and development of the wastewater treatment technology in China. In Proceedings of the Academic Annual Conference of Chinese Society for Environmental Sciences, Chengdu, China, 22–23 August 2014.

48. China Water Industry Web. Discharge Standard of Pollutants for Municipal Wastewater Treatment Plant. Available online: http://www.shuigongye.com/standard/200811/2008110612041100001.html (accessed on 6 November 2008).

49. Zhao, H. Sustainability Evaluation of Hospital Sewage Treatment System Based on Emergy Theoty-Zhuokeji Health Center Sewage Treatment System as a Case. Master's Thesis, Sichuan Agricultural University, Chengdu, China, June 2013.

50. Pulselli, R.M.; Simoncini, E.; Ridolfi, R.; Bastianoni, S. Specific emergy of cement and concrete: An energy-based appraisal of building materials and their transport. *Ecol. Indic.* **2008**, *8*, 647–656. [CrossRef]

51. Li, M. Emergy Based Research for the Sustainable Development of a Livestock Sewage Treatment Plant System. Master's Thesis, Sichuan Agricultural University, Chengdu, China, June 2013.

52. Bjorklund, J.; Geber, U.; Rydberg, T. Emergy analysis of municipal wastewater treatment and generation of electricity by digestion of sewage sludge. *Resour. Conserv. Recy.* **2001**, *31*, 293–316. [CrossRef]

53. Huang, H.H.; Zhang, J.; Wen, X.H.; Gan, Y.P.; Zhou, J. Study on energy saving methods for A2O process in wastewater treatment plants. *Chin. J. Environ. Eng.* **2009**, *3*, 35–38.

7

Sustainability Reporting in Family Firms: A Panel Data Analysis

Giovanna Gavana [1], Pietro Gottardo [2] and Anna Maria Moisello [2,*]

[1] Department of Economics, University of Insubria, 21100 Varese VA, Italy; giovanna.gavana@uninsubria.it
[2] Department of Economics and Management, University of Pavia, 27100 Pavia PV, Italy; pietro.gottardo@unipv.it
* Correspondence: annamaria.moisello@unipv.it

Academic Editor: Yongrok Choi

Abstract: We analyze the largely unexplored differences in sustainability reporting within family businesses using a sample of 230 non-financial Italian listed firms for the period 2004–2013. Drawing on legitimacy theory and stakeholder theory, integrated with the socio-emotional wealth (SEW) approach, we study how family control, influence and identification shape a firm's attitude towards disclosing its social and environmental behavior. Our results suggest that family firms are more sensitive to media exposure than their non-family counterparts and that family control enhances sustainability disclosure when it is associated to a family's direct influence on the business, by the founder's presence on the board or by having a family CEO. In cases of indirect influence, without family involvement on the board, the level of family ownership is negatively related to sustainability reporting. On the other hand, a formal identification of the family with the firm by business name does not significantly affect social disclosure.

Keywords: sustainability reporting; family firms; legitimacy; stakeholders; socioemotional wealth; Global Reporting Initiative

1. Introduction

Society's increased awareness and criticism of the environmental and social impact of corporate activities have coincided with the call for firms to effectively satisfy the need for information emerging from a broader range of constituents than shareholders and creditors [1]. Indeed, firms, since the end of 1990s, have shown a growing attitude to participate in sustainability reporting with the aim to approach in a more comprehensive and systematic way the supply of disclosures encompassing economic, environmental and social issues, in accordance with a construct of corporate sustainability based on social equity, economic and environmental integrity [2].

Empirical literature has widely addressed the issue of corporate voluntary disclosure focusing on individual areas, mostly on the environment. A key assumption behind this stream of literature is that the extent of voluntary disclosure provided by a firm shows the importance it attaches to such matters [3] and a firm would use disclosure in order to communicate its commitment to the environment or to manipulate the perception of its environmental performance [4]. Regarding research interested in the provision, by firms, of information covering multiple areas of social disclosure and, more recently, dealing with sustainability reporting, numerous studies have investigated the relationship between social disclosure and the level of pressure that a company may experience, stemming from the concern and scrutiny of the general population, pressure groups, or significant stakeholders. These studies suggest that firms with a higher visibility—in terms of nature of activity [5], environmental sensitivity of its industry or proximity to consumer [6], media exposure [7,8]—seem to exhibit greater concern to improve their image through voluntary disclosure, and that the different groups of stakeholders have different power in influencing social and environmental disclosure [9–11]. Other works have pointed out that corporate governance quality [3] and characteristics, in terms of board composition and attributes [12–14], influence social disclosure practices. Some contributions have tried to explain cross-national differences in corporate social disclosure [1,15,16]. Research has also provided some insights into the choice of the reporting media used to communicate social behavior [17,18]. Despite the fact that existing literature has dealt with a variety of corporate social and environmental disclosure issues, the reporting practices of family firms, intended as companies where a family is the ultimate controlling owner, are still relatively unexplored [19]. Voluntary disclosure literature focusing on family firms has mainly been concerned with firms' propensity to voluntarily disclose financial information or to provide voluntary disclosures related to corporate governance practices [20–22]. Firms' sustainability reporting is shaped by the relevance attached to different stakeholders, which may be internal such as employees or external such as consumers, society, and environment. In family businesses, things are more complicated because the family itself is an internal stakeholder. Given family businesses' relevance across all world economies [19] it is of interest to understand their sustainability disclosure behavior. Therefore, a great deal of effort is needed in order to point out the differences in this form of reporting within family firms, as they cannot be considered a homogeneous group [23]. Iyer and Lulseged [24] take into account discrete levels of sustainability reporting only for one year, without analyzing the effect of visibility on family businesses' sustainability disclosure over time. Moreover, they address family firms as a homogeneous group without capturing their peculiarities in terms of family influence on management, and they do not find significant differences between family and non-family firms. Campopiano and De Massis [19] explore a wide range of disclosure modes for 98'private and listed family and non-family firms using a cross-sectional analysis. They take into account family-owned firms with at least one family member in management and find differences in behavior between family and non-family firms. What has not yet been investigated is the effect of indirect family influence—that is, when there are no family members on the board, but the family appoints the members of the board by the means of ownership control [25]—and direct influence by specific forms of owner family involvement. Moreover, Campopiano and De Massis [19] call for large-scale, longitudinal studies with accurate measures of family involvement in order to capture the cause-effect and temporal relationships between sustainability report content and its evolution over time.

Therefore our research questions are:

RQ1: How visibility affects family firms sustainability disclosure behavior in a longitudinal framework?
RQ2: How family influence shapes this behavior?

We answer these research questions by analyzing how the level of family ownership, and different forms of family involvement, shape family influence in a firm's disclosure behavior taking into account different forms of visibility for a longitudinal dataset of 230 non-financial Italian listed companies during the period 2004–2013. We address this topic by drawing on legitimacy theory and stakeholder theory, integrated with the socioemotional wealth (SEW) approach. Within the legitimacy theory framework [26], sustainability disclosure is meant as a response to public pressure, in the attempt to prove that firms' behavior is in line with societal norms and values [27]. Legitimacy theory shares these arguments with stakeholder theory [28]—the latter provides a useful framework to examine social and environmental disclosure as a response to the expectations (or as a means to change perceptions) of particular stakeholder groups [29].

We argue that the relationship between family firms and their stakeholders is strongly influenced by how family businesses pursue and preserve SEW, conceived as the non-financial values that a family derives from its controlling position in the firm [30]. Accordingly, we explore the relationship between the extent of sustainability disclosure and two major dimensions of SEW, namely family control and influence on the business and identification of family members with the firm.

Overall, our study suggests that family control positively affects sustainability disclosure when it is associated to direct influence of the family on the business, by the founder's presence on the board or by having a family CEO, but not when associated to indirect influence. When a family exerts indirect influence by mere ownership control, without involvement in the business, the extent of sustainability disclosure is negatively associated with the level of family ownership. Family identification with the firm does not have a significant effect on disclosure when it is formal, that is, the business carries the family name. Our results confirm that a firm's visibility, in terms of media exposure, affects its attitude to disclosure, and this effect is particularly significant for family businesses.

We contribute to prior studies on sustainability disclosure by focusing on family firms and the differences between them. We answer the call from family business literature to better explore the differences within family firms [31,32] by analyzing the extent and type of their social and environmental disclosure. We expand the theoretical framework adopted by many studies in the field of corporate sustainability disclosure, as we use legitimacy theory coupled with stakeholder theory integrated with the SEW approach. We contribute to SEW literature by empirically testing the effect of two main dimensions and pointing out that "family control and influence" is not a homogeneous dimension with a unique effect as it differs depending on whether there is direct or indirect influence and how direct influence is exerted.

The remainder of the paper is organized as follows: Section 2 provides the theoretical background and reviews related literature; Section 3 develops the research hypotheses; Section 4 describes the methodology and data; Sections 5 and 6, respectively, report and discuss the empirical results; and Section 7 concludes, underlines the study's limitations and implications for practice and offers some suggestions for future research.

2. Theoretical Background and Literature Review

Literature provides varied interpretations as to why companies become involved in social and environmental responsible activities and in voluntary non-financial disclosure. Legitimacy theory offers a useful theoretical framework in this field as it recognizes that firms operate within a social contract that links the approval of their objectives to a behavior consistent with social values [33]. Legitimacy is defined as "a generalized perception or assumption that the actions of an entity are desirable, proper, or appropriate within some socially-constructed system of norms, values, beliefs and definitions" [34] (p. 574).

Firms try to align their actions to the values of their general and relevant publics and stakeholders [12], as actual or potential inconsistency between social values and the values of the enterprise generates a legitimacy gap [35]. Failure to conform to institutionalized norms of acceptability jeopardizes a firm's organizational legitimacy, resources and durability [36,37].

Firms may use a range of strategies in order to achieve, maintain and repair legitimacy [18,34]. They may inform stakeholders about intended improvements in social performance or shift attention from sensitive issues [38], as well as try to adapt the company's goals, actions and outputs to be consistent with the definition of legitimacy, or change their perceptions through communication [4,26]. They often use symbolic actions which "form part of an organization's public image" [14] (p. 481). Communication plays a fundamental role in recognizing a firm's legitimacy as it informs stakeholders, and society, that organizational behavior is congruent with its values, norms and expectations; non-communication may threaten performance, resource availability and survival. Society attaches great importance to companies that engage in reporting their socially-responsible behavior [39]. There is some evidence that companies with a high cost of equity capital tend to initiate sustainability reporting and that initiating firms with superior social responsibility performance attract dedicated institutional investors and analyst coverage, obtaining a reduction in the cost of equity capital [40]. A social reporting commitment is a strategy to change the public perception of a firm's legitimacy [41] and there is increasing evidence of businesses seeking to support their competitive advantage through voluntary disclosure [27].

Stakeholder theory, as well as legitimacy theory, "conceptualize[s] the organization as part of a broader social system wherein the organization impacts, and is impacted by, other groups within society", but while legitimacy theory addresses the expectations of society in general, stakeholder theory recognizes that society is made up of various groups with different views about how an organization should behave and with different abilities to influence an organization [1,7,33]. Legitimacy and stakeholder theories should be seen as two perspectives of the issue [42] as stakeholder theory offers important insights in order to identify which expectations an organization should meet and what groups of stakeholders might be relevant to its behavior and in fulfilling its objectives. In this perspective, sustainability reporting can be seen as an aspect of the dialog that the company holds with its stakeholders [42], through which it informs them of its good practices [43].

Stakeholders' salience, i.e., the degree to which management prioritizes competing stakeholders demands, is different, and more complex, in a family than in a non-family company as the family itself is a pivotal stakeholder with peculiar claims and concerns [44]. Recent studies underline that the main concern of family firms is not limited to economic performance and argue that their reference point is the preservation of socioemotional wealth [30,45–47]. The concept of socioemotional wealth refers to "the stock of affect-related value that a family derives from its controlling position in a particular firm" [48] (p. 259). SEW draws on several dimensions which make family different from non-family firms and help to explain family businesses' heterogeneity [49,50]. Prior literature identifies five main dimensions: family control and influence on the business, identification of family members with the firm, binding social ties, emotional attachment of family members and renewal of family bonds to the firm through dynastic succession [48]. Family firms show a greater preference for control than non-family businesses in order to preserve family endowment in the business [30,45]. Family members identify more strongly with a family business than non-family owners do with a firm. More favorable firm reputation is associated with having the family name as part of the corporate name, the level of family ownership and a presence on the board [51] as internal and external stakeholders see the firm as an extension of the family [48]. Empirical evidence shows that family firms have more socially-responsible behavior [52,53] as they are particularly concerned with family reputation and image [54]. The sense of identification and belonging is often shared with non-family employees; the ties among the members of this extended family help the development of strong social links with the community and the reciprocal bonds which characterize the family business involve a wide set of constituencies [48]. This social capital is itself a source of wealth for the family that behaves in order to preserve the bonds with its internal and external stakeholders, so family members with deep roots in the community tend to be more sensitive to social and environmental problems [55]. The bonds that family members experience through the business satisfy their needs in terms of belonging and affect, providing them with emotional returns [56], which engender the family's emotional attachment to the firm. The "renewal of family bonds" dimension refers to the family firm principal's purpose to transfer the business to future generations and meet the affective need to perpetuate the family dynasty [57]. For this reason, family owners and managers are concerned with the preservation of the stock of values related to the firm, among them image, reputation and the long-term relationships with internal and external stakeholders [55].

3. Hypothesis Development

Organizations try to assure congruence between their value system and the value system shared by the community, and try to display an image that matches with the expectations of their stakeholders. In so doing, they try to attain legitimacy by social and environmental disclosure [58] as providing adequate and verifiable information on their commitment to social and environmental responsible activities consolidates their reputation and stakeholders' trust [59].

Firms which have a greater visibility in terms of size or media exposure are subject to the judgment of a broader community, attract more attention from stakeholders, are more susceptible to political actions which could affect their performance and, hence, are more committed to social disclosure [3,10,60]. Social visibility also depends on industry affiliation: firms which operate in businesses with a high potential environmental impact, or companies that are better known to the end consumer, tend to exhibit a higher engagement in social disclosure [6,61]. There is some evidence that companies facing the risk of a tightening of environmental laws engage in sustainability disclosure in order to moderate the extent of public policy pressure [7,16].

Empirical research [53] shows that family businesses are more committed to the prevention of social concerns, in terms of damage to interest groups, than their non-family counterparts. This suggests that these firms are very concerned with reputation and social legitimation. Family firms care greatly about community scrutiny because it could harm the socioemotional wealth, not only prejudicing their social ties but also damaging the image of the company and, consequently, family reputation, so they try to meet stakeholders' expectations in terms of environment preservation [52]. Therefore, the affective endowment in a firm would be a prosocial stimulus to behave responsibly towards external stakeholders [62]. Family businesses are more prone to adopt initiatives visible to the community such as financing sports teams or sports facilities [63] and charitable giving [64]. More visible firms would be more concerned with social legitimacy and should engage more in social and environmental disclosure [8]. Overall, family firms' literature suggests that family businesses are especially prone to seeking legitimation for their actions, thus enhancing their reputation with stakeholders [65]. We expect that they would more actively respond to visibility pressure by engaging in social reporting in order to influence stakeholder and society perceptions and to be perceived as good corporate citizens [66].

H1: *The effect of firm visibility on sustainability disclosure is higher for family than for non-family firms.*

Some studies on social and environmental reporting propose that a firm's internal context affects disclosure [67,68]. Campopiano and De Massis [19] find that family firms are more engaged in social and environmental reporting than their non-family counterparts. All the family firms in their sample have at least one family member in the firm's top management team. This suggests that these findings could be affected by the family involvement in active management; therefore, it is of interest to estimate the effect of different levels of involvement. Several authors suggest that the extent of family involvement shape a firm's attitude towards social and environmental responsible activities and generate heterogeneity among family businesses [69–71]. Bingham et al. [69] assert that increased levels of family involvement bring a collectivistic orientation towards stakeholders as non-economic objectives, such as endorsing the ethical values shared and perpetuated by the family members [71], become more important [30,45]. This transgenerational perspective is enhanced by the presence of the founder in active management [72] who plays a relevant role in the adoption of this collectivistic orientation, enhancing the firm's activism towards the community, employees and the supplying of quality products and services to consumers [69]. According to literature, family businesses, as non-family businesses [73], are more likely to engage in social activities when top management support is high and there is some evidence that the presence of a family CEO enhances social and environmental commitment and communication [70].

The involvement of the family in the business in terms of the presence of multiple family members on the board, family CEO or its founder being still active in the business governance increases the sense of identification between the firm and the family. This implies concern for corporate reputation and the satisfaction of non-family internal stakeholders, resulting in the pursuit of responsible work practices [74]. The presence of multiple family members on the board causes the business to be perceived, by internal and external stakeholders, as an extension of the family and family members would tend to proactively take care of the business' external image [75]. Family firm owner-managers are worried that a firm's bad reputation may damage the "good name" of their family and, in turn, reflect on them as individuals [53] and family firms' founders are more likely to regard their businesses "as an extension of themselves-their identity, or self-view", a mirror of the personal values that they share with the present generations and will transmit to the future ones [49]. A closer identification of the family with the business makes family members more sensitive about the firm's reputation among external stakeholders and they tend to proactively take care of the business' external image [75].

Family firms are characterized by the "embeddedness" of the business within a family [76]. The presence of family members in a firm's management enhances family identification, influence and personal investment in the business, increasing their endowment in the firm and the need to protect it [49,50]. Therefore, we expect that firms directly influenced by the family—i.e., with family members involved in business management—are more prone to use sustainability disclosure in order to demonstrate that they respect societal values and that they are legitimate to continue their operations.

H2: Family involvement positively affects sustainability disclosure.

Carrying the family name over to the firm strengthens the integration between the family and the firm, inspiring family members to uphold the values of the family firm [77]. Family-owners having their "name on the building" are more conscious of their standing in the firm [78], they may have greater difficulty distancing themselves from the firm they control and, in so doing, would be more concerned with the effect the firm's behavior has on family's reputation [53]. Concern for the environment is more evident in businesses bearing the family name as "soiling the environment reflects badly on family name" [63] (p. 140). In cases where the family and business name is the same, the visibility of the family as controlling coalition of the firm is higher, social monitoring and expectations stronger [74] and, given that public opinion would have heavy emotional effects on family members [79], the family would be more likely to project a positive image of the business through sustainability disclosure.

H3: Family name in business name positively affects sustainability disclosure.

4. Methodology

4.1. Sample

Our sample comprises publicly-traded firms listed on the Italian Stock Exchange, after checking for the availability of accounting data and excluding financial firms. Some of our explanatory variables are based on accounting data, so, consistently with previous literature, we have excluded financial companies because of the peculiarities of their accounting system [10]. The final sample is represented by an unbalanced panel of 230 firms with data available for the period 2004–2013. We collected the financial, accounting and ownership data from AIDA (Italian Digital Database of Companies), the Italian provider of the Bureau van Dijk European Database. Ownership data was hand-collected cross-checking the information provided by AIDA, the Italian Stock Exchange and CONSOB. Social responsibility disclosure information was hand-collected by the content analysis of stand-alone sustainability reports for each year in the analysis period. The sustainability report is a suitable information recipient also for stakeholders who are less likely to consult the annual report for information but are highly concerned with social issues. The rise in the propensity for firms to publish stand-alone sustainability reports [80] is internationally encouraged and supported by the issue of several sustainability reporting frameworks, among which the Global Reporting Initiative (GRI) Sustainability Reporting Guidelines [81] has become the most adopted standard worldwide.

We define a family firm as one where a family is the ultimate owner, assuming a minimum control threshold of 20%. This definition allows us to highlight the effect of different degrees of family involvement based on ownership, board membership, management, founder and firm-family name identity on voluntary disclosure. The threshold is consistent with those used in the literature [82–85]. Assuming this threshold the family firms in our sample represent the 48.26% of the Italian non-financial listed companies.

4.2. Measures

4.2.1. Dependent Variables

We perform a content analysis of stand-alone sustainability reports as they represent one of the more advanced forms of conveying social responsibility disclosures, covering systematically a large spectrum of topics relevant to a wide variety of different stakeholder groups. Most importantly, a sustainability report is a suitable source of information for those individuals and groups less accustomed to reading of the annual report, and perhaps even more interested in social matters than shareholders. Moreover, some studies provide evidence that the adoption of a set of alternative reporting media leads to lower coverage of sustainability issues in the annual report [17,86].

Iyer and Lulseged [24] analyzing family firms sustainability reporting used a discrete independent variable. They assigned a value of 4, 3, 2, and 1 when a company issued a sustainability report in 2010, and the report level (self-declared and checked by GRI) was level A, B, C, and below C, respectively; or 0 when a firm did not issue a sustainability report in that year. When a firm did not declare the level, the authors attributed a value in accordance with the GRI G3 guidelines.

In order to avoid the weaknesses of discrete variables, we opted for a continuous index. Based on the GRI Sustainability Reporting Guidelines G3.1 [81], we derived a list of 86 items related to environmental information, labor practices, society respect, product responsibility, human rights, stakeholder engagement, economic performance and market presence. The comparison of sustainability reporting practices with key indicators outlined in the GRI framework is consistent with previous literature on the extent of sustainability disclosure [15,17]. We then performed a content analysis of the sustainability reports in order to assign each firm a numerical rating (ranging from 0 to 1) depending on the number of items actually disclosed. The content analysis we use detects only the presence or absence of the relevant information [6,7,12] given the number of items relative to several different topics it is a good measure of management propensity to provide voluntary social disclosure. For each sample firm, and for each period analyzed, the general disclosure index is calculated as $I = \Sigma dj/M$, where M is the maximum number of items a firm may disclose, thus excluding the items considered as not-relevant to the specific firm, and dj takes the value 1 if the item j is disclosed, and 0 if the sustainability report does not present such an item [13]. Further, we compute five sub-indexes, respectively for environment, labor practices, society, human rights respect and product responsibility-related information in order to assess the different emphasis firms place on these themes and, in turn, to understand to which stakeholders they are paying special attention. The related checklists consist of 30 items for environmental information: 15 for labor practices, 10 for human rights respect, 10 for society respect, and 9 for product responsibility.

4.2.2. Independent Variables

SEW Related Variables

Our first proxy for the family control and influence dimension of SEW is family ownership. Moreover, to capture in full the impact of family involvement and influence on sustainability disclosure we use, in the regression models, several other indicators [30,53,64,78,87]: family CEO, presence of multiple family members on the board, presence of the founder in the firm and business bearing the family's name. Family ownership is the sum of equity stakes (%) that family members have in the firm, directly or indirectly. Family CEO and multiple family members are dummy variables that take value 1 if the CEO is a family member and if multiple family members sit on the board, respectively. The presence of the founder in the firm and the family name are dummy variables that identify the family firms where the founder is involved in the business and where the business carries the family name, respectively.

Control Variables and Interaction Variables

We include a set of variables that, as found in many previous studies, may correlate with firm disclosure, notably size, profitability and leverage [7,8,10–13,66,88–91]. As a measure of size we use the natural logarithm of sales. Profitability is measured by ROA. Leverage is measured by the gearing ratio, defined as the ratio of long-term and short-term financial debts to equity. We also use Public control [90] defined as a dummy variable that takes value 1 if the firm is owned by public bodies. To control for residual effects related to a specific industry affiliation not captured by the variables "consumer proximity" and "environmental sensitivity", we include in the models sector dummies at the two-digit ATECO level. Consumer proximity [6,61] is measured as a dummy variable that takes value 1 for the firms that are better known to the public (high profile) and 0 otherwise (low profile). Based on prior research, high-profile firms are identified as those in the sectors of household goods and textiles, beverages, food and drug retailers, telecommunications, electricity, gas distribution and water. Environmental sensitivity [6,7] is a dummy variable that takes value 1 for firms operating in sectors with a higher risk of environmental impact and 0 otherwise. Based on prior research, the higher environmental risk sectors are mining, oil and gas, chemicals, construction and building materials, forestry and paper, steel and other metals, electricity, gas distribution and water. Media exposure [88] is measured by counting, for each year, the number of articles from the most renowned Italian financial newspaper—"Il Sole 24 Ore"—that contain the firm's name in the sample period 2004–2013. Sport is a dummy to distinguish firms funding sports activities and facilities from the others.

4.3. Empirical Models

The presence of multicollinearity is tested based on the correlation matrix and by computing the variance inflation factors. Results indicate that multicollinearity is unlikely to be a problem in our dataset. The statistical analysis includes the use of linear panel regression models to verify the hypotheses reported in the previous section. The panel analysis uses efficiently cross and time-series data, increasing the number of observations and the parameter's reliability while reducing the likelihood of multicollinearity. For the total disclosure index and the four sub-indices, two model formulations are shown that take the following general form:

Model 1: Disclosure scorej = f(SEW variables, control variables)
Model 2: Disclosure scorej = f(SEW variables, family ownership interactions, control variables)

5. Results

Table 1 reports the descriptive statistics for the dependent and independent variables, distinguishing family and non-family firms. Family firms, as a mean, show a significantly higher propensity to engage in sustainability disclosure than non-family businesses. This attitude is confirmed when focusing on each particular area of sustainability disclosure, namely environment, community, human rights, labor and product. The results of the *t*-tests of differences in the disclosure indexes means are always significant at the conventional level. Our findings show that family firms are significantly more likely to support local sports than their non-family counterparts. Our sample family firms are older, bigger and more profitable than non-family firms. A consistent number of family firms has a family CEO (73.9%), 51.4% has the founder on the board and 73% shows multiple presence of family members on board. On average, families control the company through the absolute majority of ownership stakes. The family lends its name to the company in 25.2% of the family businesses.

Table 1. Descriptive statistics.

	Family Firms		Non-Family Firms		*t*
	Mean	SD	Mean	SD	
SUS	0.056	0.188	0.024	0.117	−4.71 ***
ENV	0.058	0.196	0.025	0.124	−4.74 ***
SOC	0.048	0.177	0.022	0.117	−3.96 ***
LAB	0.066	0.220	0.033	0.154	−4.09 ***
PROD	0.046	0.176	0.015	0.091	−5.12 ***
HUM	0.041	0.160	0.014	0.088	−4.83 ***
Size	12.349	1.711	11.395	2.108	−11.03 ***
Roa	2.196	9.683	0.193	18.688	−4.87 ***
Gearing ratio	0.573	4.435	0.529	13.515	1.39
Media Exposure	12.707	39.069	9.035	29.466	1.50
Consumer Prox	0.351	0.478	0.230	0.421	−6.45 ***
Environmental Sens	0.541	0.499	0.340	0.474	−9.42 ***
Sport	0.270	0.444	0.200	0.400	−4.05 ***
qFamily	58.971	14.764	-	-	-
FamilyCEO	0.739	-	-	-	-
Founder	0.514	-	-	-	-
FMulty	0.730	-	-	-	-
FamilyName	0.252	-	-	-	-

Note: *** *p*-value significant at the 1% level.

Table 2 presents the results of the correlation analysis. Overall, these results suggest that multicollinearity will not be a major concern in the following regression analysis.

Table 2. Correlation analysis.

	0	1	2	3	4	5	6	7	8	9	10	11	12	13	14	15	16	17
0 SUS																		
1 ENV	0.986																	
2 SOC	0.975	0.949																
3 LAB	0.977	0.958	0.938															
4 PROD	0.938	0.911	0.930	0.893														
5 HUM	0.913	0.875	0.894	0.859	0.859													
6 Size	0.466	0.459	0.454	0.455	0.438	0.425												
7 Roa	0.020	0.024	0.018	0.017	0.005	0.020	0.058											
8 Gearing ratio	-0.027	-0.028	-0.025	-0.028	-0.024	-0.022	-0.022	-0.006										
9 Public	0.430	0.422	0.420	0.420	0.416	0.385	0.378	0.014	-0.026									
10 DConsumer Prox	0.049	0.042	0.052	0.039	0.070	0.062	0.056	0.024	-0.022	0.168								
11 DEnvironmental Sens	0.165	0.169	0.159	0.151	0.128	0.182	0.182	0.001	0.021	0.192	-0.230							
12 Media Exposure	0.442	0.424	0.425	0.420	0.441	0.476	0.457	0.015	-0.014	0.254	0.070	0.105						
13 DSport	0.182	0.193	0.157	0.166	0.152	0.175	0.205	0.025	-0.025	0.116	0.056	0.116	0.156					
14 DFounder	-0.058	-0.048	-0.075	-0.062	-0.063	-0.054	-0.012	0.079	-0.019	-0.183	-0.070	-0.090	-0.125	-0.053				
15 qFamily	-0.079	-0.075	-0.086	-0.081	-0.071	-0.070	0.058	0.105	-0.020	-0.253	0.084	0.113	-0.071	0.094	0.547			
16 DFamilyCEO	-0.030	-0.024	-0.042	-0.027	-0.035	-0.038	0.089	0.113	-0.017	-0.192	0.058	0.100	-0.090	0.049	0.488	0.736		
17 DFMulty	-0.001	0.004	-0.018	-0.011	0.004	0.010	0.119	0.077	-0.012	-0.189	0.084	0.109	0.016	0.054	0.495	0.726	0.629	
18 DFamilyName	-0.047	-0.047	-0.037	-0.057	-0.041	-0.035	0.050	0.048	0.008	-0.112	0.032	0.076	-0.033	-0.002	0.148	0.401	0.306	0.366

Note: correlations in bold are significant at the 5% level.

Table 3 shows the results of the Ordinary Least Squares (OLS) regression analysis performed through a basic model (model 1) and a model with interactions (model 2) using the sustainability index as dependent variable.

Table 3. General disclosure index—Panel OLS regressions.

Variables	Model 1	Model 2
Interc.	−0.238 ***	−0.240 ***
Controls Variables		
Size	0.021 ***	0.020 ***
Roa	−0.000	−0.000
Gearing ratio	−0.000	−0.000
Public	0.207 ***	0.211 ***
Consumer Proximity	0.050 **	0.052 **
Environmental Sensitivity	−0.002	0.005
Media Exposure	0.002 ***	0.001 ***
Sport	0.033 ***	0.054 ***
Industry	Yes	Yes
SEW Variables		
qFamily	−0.001 ***	−0.000
FamilyCEO	0.027 **	0.025 **
Founder	0.024 **	0.031 ***
FMulty	0.013	0.007
FamilyName	−0.013	−0.013
Interactions Variables		
qFamily × Media Exposure		0.000 ***
qFamily × Sport		−0.001 **
qFamily × Consumer Proximity		−0.000
qFamily × Environmental Sensitivity		−0.001 **
R^2	0.382	0.386

Note: *, **, *** *p*-value significant at the 10%, 5% and 1% level.

The F test for both models is significant at the 1% level. Model 1 presents an adjusted R^2 of 0.382, the explicative power of the model rises to 0.386 adding the interaction variables. Both models highlight that general sustainability disclosure score is significantly and positively affected by firm size, consistent with prior research in the UK [12,24,92]. Our findings confirm the positive and significant effect of media exposure highlighted in literature [6,88]. We find a significant positive relation between the proximity of the business to the consumer and the sustainability score, coherent with the results of Branco and Rodrigues [4] for Portuguese companies, although for these firms the effect was not significant. Consistent with this study, environmental sensitivity is positively related to disclosure but never significant. In line with empirical literature [6,88], we find that profitability does not influence the attitude to disclose a firm's social behavior. We confirm that a firm's capital structure does not affect sustainability disclosure as we find, consistent with the literature [12,93] a negative but not significant relation with the gearing ratio. Our results show that firms that engage in sports funding are significantly more sensitive to corporate sustainability disclosure and this is coherent with the suggestions of Smith and Westerbeek [94] on the role that sport can play as a vehicle for deploying a social responsible behavior. Like Cheng and Courtenay [89], who find that governmental ownership has a positive—but not significant—relation with disclosure, we find a strong and significant positive effect of public control on the business.

As regards the variables that characterize specifically family businesses, in model 1 we find that family ownership has a significant negative effect on sustainability although family business are more engaged in sustainability disclosure than non-family businesses. According to H2, family CEO and the presence of the founder on the board have, in both models, a positive and significant effect. Unlike the evidence of Ho and Wong [93] on financial voluntary disclosure, we find that the presence of multiple family members on the board has a positive, but not significant, effect on sustainability reporting. Unlike the predictions hypothesized in H3, family name does not affect sustainability disclosure.

When we add the interactions [95] in model 2, the signs and significance of the control variables, and of the SEW variables, do not change, except for family ownership. The interaction between family ownership and media exposure is significantly positive, that is to say that visibility reinforces family firms' disclosure propensity and that family-controlled firms are more sensitive to media exposure than non-family businesses. The interactions of family ownership with consumer proximity do not show any differential effect in family-controlled businesses. The interaction between family ownership and environmental sensitivity shows that the positive effect of the latter on disclosure is mitigated by family control. These results suggest that H1 is confirmed when a family firm's visibility is related to media exposure.

The interaction between family ownership and sports funding is negative, that is the relation between the propensity to engage in sustainability reporting, and sport supporting, is lower for family than for non-family firms.

Table 4 reports the results of the pooling regressions for the specific indexes related to environment, society, human rights, labor and product responsibility disclosure.

Overall, the control variables maintain the sign and significance pointed out using the sustainability index, except for the presence of a family CEO, which is not significant for product responsibility, human rights and society. The regression results show that the presence of multiple family members on the board has a positive, but never significant, effect for the specific disclosure indexes, as it does for the general sustainability index. The interaction variables maintain the same sign and significance shown in the case of the general sustainability index, with the exception of the interaction between family ownership and environmental sensitivity; this loses significance for product responsibility and human rights respect disclosure.

The explicative power of model 2, using the specific disclosure indexes, ranges from 0.356 to 0.373.

Table 4. Disclosure sub-indexes—Panel OLS regressions.

	PANEL A—Model 1				
Variables	Environment	Society	Labour	Prod. Responsibility	Hum. Rights
Interc.	−0.249 ***	−0.242 ***	−0.277 ***	−0.199 ***	−0.151 ***
Controls					
Size	0.021 ***	0.021 ***	0.024 ***	0.017 ***	0.013 ***
Roa	−0.000	−0.000	−0.000	−0.000	−0.000
Gearing ratio	−0.000	−0.000	−0.000	−0.000	−0.000
Public	0.210 ***	0.190 ***	0.237 ***	0.195 ***	0.148 ***
Cons Proximity	0.054 **	0.055 ***	0.058 **	0.055 ***	0.022
Env. Sensitivity	0.001	−0.004	−0.001	−0.006	−0.004
Media Exposure	0.002 ***	0.001 ***	0.002 ***	0.002 ***	0.002 ***
Sport	0.042 ***	0.021 **	0.029 ***	0.023 **	0.025 ***
Industry	Yes	Yes	Yes	Yes	Yes
SEW Variables					
qFamily	−0.001 ***	−0.001 ***	−0.001 ***	−0.000 **	−0.001 ***
FamilyCEO	0.029 **	0.020	0.036 **	0.015	0.013
Founder	0.030 ***	0.016	0.025 *	0.016	0.018 *
FMulty	0.017	0.004	0.007	0.013	0.013
FamilyName	−0.017	−0.003	−0.024	−0.011	−0.005
R^2	0.370	0.355	0.357	0.352	0.356

Table 4. *Cont.*

Variables	Environment	Society	Labour	Prod. Responsibility	Hum. Rights
		PANEL B—Model 2			
Interc.	−0.252 ***	−0.244 ***	−0.277 ***	−0.201 ***	−0.151 ***
Controls					
Size	0.021 ***	0.021 ***	0.024 ***	0.017 ***	0.012 ***
Roa	−0.000	−0.000	−0.000	−0.000	−0.000
Gearing ratio	−0.000	−0.000	−0.000	−0.000	−0.000
Public	0.208 ***	0.190 ***	0.243 ***	0.202 ***	0.150 ***
Cons Proximity	0.056 **	0.055 **	0.055 **	0.070 ***	0.026
Env. Sensitivity	0.007	0.004	0.011	−0.004	0.001
Media Exposure	0.001 ***	0.001 ***	0.001 ***	0.001 ***	0.001 ***
Sport	0.068 ***	0.045 ***	0.051 ***	0.043 ***	0.037 ***
Industry	Yes	Yes	Yes	Yes	Yes
SEW Variables					
qFamily	−0.000	0.000	−0.000	−0.000	−0.000
FamilyCEO	0.027 **	0.018	0.034 **	0.013	0.012
Founder	0.035 ***	0.022 **	0.035 ***	0.023 **	0.022 **
FMulty	0.013	−0.001	0.001	0.005	0.010
FamilyName	−0.017	−0.002	−0.023	−0.012	−0.005
SEW Interactions					
qFamily × Media Exp.	0.000 *	0.000 ***	0.000 ***	0.000 ***	0.000 **
qFamily × Sport	−0.001 **	−0.001 **	−0.001 *	−0.001 **	−0.000
qFamily × Cons. Prox.	−0.000	−0.000	−0.000	−0.000	−0.000
qFamily × Env. Sensitivity	−0.001 *	−0.001 **	−0.001 ***	−0.000	−0.000
R^2	0.373	0.359	0.364	0.356	0.358

Note: *, **, *** *p*-value significant at the 10%, 5% and 1% level.

6. Discussion

Firms seek legitimacy by converging values pursued by the company, with the values shared by society, and making this convergence evident in the eyes of society [12,34,35]. Society is not a homogeneous group of individuals with identical expectations and a firm's different groups of stakeholders are members of society [96]. Businesses seek legitimacy by different stakeholder groups depending on how they help to achieve the firm's goals. Family businesses' behavior is strongly influenced by non-economic objectives, i.e., family members' socioemotional wealth preservation [30,45]. Sustainability disclosure is a means of demonstrating to stakeholders that a firm's actions are in accordance with the system of values shared by society and thus preserve the SEW. The business is seen as an extension of the family itself because of a strong sense of identification between family members and the firm. Through the business, a family binds important social ties—with vendors, suppliers, employees and the community—which give family members a relevant emotional and reputational return. Within the firm family members can satisfy their needs for affection and belonging which may also involve non-family employees as members of the extended family [48]. A family business is strongly legitimated to exist for an extended period as it allows the objectives of the internal stakeholders, both monetary and non-monetary, to be achieved. It will renew the family bond and provide socioemotional wealth to the future generations through dynastic succession. In general, family businesses are concerned with both internal and external stakeholders and they are more prone to disclose their corporate social responsible behavior.

Our findings confirm that visibility increases a firm's voluntary disclosure and demonstrate that media exposure significantly affects family businesses sustainability reporting. This is likely to be due to the dual effect of the media: the media exposes the family to the scrutiny of "wider society", of groups with which the company has no social bonds, and it can also influence community perceptions of a business [97]. The image of the family is the mirror image of the company so firms are particularly concerned with this type of visibility and, as a result, are more prone to disclose their social behavior, as suggested by H1. Our results show that the positive relation between sports funding and disclosure, which emerged for all firms, changes sign in the case of family firms; it could be explained as a substitution effect between funding sport and voluntary disclosure because the former is a means of building a stronger corporate image in the community where a firm operates.

As mentioned above, socioemotional wealth comprises different dimensions and, on the basis of our results, "family control and influence" and "identification" do not seem to have the same effect on sustainability disclosure. Moreover, "family control and influence" is not a homogeneous dimension because its effect depends on how a family exerts its influence on the firm. An owning family may exert its control on the business indirectly, by appointing the CEO and the board, or directly by having a family CEO and by having a presence on the board. Our study suggests that indirect control, i.e., family ownership without the presence of the family in management, does not affect disclosure and that what really matters is the family's direct influence by its involvement in the business. Our study shows that when family ownership increases, in absence of family involvement, we have a negative effect on disclosure. In this case, the relevance of external shareholders decreases, the absence of the family in management limits the affective endowments in the firm so concern for external stakeholders falls as the family prefers to devote resources to satisfying the groups that would reinforce family control [55].

Thus, our empirical finding shed lights on the effect of a second important dimension of SEW, namely family members' identification with the firm. The sense of identification depends on the family's proximity to the business. We may have a considerable proximity when the family is actively involved in business governance and management and a formal proximity when the firm bears the family name. Our results confirm H2 and suggest that the sense of identification is stronger, and the attitude to preserve/enhance family image and reputation through disclosure higher, when the family is considerably close to the business. Family involvement through the CEO, or by having the founder on the board, has a significant positive effect on disclosure while the multiple presence of board members does not affect sustainability disclosure. Multiple family members on the board may belong to different branches of the family, rather than the founder's nuclear family, and they may even be in conflict with each other. In this case, the SEW and the sense of identification tend to lessen. The founder's presence and a family CEO strongly affect environmental and labor respect disclosure. Previous research points out that family firms behave consistently with respect for the environment [52] and for their employees, by improving their quality of life [98] and more stable employment [99]; our results go further by suggesting that family firms choose to actively engage in formal communication—i.e., sustainability reporting—to show their interest in these stakeholders.

Our findings do not support H3 and suggest that if a family is formally, by the means of name, but not considerably close to the business we do not have an effect on sustainability disclosure.

7. Conclusions

This paper focuses on a sample of 230 Italian non-financial listed firms. It studies sustainability disclosure, by the means of a continuous index determined for each year of the period 2004–2013. It highlights that the way family ownership affects sustainability reporting depends on how the family exerts its influence on the business. The presence of the founder on the board or a family CEO have a significant positive effect, in particular on environment and labor disclosure, because of the family's closer identification with the business through its active involvement in the firm. Moreover, it shows that a formal identification of the family and the firm by the means of the name of the business does not affect disclosure. Our study confirms the relevant effect on disclosure of some industry and firm characteristics suggested by prior studies, such as consumer proximity, firm size and media exposure. We extend prior research by pointing out a higher effect of media pressure for family firms and the positive relation between a firm's propensity to finance sport and sustainability reporting. However, this is lower for family firms. Further, our empirical evidence points out that family firms operating in environmentally-sensitive industries are less prone to disclose their behavior toward sustainability than non-family businesses.

This study has some practical implications. It provides valuable information for public policy makers which may address more effectively their regulatory activity taking into account the different motivations which affect family firms' attitude towards sustainability disclosure in different aspects, namely environment, society, human rights, labor and product responsibility. The information on the motivations and incentives underlying the decision to disclose on the various aspects of corporate sustainability has implications also for investors, employees and consumers, as it can help them in selecting which companies to invest their money in, their work or in choosing whose products to buy. It has practical implications for family firms, especially those operating in environmentally-sensitive industries, suggesting better management of voluntary disclosure as it has a strategic function as well as a communicative one to those outside the company. Our findings indicate, for the family-owned firms, a possible need for improvement in the areas of the impact on product responsibility, providing accounting practitioners with useful information in order to assist effectively these businesses in the sustainability-reporting process.

This study presents some limitations. As a proxy for SEW, we use the percentage of shares controlled by a family as it is the only valuable alternative for research based on large archival databases [48]. This variable measures the "family control and influence" dimension but is not able to capture other aspects such as family members' sense of identification with the business and willingness to transfer the business to future generations, relations within the owning family and between the family and other stakeholders. Further studies could overcome these limitations by using different research methods such as surveys and case studies, in order to develop a multidimensional measurement of family-affective endowment in the firm and evaluate the impact of all SEW dimensions on sustainability disclosure.

Another limit of this study, which could be addressed by further research, is that it takes into account only three moderators of family endowment, namely family CEO, founder and multiple family members on the board. However, other factors affect endowment, modifying the relationship with stakeholders and the need to gain legitimacy, such as a qualified presence of non-family shareholders, the family's generational stage, family conflicts and litigation. Moreover, this study is single-country focused and it would be useful to address its topics on an international sample in order to evaluate the effect of different institutional settings and environmental issues on family and non-family firms' sustainability disclosure.

Acknowledgments: The authors are grateful to the Editor and to the anonymous reviewers for their valuable comments.

Author Contributions: Giovanna Gavana, Pietro Gottardo and Anna Maria Moisello conceived and designed the study; Giovanna Gavana, Pietro Gottardo and Anna Maria Moisello constructed the database and analyzed the data; Giovanna Gavana, Pietro Gottardo and Anna Maria Moisello contributed analyses tools; Giovanna Gavana, Pietro Gottardo and Anna Maria Moisello wrote the paper.

Conflicts of Interest: The authors declare no conflict of interest.

References

1.   Van der Laan Smith, J.; Adhikari, A.; Tondkar, R.H. Exploring differences in social disclosures internationally: A stakeholder perspective. *J. Account. Public Policy* **2005**, *24*, 123–151. [CrossRef]

2.   Bansal, P. Evolving sustainably: A longitudinal study of corporate sustainable development. *Strat. Manag. J.* **2005**, *26*, 197–218. [CrossRef]

3.   Chan, M.C.; Watson, J.; Woodliff, D. Corporate governance quality and CSR disclosures. *Bus. Ethics* **2014**, *125*, 59–73. [CrossRef]

4.   Ling, Q.; Mowen, M.M. Competitive strategy and voluntary environmental disclosure: Evidence from the chemical industry. *J. Account. Public Policy* **2013**, *13*, 55–84. [CrossRef]

5.   Brammer, S.; Pavelin, S. Voluntary social disclosures by large UK Companies. *Bus. Ethics* **2004**, *13*, 86–99. [CrossRef]

6.   Branco, M.C.; Rodrigues, L.L. Factors influencing social responsibility disclosure by Portuguese companies. *Bus. Ethics* **2008**, *83*, 685–701. [CrossRef]

7.   Fernandez-Feijoo, B.; Romero, S.; Ruiz, S. Effect of Stakeholders' Pressure on Transparency of Sustainability Reports within the GRI Framework. *Bus. Ethics* **2014**, *122*, 53–63. [CrossRef]

8.   Gamerschlag, R.; Möller, K.; Verbeeten, F. Determinants of voluntary CSR disclosure: Empirical evidence from Germany. *Rev. Manag. Sci.* **2011**, *5*, 233–262. [CrossRef]

9.   Patten, D.M. Media Exposure, Public Policy Pressure, and Environmental Disclosure: An Examination of the Impact of Tri Data Availability. *Account. For.* **2002**, *26*, 152–171. [CrossRef]

10. Reverte, C. Determinants of corporate social responsibility disclosure ratings by Spanish listed firms. *Bus. Ethics* **2009**, *88*, 351–366. [CrossRef]

11. Roberts, R.W. Determinants of corporate social responsibility disclosure: An application of stakeholder theory. *Account. Organ. Soc.* **1992**, *17*, 595–612. [CrossRef]

12. Haniffa, R.M.; Cooke, T.E. The impact of culture and governance on corporate social reporting. *J. Account. Public Policy* **2005**, *24*, 391–430. [CrossRef]

13. Khan, A.; Muttakin, M.B.; Siddiqui, J. Corporate governance and corporate social responsibility disclosures: Evidence from an emerging economy. *Bus. Ethics* **2013**, *114*, 207–223. [CrossRef]

14. Michelon, G.; Parbonetti, A. The effect of corporate governance on sustainability disclosure. *J. Manag. Gov.* **2012**, *16*, 477–509. [CrossRef]

15. Fifka, M.S.; Drabble, M. Focus on Standardization of Sustainability Reporting—A Comparative Study of the United Kingdom and Finland. *Bus. Strategy Environ.* **2012**, *21*, 455–474. [CrossRef]

16. Guthrie, J.; Parker, L.D. Corporate social disclosure practice: A comparative international analysis. *Adv. Public Interest Account.* **1990**, *3*, 159–175.

17. Frost, G.; Jones, S.; Loftus, J.; van der Laan, S. A survey of sustainability reporting practice of Australian reporting entities. *Austral. Account. Rev.* **2005**, *15*, 89–96. [CrossRef]

18. Holder-Webb, L.; Cohen, J.R.; Nath, L.; Wood, D. The supply of Corporate Social Responsibility Disclosures among U.S. Firms. *Bus. Ethics* **2009**, *84*, 497–527. [CrossRef]

19. Campopiano, G.; de Massis, A. Corporate social responsibility reporting: A content analysis in family and non-family firms. *Bus. Ethics* **2015**, *129*, 511–534. [CrossRef]

20. Ali, A.; Chen, T.Y.; Radhakrishnan, S. Corporate disclosures by family firms. *J. Account. Econ.* **2007**, *44*, 238–286. [CrossRef]

21. Chen, S.; Chen, X.; Cheng, Q. Do family firms provide more or less voluntary disclosure? *J. Account. Res.* **2008**, *46*, 499–536. [CrossRef]

22. Hutton, A.P. A discussion of 'corporate disclosure by family firms'. *J. Account. Econ.* **2007**, *44*, 287–297. [CrossRef]

23. Chua, J.H.; Chrisman, J.J.; Steier, L.P.; Rau, S.B. Sources of heterogeneity in family firms: An introduction. *Entrepreneurship* **2012**, *36*, 1103–1113. [CrossRef]

24. Iyer, V.; Lulseged, A. Does family status impact US firms' sustainability reporting? *Sustain. Acc. Manag. Pol. J.* **2013**, *4*, 163–189. [CrossRef]

25. Dowling, J.; Pfeffer, J. Organizational legitimacy: Social values and organizational behavior. *Pac. Sociol. Rev.* **1975**, *18*, 122–136. [CrossRef]

26. Hooghiemstra, R. Corporate communication and impression management: New perspective why companies engage in Corporate Social Reporting. *Bus. Ethics* **2000**, *27*, 55–68. [CrossRef]

27. Astrachan, J.H.; Klein, S.B.; Smyrnios, K.X. The F-PEC scale of family influence: A proposal for solving the family business definition problem. *Fam. Bus. Rev.* **2002**, *15*, 45–58. [CrossRef]

28. Freeman, R.E. Strategic management. In *A Stakeholder Approach*; Pitman: Boston, MA, USA, 1984.

29. Deegan, C.; Blomquist, C. Stakeholder influence on corporate reporting: An exploration of the interaction between WWF-Australia and the Australian minerals industry. *Account. Org. Soc.* **2006**, *31*, 343–372. [CrossRef]

30. Gomez-Mejia, L.R.; Haynes, K.T.; Nunez-Nickel, M.; Jacobson, K.J.L.; Moyano-Fuentes, J. Socioemotional wealth and business risks in family-controlled firms: Evidence from Spanish olive oil mills. *Adm. Sc. Q.* **2007**, *52*, 106–137.

31. Salvato, C.; Moores, K. Research on accounting in family firms: Past accomplishments and future challenges. *Fam. Bus. Rev.* **2010**, *23*, 193–215. [CrossRef]

32. Uhlaner, L.M.; Kellermanns, F.; Eddeleston, K.; Hoy, F. The entrepreneuring family: A new paradigm for family business research. *Small Bus. Econ.* **2012**, *38*, 1–11. [CrossRef]

33. Deegan, C. Introduction: The legitimizing effect of social and environmental disclosures-a theoretical foundation. *Account. Aud. Account. J.* **2002**, *15*, 282–311. [CrossRef]

34. Suchman, M.C. Managing legitimacy: Strategic and institutional approaches. *Acad. Manag. J.* **1995**, *20*, 571–610.

35. Sethi, S.P. A conceptual framework for environmental analysis of social issues and evaluation of business response patterns. *Acad. Manag. J.* **1979**, *4*, 63–74.

36. Oliver, C. Strategic responses to institutional processes. *Acad. Manag. J.* **1991**, *16*, 145–179. [CrossRef]

37. Scott, W.R. The adolescence of institutional theory. *Adm. Sci. Q.* **1987**, 493–511. [CrossRef]

38. Lindblom, C.K. The implications of organizational legitimacy for corporate social performance and disclosure. In Proceedings of the Critical Perspectives on Accounting Conference, New York, NY, USA, 20 April 1994.

39. Fisher, J.; Gunz, S.; McCutcheon, J. Private/public interest and the enforcement of a code of professional conduct. *Bus. Ethics* **2001**, *31*, 191–207. [CrossRef]

40. Dhaliwal, D.S.; Li, O.Z.; Tsang, A.; Yang, Y.G. Voluntary nonfinancial disclosure and the cost of equity capital: The initiation of corporate social responsibility reporting. *Acc. Rev.* **2011**, *86*, 59–100. [CrossRef]

41. Deegan, C.; Rankin, M.; Voght, P. Firms' disclosure reactions to major social incidents: Australian evidence. *Account. For.* **2000**, *24*, 101–130. [CrossRef]

42. Gray, R.; Kouhy, R.; Lavers, S. Corporate social and environmental reporting: A review of the literature and a longitudinal study of UK disclosure. *Account. Aud. Account. J.* **1995**, *8*, 47–77. [CrossRef]

43. Clatworthy, M.; Jones, M.J. The effect of thematic structure on the variability of annual report readability. *Account. Aud. Account. J.* **2001**, *14*, 311–326. [CrossRef]

44. Mitchell, R.K.; Agle, B.R.; Chrisman, J.J.; Spence, L.J. Toward a theory of stakeholder salience in family firms. *Bus. Ethics Q.* **2011**, *21*, 235–255. [CrossRef]

45. Kalm, M.; Gomez-Mejia, L.R. Socioemotional wealth preservation in family firms. *Rev. Adm.* **2016**, *51*, 409–411. [CrossRef]

46. Cruz, C.; Justo, R.; de Castro, J.O. Does family employment enhance MSEs performance?: Integrating socioemotional wealth and family embeddedness perspectives. *J. Bus. Vent.* **2012**, *27*, 62–76. [CrossRef]

47. Leitterstorf, M.P.; Rau, S.B. Socioemotional wealth and IPO underpricing of family firms. *Strateg. Manag. J.* **2014**, *35*, 751–760. [CrossRef]

48. Berrone, P.; Cruz, C.; Gomez-Mejia, L.R. Socioemotional wealth in family firms theoretical dimensions, assessment approaches, and agenda for future research. *Fam. Bus. Rev.* **2012**, *25*, 258–279. [CrossRef]

49. Gomez-Mejia, L.R.; Cruz, C.; Berrone, P.; de Castro, J. The bind that ties: Socioemotional wealth preservation in family firms. *Acad. Manag. Ann.* **2011**, *5*, 653–707. [CrossRef]

50. Vandekerkhof, P.; Steijvers, T.; Hendriks, W.; Voordeckers, W. The effect of organizational characteristics on the appointment of nonfamily managers in private family firms: The moderating role of socioemotional wealth. *Fam. Bus. Rev.* **2015**, *28*, 104–122. [CrossRef]

51. Deephouse, D.L.; Jaskiewicz, P. Do Family Firms Have Better Reputations than Non-Family Firms? An Integration of Socioemotional Wealth and Social Identity Theories. *J. Manag. Stud.* **2013**, *50*, 337–360. [CrossRef]

52. Berrone, P.; Cruz, C.; Gomez-Mejia, L.R.; Larraza-Kintana, M. Socioemotional wealth and corporate responses to institutional pressures: Do family-controlled firms pollute less? *Adm. Sci. Q.* **2010**, *55*, 82–113. [CrossRef]

53. Dyer, W.G.; Whetten, D.A. Family firms and social responsibility: Preliminary evidence from the S&P 500. *Entrep. Theory Parct.* **2006**, *30*, 785–802.

54. Sharma, P.; Manikutty, S. Strategic divestments in family firms: Role of family structure and community culture. *Entrep. Theory Parct.* **2005**, *29*, 293–311. [CrossRef]

55. Cennamo, C.; Berrone, P.; Cruz, C.; Gomez-Mejia, L.R. Socioemotional Wealth and Proactive Stakeholder Engagement: Why Family-Controlled Firms Care More about Their Stakeholders. *Entrep. Theory Parct.* **2012**, *36*, 1153–1173. [CrossRef]

56. Astrachan, J.H.; Jaskiewicz, P. Emotional returns and emotional costs in privately held family businesses: Advancing traditional business valuation. *Fam. Bus. Rev.* **2008**, *21*, 139–149. [CrossRef]

57. Zellweger, T.M.; Kellermanns, F.W.; Chrisman, J.J.; Chua, J.H. Family control and family firm valuation by family CEOs: The importance of intentions for transgenerational control. *Org. Sci.* **2012**, *23*, 851–868. [CrossRef]

58. Deegan, C. Organisational legitimacy as a motive for sustainability reporting. In *Sustainability Accounting and Accountability*; Unerman, J., Bebbington, J., O'Dwyer, B., Eds.; Routledge: London, UK, 2007; pp. 127–149.

59. Perrini, F. Building a European portrait of corporate social responsibility reporting. *Europ. Manag. J.* **2005**, *23*, 611–627. [CrossRef]

60. Morhardt, J.E. Corporate social responsibility and sustainability reporting on the internet. *Bus Strategy Environ.* **2010**, *19*, 436–452. [CrossRef]

61. Clarke, J.; Gibson-Sweet, M. The use of corporate social disclosures in the management of reputation and legitimacy: A cross sectorial analysis of UK top 100 companies. *Bus. Ethnics Eur. Rev.* **1999**, *8*, 5–13. [CrossRef]

62. Cruz, C.; Larraza-Kintana, M.; Garcés-Galdeano, L.; Berrone, P. Are Family Firms Really More Socially Responsible? *Entrep. Theory Parct.* **2014**, *38*, 1295–1316. [CrossRef]

63. Uhlaner, L.M. Business family as a team: Underlying force for sustained competitive advantage. *Hand. Res. Fam. Bus.* **2006**, 125–144.

64. Déniz, M.D.L.C.D.; Suárez, M.K.C. Corporate social responsibility and family business in Spain. *Bus. Ethnics* **2005**, *56*, 27–41. [CrossRef]

65. Dunn, B. Family enterprises in the UK: A special sector. *Fam. Bus. Rev.* **1996**, *9*, 139–155. [CrossRef]

66. Neu, D.; Warsame, H.; Pedwell, K. Managing public impressions: Environmental disclosures in annual reports. *Account. Org. Soc.* **1998**, *23*, 265–282. [CrossRef]

67. Adams, C.A. Internal organisational factors influencing corporate social and ethical reporting: Beyond current theorizing. *Account. Aud. Account. J.* **2002**, *15*, 223–250. [CrossRef]

68. Prado-Lorenzo, J.M.; Gallego-Alvarez, I.; Garcia-Sanchez, I.M. Stakeholder engagement and corporate social responsibility reporting: The ownership structure effect. *Corp. Soc. Resp. Environ. Manag.* **2009**, *16*, 94–107. [CrossRef]

69. Bingham, J.B.; Dyer, W.G., Jr.; Smith, I.; Adams, G.L. A stakeholder identity orientation approach to corporate social performance in family firms. *Bus. Ethics* **2011**, *99*, 565–585. [CrossRef]

70. Marques, P.; Presas, P.; Simon, A. The Heterogeneity of Family Firms in CSR Engagement: The Role of Values. *Fam. Bus. Rev.* **2014**, *27*, 206–227. [CrossRef]

71. O'Boyle, E.H.; Rutherford, M.W.; Pollack, J.M. Examining the relation between ethical focus and financial performance in family firms: An exploratory study. *Fam. Bus. Rev.* **2010**, *23*, 310–326. [CrossRef]

72. Le Breton-Miller, I.; Miller, D. Socioemotional wealth across the family firm life cycle: A commentary on "Family Business Survival and the Role of Boards". *Entrep. Theory Parct.* **2013**, *37*, 1391–1397. [CrossRef]

73. Dobele, A.R.; Westberg, K.; Steel, M.; Flowers, K. An examination of corporate social responsibility implementation and stakeholder engagement: A case study in the Australian mining industry. *Bus. Strategy Environ.* **2014**, *23*, 145–159. [CrossRef]

74. Zellweger, T.M.; Nason, R.S.; Nordqvist, M.; Brush, C.G. Why do family firms strive for nonfinancial goals? An organizational identity perspective. *Entrep. Theory Parct.* **2013**, *37*, 229–248.

75. Micelotta, E.R.; Raynard, M. Concealing or revealing the family? Corporate brand identity strategies in family firms. *Fam. Bus. Rev.* **2011**, *24*, 197–216. [CrossRef]

76. Le Breton-Miller, L.; Miller, D. Agency vs. stewardship in public family firms: A social embeddedness reconciliation. *Entrep. Theory Parct.* **2009**, *33*, 1169–1191. [CrossRef]

77. Sundaramurthy, C.; Kreiner, G.E. Governing by managing identity boundaries: The case of family businesses. *Entrep. Theory Parct.* **2008**, *32*, 415–436. [CrossRef]

78. De Vries, M.F.K. The dynamics of family controlled firms: The good and the bad news. *Organ. Dyn.* **1994**, *21*, 59–71. [CrossRef]

79. Westhead, P.; Cowling, M.; Howorth, C. The development of family companies: Management and ownership imperatives. *Fam. Bus. Rev.* **2001**, *14*, 369–385. [CrossRef]

80. Bebbington, J.; Larrinaga, C.; Moneva, J.M. Legitimating reputation/the reputation of legitimacy theory. *Account. Aud. Account. J.* **2008**, *21*, 371–374. [CrossRef]

81. Global Reporting Initiative. *Sustainability Reporting Guidelines, G3.1*; Global Reporting Initiative: Amsterdam, The Netherlands, 2011.

82. Anderson, R.C.; Reeb, D.M. Founding-family ownership and firm performance: Evidence from the S&P 500. *J. Fin.* **2003**, *58*, 1301–1328.

83. Croci, E.; Doukas, J.A.; Gonec, H. Family control and financing decisions. *Europ. Financ. Manag.* **2011**, *17*, 860–897. [CrossRef]

84. Faccio, M.; Lang, L.H.P. The ultimate ownership of Western Europe corporations. *J. Financ. Econ.* **2002**, *65*, 365–395. [CrossRef]

85. Villalonga, B.; Amit, R. Family control of firms and industries. *Financ. Manag.* **2010**, *39*, 863–904. [CrossRef]

86. Guthrie, J.; Cuganesan, S.; Ward, L. Industry specific social and environmental reporting: The Australian Food and Beverage Industry. *Account. For.* **2008**, *32*, 1–15. [CrossRef]

87. Stavrou, E.; Kassinis, G.; Filotheou, A. Downsizing and Stakeholder Orientation among the Fortune 500: Does Family Ownership Matter? *J. Bus. Ethics* **2008**, *72*, 149–162. [CrossRef]

88. Brammer, S.; Pavelin, S. Factors influencing the quality of corporate environmental disclosure. *Bus. Strategy Environ.* **2008**, *17*, 120–136. [CrossRef]

89. Cheng, E.C.; Courtenay, S.M. Board composition, regulatory regime and voluntary disclosure. *Int. J. Account.* **2006**, *41*, 262–289. [CrossRef]

90. Naser, K.; Al-Hussaini, A.; Al-Kwari, D.; Nuseibeh, R. Determinants of corporate social disclosure in developing countries: The case of Qatar. *Adv. Int. Account.* **2006**, *19*, 1–23. [CrossRef]

91. Cormier, D.; Gordon, I.M.; Magnan, M. Corporate environmental disclosure: Contrasting management's perceptions with reality. *Bus. Ethics* **2004**, *49*, 143–165. [CrossRef]

92. Mio, C.; Venturelli, A. Non-financial information about sustainable development and environmental policy in the annual reports of listed companies: Evidence from Italy and the UK. *Corp. Soc. Resp. Environ. Manag.* **2013**, *20*, 340–358. [CrossRef]

93. Ho, S.S.; Wong, K.S. A study of the relationship between corporate governance structures and the extent of voluntary disclosure. *J. Int. Account. Audit. Tax.* **2001**, *10*, 139–156. [CrossRef]

94. Smith, A.C.; Westerbeek, H.M. Sport as a vehicle for deploying corporate social responsibility. *J. Corp. Citiz.* **2007**, *25*, 43–54. [CrossRef]

95. Namazi, M.; Namazi, N.R. Conceptual analysis of moderator and mediator variables in business research. *Proc. Ecol. Financ.* **2016**, *36*, 540–554. [CrossRef]

96. Zientara, P. Socioemotional wealth and corporate social responsibility: A critical analysis. *Bus. Ethics* **2015**. [CrossRef]

97. Brown, N.; Deegan, C. The public disclosure of environmental performance information—A dual test of media agenda setting theory and legitimacy theory. *Account. Bus. Res.* **1998**, *29*, 21–41. [CrossRef]

98. Stavrou, E.T.; Swiercz, P.M. Securing the future of the family enterprise: A model of offspring intentions to join the business. *Entrep. Theory Parct.* **1998**, *23*, 19–21.

99. Block, J. Family management, family ownership, and downsizing: Evidence from S&P 500 firms. *Fam. Bus. Rev.* **2010**, *23*, 109–130.

8

Towards Transgressive Learning through Ontological Politics: Answering the "Call of the Mountain" in a Colombian Network of Sustainability

Martha Chaves [1], Thomas Macintyre [2,*], Gerard Verschoor [3] and Arjen E. J. Wals [4]

[1] Sociology of Development and Change Group, Wageningen University, P.O. Box 8130, 6706 KN Wageningen, The Netherlands; marthacecilia.chaves@gmail.com

[2] MINGAS in Transition Research Group, Calle 8 # 16-218 Rozo, Palmira, Colombia

[3] Sociology of Development and Change Group, Wageningen University, Hollandseweg 1, 6706 KN Wageningen, The Netherlands; Gerard.Verschoor@wur.nl

[4] Education and Competence Studies Group (ECS), Wageningen University, Hollandseweg 1, 6706 KN Wageningen, The Netherlands; arjen.wals@wur.nl

* Correspondence: Thomas.macintyre@gmail.com

Academic Editor: Helmut Haberl

Abstract: In line with the increasing calls for more transformative and transgressive learning in the context of sustainability studies, this article explores how encounters between different ontologies can lead to socio-ecological sustainability. With the dominant one-world universe increasingly being questioned by those who advocate the existence of many worlds—a so-called pluriverse—there lays the possibility of not only imagining other human–nature realities, but also engaging with them in practice. Moving towards an understanding of what happens when a multiplicity of worlds encounter one another, however, entails a sensitivity to the negotiations between often competing ontologies—or ontological politics. Based on an ethnographic methodology and narrative methods, data were collected from two consecutive intercultural gatherings called *El Llamado de la Montaña* (The Call of the Mountain), which take place for five days every year in different parts of Colombia. By actively participating in these gatherings of multiplicity, which address complex socio-ecological challenges such as food sovereignty and defence of territory, results show how encounters between different ontologies can result in transformative and potentially transgressive learning in terms of disrupting stubborn routines, norms and hegemonic powers which tend to accelerate *un*sustainability. Although we argue that a fundamental part of the wicked sustainability puzzle lies in supporting more relational ontologies, we note that such learning environments also lead to conflicts through inflexibility and (ab)use of power which must be addressed if sustained socio-ecological learning is to take place.

Keywords: ontological politics; transformative learning; transgressive learning; sustainability; Colombia; narrative methods

---

1. Introduction to Other Realities

"We are the new seeds that sprout from the earth. We have been called upon to restore the times of our peoples, and we are going towards the call of the mountain, from whose veins sprout great memories of new dawns in which to live. And we stand up in a silent way, because we recognize the silence of the sun and we know how to listen to the moon. It is the time of the new beings, and the air will give us the strength to carry this great message." (First part of a song written by Lorenzo Muelas Tombé, an Indigenous Misak youth who helped organize the 2015 gathering of the Call of the Mountain)

High up in the Andean mountains of Southern Colombia, in the Indigenous territory of Misak, a sacred walk is taking place. Led by the Misak people, and followed by *abuelos* (elders), ecovillagers, Hare Krishna, urban intellectuals, and foreigners, this diverse collection of people are walking towards the pueblo of Silvia to *"activate and heal the bond with one's own territory . . . and as a collective prayer for life and for peace"*. (Ana María, Council of Women Elders of Colombia [1]). Everyone is holding a seedling in their hands, and walking to the rhythm of the traditional Misak drums and flutes. Arriving to the pueblo, everyone boards a colorful (if dilapidated) *chiva* bus, which carries its motley crew to the agricultural development land of the Indigenous Misak University. After being received with the traditional fermented maize drink *chicha*, the group of over 300 people prepares to *sembrar agua* (plant water) through the reforestation of 2000 trees in a neighboring wetland. While the *mamitas* (women elders) are cooking a traditional soup called *sancocho* over an open fire (and remarking on the difficulty of making a tasty *sancocho* without meat as requested by the organizers), an animated group of *abuelas*, children, Hare Krishnas and ecovillagers are singing and chanting to the soup, while a human chain forms to efficiently move the seedlings to where they are to be planted. Under the animated discussion between Misak youth and permacultural experts about which variety of seedling should be planted where, and accompanied by a steady drizzle of rain, holes are dug, hands reach into the soil, and seedlings are given a new home. When the last seedling is rooted, everyone trudges back to the farmhouse to eat the *sancocho*, drink an *agua panela* (sugar cane tea) and celebrate. *"Hermanos, thank you for helping us plant our water and for sharing in this collective effort"* a Misak organizer cries aloud to everyone, who respond with a cheer. In all this excitement a little girl asks her mom: *"why are we planting water? why are we singing to the soup?"*

This research addresses ontological encounters entailed in bringing about socio-ecological sustainability. The addition of "socio-ecological" to sustainability is intentional, as much work done on sustainability nowadays tends to focus on economic sustainability, often without people and planet in mind. Adrian Parr [2] even suggests that sustainability has gradually been hijacked and neutered by neo-liberal economic interests. While economics inevitably is part of the sustainability puzzle, the need to pay full attention to the ecological boundaries within which both humans and non-humans will have to live together requires taking on board issues that vastly overflow the economic undercurrent that dominates the on-going sustainability discourse [3–5].

This emphasis is particularly prudent in a time when the hegemony of the development and globalization "projects" [6] are in crisis, with new narratives of human–nature relations emerging which propose fundamental changes in how we understand the world and its relations. This is nowhere more apparent than in the region of Latin America where counter-hegemonic movement at the political and social level are being witnessed. The new constitutions of Bolivia and Ecuador have respectively acknowledged a plurinational state and the rights of nature (both affronts to the notion of the modern state), while social movements combining Indigenous communities and environmental activists are gathering around endogenous concepts like *buen vivir* (the good life) which propose more biocentric, relational and communal relations [7].

The above examples provide a glimpse into the profound but difficult notion of different worlds living side by side in what some authors are calling the pluriverse [8–11]. Such coexistence of multiple worlds denotes a departure from the homogenizing and euphemistic idea of the "global village" [11], instead giving status to alternative ways of being in the world. These sub-altern alternatives, it has been argued, can provide a diversity of responses to the global crisis if only one would consider their knowledge to be equally valid [12]. Such a decolonial attitude fits into the greater sustainability transition discourses which call for radical cultural and institutional transformations to an "altogether different world" [13] (p. 138).

However, the question arises as to *how* alternative ways of being and knowing can contribute to addressing the sustainability challenges of our time [9]. This article is based on the premise that people learn more from each other when they are confronted by different realities—what exists, and our underlying assumptions of what is, what is not and what might be. When an urban environmentalist meets an ecovillager and an Indigenous person, all three expounding different understandings of what it means to plant a tree—or when a strictly vegetarian Hare Krishna devotee is confronted with an Indigenous Misak or Arhuaco who eats (wild) meat as part of their culture, there is the potential for clashes and conflicts, but also for new insights and understandings in what it means to live an environmentally and socially responsible life *together* with other people who do not necessarily share that reality. Such encounters have the potential to transform the way we learn from the world around us through challenging deeply held beliefs and habits [4]. This is based on the increasing recognition that more emancipatory forms of transformative learning can lead to far deeper and more meaningful engagement in sustainability issues than, for example, trying to change people's environmental behavior instrumentally through, for instance, persuasion, social marketing or by law [14].

Returning to the opening scene of this article—high up in the Andean mountains—we can see a description taken from *El Llamado de la Montaña* (The Call of the Mountain). Organised by the sustainability network C.A.S.A. Colombia [15], this yearly Colombian gathering brings together a diverse array of people, communities and projects for five days of communal living, in which participants exchange experiences on sustainable living while partaking in working councils, workshops, panel discussions, dances and other artistic pursuits. *El Llamado*, as the event is referred to, brings together Indigenous elders and businesspeople, urban permaculturalists and peasants, and Hare Krishna devotees and academics, to name just a few. The event is self-financed and self-organized, and has the aim of articulating and forging alliances between diverse grassroots movements in Colombia around pressing socio-ecological concerns such as food sovereignty, mega-mining, and post-conflict reconstruction. The interactions during the event represents an "environmentalism of everyday life" [16], where participants share social justice and environmental concerns through embodied practices during the event, often explicitly acknowledging and promoting the role of the non-human realm. From this yearly exercise of community and human–nature interaction, a central challenge has emerged in how to deal with encounters within and between a diversity of peoples, visions and knowledges. It has been recognized by event organizers that such encounters can promote innovative thinking and action when differences come together in a generative learning environment, yet it is also acknowledged as leading to conflict through misunderstandings, poor communication and other underlying issues.

2. Transgressive Learning and the Ontological Politics of the Pluriverse

Studying the challenges and opportunities of an "environmentalism of everyday life" can be done from a variety of perspectives. We approach these challenges and opportunities through a twofold strategy in which we combine an interest in ontological politics and the transgressive learning that may be obtained from this.

The concept of "ontological politics" originates from the so-called "ontological turn" in anthropology and science and technology studies [17,18]. The concept of ontology itself originally comes from philosophy, and involves the study of reality and questions related to the kind of entities that can be said to exist as well as the relations between them [19]. Importantly, the ontological turn assumes that there exist a multiplicity of realities or worlds [20–22]. Underlying this proposition is the argument that the reality we live in is one performed in a variety of practices [18], whereby reality does not *precede* the everyday practices in which we interact with "the world", but is rather shaped *within* those practices [23]. Since practices are multiple, so too are the realities they produce—hence, "if reality is *done*, then it is also *multiple*" [18] (p. 75). Therefore, multiple worlds or ontologies do not form a universe, but rather what William James called a "*pluriverse*" or a "*multiverse*" [24]. In this pluriverse, these different worlds or "ways of being" are partially connected, i.e., they are connected without implying that they share a common ontology [25].

Partial connections between different worlds or realities inevitably lead to ontological encounters. In these encounters, ontological disjunctures or misunderstandings are very likely to occur. Viveiros de Castro calls these situations occasions of "uncontrolled equivocation": "a type of communicative disjuncture where the interlocutors are not talking about the same thing, and do not know this" [26] (p. 9). Situations such as these (involving Indigenous *taitas* and *abuelas*, Hare Krishnas and ecovillagers) of course abound in the different *Llamados de la Montaña* which are the object of our study. We find these situations interesting—not for the clashes between different ways of being, but rather for the ontological politics that come with them; that is, the possibility that practices (and hence realities) might be changed for other ways of being that "could be" [17]. When deployed in the "transgressive" context of our own sustainability struggles, ontological encounters provide a treasure for learning that unsustainable realities are not destiny.

In the context of this article, then, we are interested in the dynamics that take place in the encounter and interaction between the different worlds or ontologies that meet in the micro-cosmos apparent in the *Llamado de la Montaña*. In particular, we focus on the power-saturated, "partially connected unfolding of worlds" [19] that are constantly becoming, giving rise to new ontologies through concrete relations and actions among persons, things, spirits, and deities. In these new "worlds in the making" transgressive learning results from the power-laden ontological interactions, interferences and blendings that are characteristic of complex socio-ecological settings.

Taking these affections (and the notion of the pluriverse) seriously, however, means addressing fundamental issues underlying power relations upheld through practices. Several social movements and theories of decolonialization acknowledge this and have been identified as a stream of emerging transgressive and transformative research and praxis in the sustainability sciences [27]. In this paper we will employ the emerging concept of transgressive learning in challenging the taken for granted, normalized status quo of global systemic dysfunction [28,29]. As a form of transformative learning, the concept focuses on uprooting structures of privilege and hegemonies of power through innovative strategies which foreground cognitive, epistemic, social and environmental justice, often through activism and normative interventions. While Mezirow´s theory of Transformative learning [30] is often used to frame such discussions about the changes in values and worldviews needed to move towards a more sustainable world, its theory is mainly based on cognitive change at the level of the individual [27]. The emerging concept of transgressive theory attempts to take a more decolonial and transdisciplinary stance [31] building on such work as that of the critical pedagogy of Paolo Freire [32] in Latin America, and other strands such as reflexive social learning and capabilities theory, critical phenomenology, and socio-cultural and cultural historical activity theory [27]. Important for transgressive learning theory is recognizing that socio-technical transitions to sustainability do not come about easily because of lock-in mechanisms which maintain poverty and social injustices. To address this, transgressive learning posits that radical innovations instead occur in "niches" [33], in which the cultivation and productive utilization of multiplicity are necessary for transformative disruptions to emerge [4]. Put simply, people learn more from each other when they are different from each other, as this creates more space for reflection through disruption and dissonance [34]. However, it has also been emphasized that the tensions which arise between different ways of thinking are only productive when strong affects exist within the group [35], and the upscaling of these processes are dependent on external landscape developments putting pressure on dominant regimes so as to open windows of opportunity for niches to expand [33]. Transgressive learning therefore plays an important part of this paper as it represents a *type* of learning which can disrupt normalized unsustainable habits, of which we argue ontological politics play a vital role.

In the search for "worlds and knowledges otherwise" [36], and the potential for transgressive learning which result from the politics of their encounters, this research therefore aims to explore how concrete intercultural practices can lead to insights into how to imagine and practice the pluriverse in the sustainability arena. Considering that the pluriverse can be made visible by examining ontological conflicts (the unequal encounter between worlds) [37], or what has also been referred to as political ontology [20], our aims are to explore how this pluriverse might look like in sustainability practices, and the extent to which ontological encounters can lead to transgressive learning. It is important to address ontological encounters in the sustainability debate for three reasons. The first is that there is an increasing recognition that ontological predispositions play an important role in how people understand and engage with sustainability challenges [6,38,39]. With a country like Colombia managing to atypically balance a high Human Development Index (HDI), while maintaining a low ecological footprint [40] it is important to explore the ontological basis of this. The second reason is that when different visions of sustainability and the realities that enact them are played out in practice, their encounters lead to a politics where different realities interact and compete with one another [9]. Rather than romanticizing a "pluriverse" where everybody gets along, engaging with such "ontological politics" forces us to not only acknowledge the existence of different realities, but also the power plays within and between them which adds increasing complexity to already "wicked" challenges of sustainability [41]. This leads us to the third reason, which is that meetings between ontologies have the potential to not only be transformative in how we learn about the world around us [4], but also activate "transgressive" learning processes which challenge the status quo through action-oriented interventions [27]. By responding to the Call of the Mountain—the call for a more sustainable world—we will engage in the debates of ontological politics in practice, as well as providing some examples in which ontological encounters leads to a type of learning potentially "transgressive" in kind.

3. Methodology: New Ethnography and the Voice of the Researcher

In addressing the aims above, it is vital to employ a methodology which allows us to enter (to the extent it is possible) the pluriverse, and to critically represent the interactions which take place. Central to this methodology is the acknowledgment that research is never impartial but an assemblage with its own effects on the event researched comprising of researchers, data, methods and contexts [42]. This is in line with the increasing skepticism in the postmodern world regarding the objectivity of the researcher, the generalization of knowledge claims, and the realist agendas where the researcher is put above the subject and the method over the subject matter [43]. In keeping with the ontological turn, we have tried, as Viveiros de Castro advocates [44], taking seriously the things the people we study tell us, while acknowledging, at the same time, that we inevitably bring our own ontological assumptions into the research encounter. We have hence employed methods of "new ethnography" [45] where narrative is employed to communicate findings using the tools of storytelling. Specifically, this involves elements of auto-ethnography as a way of acknowledging the role of the researcher in voicing the results. Analytic auto-ethnography [46] involves three aspects which can be seen in our methods:

(1) We are full members of the researched group or setting: Participant observation was carried out by the first two authors who immersed themselves in the network studied (C.A.S.A. Colombia), where the first author gradually became actively involved in the organizing committee of *El Llamado*. Both authors have participated in several *Llamados* since 2011 and have carried out informal and semi-structured interviews during the *Llamados* of 2014 and 2015.

(2) We are visible in the resulting texts: As witnessed by the opening paragraph of this paper, and as will be seen in the first section of the results, narrative techniques are employed to generate "creative narratives shaped out of a writer's personal experiences within a culture and addressed to academic and public audiences" (p. 9). Employing personal voice also contributes to the idea that research is performative [47]. A good example of a monologue written in this way can be seen in the work of Mario Blaser who in his book *Storytelling globalization from the Chaco and beyond* [9] engages in knowledge practice grounded in a relational ontology, advocating a pluriverse where other forms of knowledge and ontologies are accepted.

(3) We are committed to developing theoretical understandings of broader social phenomena: As researchers the focus is on the role of enacting the pluriverse in activating transformative and transgressive learning processes which are been increasingly acknowledged as imperative in sustainability process for social change.

4. Enacting a Learning Pluriverse: The Collective Process of *El Llamado*

"We are servants of this humanity, as souls we are all drops of the same sea, love is our nourishment to sprout. To be slighter than the grass and more tolerant than a tree, our heritage is humbleness."

(Yayati, Hare Krishna devotee from the community Varsana)

"We are the humanity that is sprouting, cells of the earth remembering, a joyful singing rainbow, to awaken all of mankind." (Tatiana, Ecovillager)

"We are 'taitas', 'mayores' y 'mamitas'—[names given to Indigenous elders, men and women respectively]—we receive the message from the mountain, and together we will construct a new tomorrow." (Camilo from the community Tal, Cundinamarca) (The above are verses written by participants during the 2015 event to contribute to the song written by the Misak youth Lorenzo Muelas Tombé.)

The gathering *El Llamado* began in 2006 when pioneers in the Colombian ecovillage movement gathered in the community Pachamama to share experiences in a family setting and explore ways to foster a social movement. An ecovillage can be defined as "a planetary knowledge community grounded in a holistic ontology and seeking to construct viable living systems as an alternative to the unsustainable legacy of modernity" [48] (p. 125). The gathering remained very much a small ecovillage affair until 2012, when the organizers came together with ecovillagers from different parts of Latin America and Spain to form the Council of Sustainable Settlements of Latin America (C.A.S.A.) with the vision of articulating a broad diversity of initiatives beyond the ecovillage realm around different ways of understanding and practicing sustainability.

As of today, C.A.S.A. Colombia has been building up the network with initiatives which exert *"views of sustainability different from those of 'mainstream' [including the government] who see nature as a human resource … but instead consider nature as an entity endowed with spirit, which the human being is caregiver, and with which we want to cultivate a deep spiritual relationship that involves changing our role in the world"* (C.A.S.A. institutional documentation). This endeavor has led organizers and participants to immerse themselves in a process of assembling visions and practices of sustainability across many cultures through experiential learning, thus encountering all the challenges of such processes along the way.

In this section, we first provide an auto-ethnographic narrative as a way of presenting the transgressive learning experienced by the first author during *El Llamado* of 2015, in which she was a participant as well as an organizer. In the second part, we bring specific examples of ontological politics which surfaced during the two *Llamados* of 2014 and 2015, demonstrating further possibilities for transgressive learning through encounters of difference.

4.1. A Glimpse into El Llamado de la Montaña, 2015

"It is the 5 January 2015, and I am on the bus entering the Misak territory. Looking out the window I see a beautiful landscape full of rolling hills and mountains. I have seen them several times before, but this time they look different. What I know now about this territory and how it relates to its people has changed my perception of these mountains. They have become alive to me. With a little help from my imagination I see the hills as noses, the mountains as heads, breasts and legs, the small houses as eyes and the clouds as hair or beards, moving in the wind.

After arriving to the pueblo Silvia, it is time to walk up into the mountains along a beautiful river which leads to the Misak University. Curious faces follow me from the houses I pass by, and perhaps this is not strange as before me and behind me there are other 'backpackers' walking the same path as me. This is certainly not so common in what the Colombian government describe as an 'insecure and difficult region'.

As I walk the final steps to the Misak University I reflect on my process of trying to understand the Misak peoples' relations to their territory, and on how these relations influence the way they think and act. How much will participants manage to understand in five days of co-existence?

Misunderstandings have been common in organizing El Llamado together with the Misak. After each visit I have always carried the feeling of not getting the whole picture of our relationship, and not being able to entirely explain these differences. During the few times I thought 'I got it', later events proved the contrary. It is not a question of language; it is a question of how our individual and collective worlds work and how we experience them differently. For example, while as an organizer I was more concerned about the logistics of food, accommodation, the scheduling of activities, and special guests, the main concerns of the Misak organizers were the possible harm to the energy of the territory and the fear/excitement of sharing their knowledge with us. They argued, for example, that after their medicine men consulted the territory, it was decided that the temascal (ritual sweat lodges from the Indigenous people of North American which are run by healers in every Llamado) where not possible to carry out during this version of El Llamado: "Our territory is of water and the temascal is a ritual of fire. If we carry it out here we can create an imbalance in nature and even wake up the volcanoes." (Misak organizer)

With all this in mind, I felt the pressure of my anxiety: How will these different worlds meet and connect? How are the relatively reserved Misak people going to face the numerous hugs and other physical contact that are the 'menu of the day' during every Llamado? How are other participants going to deal with all the rules and rituals surrounding the care of the Misak territory?

The first challenge I encountered was during the registration for the event when we had to ask women if they were going to be in their menstrual cycle during El Llamado. This was a shocking question for many, and difficult for me to ask. Even more shocking was the information that if they were in their cycle then they could not help in the kitchen or cross any streams of water, they had to have a special cleansing and harmonization ritual and they could not attend the highly anticipated Minga—the collective work party. The look on the faces of several of these urban-based women was of utter disbelief. Equality? Segregation? Discrimination? I tried to explain the complexity of this 'territorio despierto' (awakened territory) of the Misak people and make them feel positive about their situation. The network of 'women circles' had organized a special event just for the women with 'the moon' (as they call women in their menstrual cycles) where they could sit with the women elders from the territory to better understand the relation between their 'state' and the territory. Though perhaps not convinced, several of these women accepted the invitation and followed the instructions. It was satisfying when one of these women came to me in the lasts days of El Llamado to tell me that the highlight of her experience had been the circle of women. Through this experience she had been able to connect with her inner-self through the stories of other women. She dealt with personal issues and was able to grasp the opportunities and the magic of being a 'woman with the moon'. She also told me of learning practical things about sustainability, such as the option of using cloth pads or women cups instead of disposable pads, and about the importance of bringing feminine qualities (such as caring, emotions, and flexibility) into her job and everyday life.

Late in the afternoon, the cleansing of participants was programed. Participants were asked to sit in a horizontal line behind a group of medicine men who were 'speaking' to the territory. Without explanation, people accepted the situation and waited for something to happen. After two hours nothing had 'happened'. The medicine men stood gazing out into the mountains, whispering softly to the wind, while participants began to feel cold, tired and bored. Hare Krishna devotees started singing and people quickly joined in—songs about taking care of mother earth—as a means of lightening the mood as well as a means to join in the message of harmony with the territory. Well, the medicine men were not happy about this; they informed me that the noise and energy was making their job more difficult. Yet how could we stop the singing if this was the way people were dealing with such a foreign situation? Eventually a thick mist began descending from the mountains enveloping everyone, further decreasing the temperature and silencing the singing. Many participants, tired and cold, started leaving for their tents. At last, a medicine man explained that the mist was the actual cleansing. He brushed each person front and back with a branch soaked in water infused with medicinal plants. The ritual was over; the remaining people stumbled back to their tents in the dark, some confused, some contented, and most just ready for bed.

The next day, during the programed panel discussion on buen vivir (the good life), in which leaders of several Indigenous communities told their story of struggling to establish their good life against the 'development machine', I was thinking how far removed my personal story was from those experiences. When it was the turn for the ecovillage representative to speak I felt uneasy. I had lived for a period of time in the ecovillage of this representative, and after listening to all the stories of indigenous repression and resistance I wondered what the ecovillage world could contribute? Jorge, the ecovillager speaker, began by explaining the role of the mestizo of urban origin in bringing back sustainability in practice to mainstream society. He talked about our need to learn from native communities how to reconnect with place and nature, but also about day-to-day struggles of dealing with waste management, and the loss of spirituality and hope in youth. This was confirmed by the next speaker, the Colombian ex-senator Gloria Cuartas stating that we (the audience) need it to become 'the geography of hope' to heal this country in conflict, Colombia. I felt inspired again and more secure of what I could offer to others.

Part of my volunteer duties was to help in the kitchen, but when I arrived there was chaos. People were passing through the kitchen on the way to their tents; one group of people were chopping carrots, while others were singing and talking around the fire where the food was being cooked. I felt sorry for the woman in charge of the kitchen who appeared stressed with all the bustle. Together with other C.A.S.A. organizers we tried to organize the flow of people through the kitchen and asked the singers and those just hanging around to leave the kitchen. Suddenly I was stopped by the Misak mamita in charge of the kitchen who asked what I was doing. "I'm helping you organize the kitchen—you looked stressed" I replied. "No, no, no", she told me. "My kitchen is open for everyone, I like having people around; this is the Misak way. Everybody is welcome, all activity and learning starts from the 'fogón' (wood stove)—it is the heart of the community. I'm not stressed because of all the people, but because those who were supposed to come and help prepare the food have not arrived. I'm running late with the lunch!" I felt ashamed to have imposed my perception of a chaotic kitchen by assuming it was the same for everyone. And also frustrated with so much talk about showing love through serving the other, but so few people actually helping in the kitchen. So I stepped out of my organizing role, and joined a gossiping group of Misak women peeling potatoes and good naturedly teasing a Norwegian participant who was helping them and trying out words in their language. I sat down next to Gobinda, a Hare Krishna woman, who shared this reflection with me: "The Misak territory teaches us unity, order, expansion, and brotherhood. Look at the way they organize the kitchen tasks. I generally organize the kitchen of my community by having everything prepared at the same time . . . five people are in charge of peeling potatoes, while three are in charge of chopping onions, and so on. Here it is different: we all sit and peel all the potatoes, then we all chop the onions . . . I interpret this as a teaching: is not only about getting the job done efficiently but being together—united in everything we do."

The day ended with the magical night in which Anthakarana, a family ecovillage of artists, performed what they call the ritual-theater, where they combine sustainability issues with rituals of different origins, which are carried out in interaction with the audience. They asked people to light the candle of the person beside them and to move together as a united group around the fire in the middle of the Misak University hall, calling out visions of how a new humanity could be. Faces of people began to light up with each candle, I saw Indigenous, afro, white, mestizo faces; I saw foreigners, locals, youth, children, elders, women and men. This was a powerful experience with over 300 candle-lit people moving together. At the beginning I felt concern with so many candles around me: my hair!—I thought. I started to think of all the things that could go wrong. But after a while, my mind relaxed and I concentrated in following the movement of people around me . . . I started to feel trust, 'we are all taking care of each other' I told myself. In that moment I felt like the room became the whole world with all its diversity, I felt that the fire was also dancing amongst us . . . I started to feel joy in my heart as I was able to feel connected with all forms of beings, all holding one similar wish: to live in a more sustainable and harmonious world. I felt like I a large family surrounded me. Perhaps the collective process of acknowledging multiplicity is the first step in building any long-term sustainability, I thought to myself, and trusting the process the second step. This new insight transformed the way I now see sustainability issues. [49]

4.2. Ontological Hotspots and Their Dissonance

As we have seen in the previous section, the diversity of participants and types of activities carried out during *El Llamado* bring about a number of encounters of difference. What we will now look into are two "ontological hotspots," which we use to denote a situation or activity which was commonly approached very differently by participants of *El Llamado*. Such hotspots created a high degree of dissonance between participants with different ontologies, providing the potential for ingrained ways of "being" in world to be shaken up a little.

4.2.1. Hare Krishna "House Rules" and the Rigidity of Ontologies

El Llamado of 2014 was held in the ecovillage Varsana, Cundinamarca, which is a sacred Hare Krishna monastery along the Vrinda line of practice. Under the hierarchical guidance of their guru Swami B.A. Paramadvaiti, devotees follow the teachings of the sacred Vedic scriptures, in which the ultimate goal is service to God. This all-encompassing philosophy is witnessed through the practices of simple living, serving others and constant prayer and meditation through the repetition of the Hare Krishna mantra. The Hare Krishna community has a hierarchical organization, in which the spiritual guide plays a strong role in teaching Vedic scriptures to its followers and in making decisions. Consequently, the "house rules" of the monastery are well defined and strict: nudity, sexual relations, consumption of meat, the use of any kind of stimulants (including coffee, alcohol and cigarettes) and even campfires are strictly prohibited in the sacred grounds of the monastery. Although nudity and the use of drugs and alcohol are also prohibited by the organization of *El Llamado*, "power plants" (plants used for rituals which have the property to induce other states of consciousness) such as tobacco and coca leaves used for rituals and sexual relations in private places are allowed, as well as consumption of coffee. Furthermore, fire is a strong symbol for several C.A.S.A. members, who consider it a sacred living entity who plays a vital role in *mambeos*—ceremonial conversations around the fire. Elders light a sacred campfire during the opening ceremony of every *Llamado*, and is kept going during the entire event, day and night.

During the organization of the event it became clear that emphasis would have to be placed on respecting the "house rules" in Varsana, while negotiating ways for participants to express their own ideas of sacredness. At the early stage of the event Hare Krishna organizers and participants found ways of negotiating some of these rules. The Hare Krishna hosts compromised by allowing the *mambeo* campfire in the monastery grounds, as they understood its spiritual purpose, but under the condition that it was only to be lit and maintained by Indigenous elders. A group of coffee lovers met every morning outside the property grounds of Varsana to share coffee prepared on a camping stove, while another group hiked to a nearby hill to smoke cigarettes. These early morning activities were for some the means to joyfully bear the rigidity of the house rules and express freedom of choice, as some conversations around the coffee conveyed. Nonetheless, a few participants were not able to follow the restriction on campfires and decided to start their own campfire in the middle of social activities to the surprise of devotees and C.A.S.A. organizers. Arguing that the fire was their means of connecting to the earth and of celebrating life, they refused to put it out after a polite request from a Hare Krishna organizer, and it took the heated words of C.A.S.A. organizers and other participants for the fire to be removed.

Further discord grew out of the Hare Krishna philosophy and practices around food, and what is to them the sacred art of cooking. As Nitia from Varsana argued, *"Food for us, as we offer it, is more a spiritual food. Of course is also for the body but is more to feed the spirit."* The preparation and consumption of food is based on the *ahimsa* principle, which in Sanskrit denotes non-violence and compassion, and is related to karma. In accordance with strict vegetarianism, Hare Krishnas have a strict etiquette in the kitchen and for serving food. One needs to be clean, bathed, have good oral hygiene and one cannot taste food or eat anything during cooking or serving (this includes tasting the food before it is served). An important part of *El Llamado*, however, is the "loving service" in which participants are asked to volunteer in communal tasks, such as helping in the kitchen. Unlike the Misak who see the kitchen and the fire as a meeting and learning point, the Hare Krishnas found it challenging to have participants not of the faith aid in food preparation and cooking. How could they be sure people were properly bathed and clean? Would they be imparting the right reverence to the food? Problematic situations arose when some participant arrived to the kitchen chewing coca leaves, which was perceived as eating by devotees and hence the participants were initially refused entry. Participants argued, however, that in their ontology chewing coca leaves was sacred and a form of meditation and connecting with the "great spirit" and were very disappointed at not being allowed to help. In another example, a participant who was a professional chef was most disappointed when his specially brought cooking knives were refused entry to the kitchen as they had previously been in contact with meat, and thus defied the principle of *ahimsa*. This participant expressed his disappointment, as according to him food making was one of the few activities he saw the possibility of sharing and connecting with Hare Krishnas. (In a later conversation with a Hare Krishna devotee it was remarked that there are in fact cleansing ceremonies which can be undertaken to cleanse utensils which have been in contact with meat. For whatever reason, such a ceremony was not suggested or carried out at the time.)

Adding to this discomfort of the rules of the kitchen was the actual practice of vegetarianism. The organization of C.A.S.A. programs vegetarian menus for *El Llamado* as many members are vegetarian and to lower costs of the event, while at the same time providing opportunities for meat-eaters to experience different types of foods. However, many participants do not consider themselves vegetarian. Several claim that animals are part of the web of life, together with humans and plants, thus respect towards them is shown by the way you breed them and by the compassionate way you kill them, thanking the animals for their sacrifice and ensuring no waste. On the other hand, some strict vegetarians (which were not only devotees) argued that what one eats is directly related to one's level of spirituality, implying that vegetarians have a higher level of conscience than those who consume meat. This ontological difference became a dissonance during *El Llamado* and got as far as the awkward situation of having some *Mamos* (highly acclaimed spiritual leaders from Indigenous groups of the region Sierra Nevada of Colombia), excusing themselves for consuming wild meat which they argue is needed to enrich their diets high up in the mountains where they live. Moreover, when writing the final declaration which would represent the voice of all participants of the gathering, the C.A.S.A. organization refused to include vegetarianism in the text as it did not represent all C.A.S.A. members. This created tensions and until today a joint declaration has not been signed.

Unable to resolve these tensions the Hare Krishna organizers became increasingly stressed. Devotees called an urgent meeting with their spiritual leader Swami B.A. Paramadvaiti who arrived during the event, to ask for guidance. A devotee shared the defining outcome of the meeting: *"Swami Paramadvaiti told us that the main idea was that people should feel good and accepted in Varsana, for them not to feel in such a strange place that they would not want to get involved. He told us to be flexible if the intentions of others were good and were carried out in a loving manner."* This message relaxed devotees who became more flexible and tolerant to the transgressions of participants. Since the 2014 *Llamado*, the involvement of the Hare Krishnas has grown, with Hare Krishna facilitators and participants in the 2015 *Llamado* in the Misak University. Although devotees helped in the kitchen and joined all activities, they also quietly prepared their own food in their tents when conditions in the kitchen did not meet their requirements.

4.2.2. Learning about Power Relations through Ontological Encounters

El Llamado of 2015 was held at the Misak University in the territory of Guambía. As a non-formal university, its purpose is to prepare young leaders to work in their territory in accordance to their ancient wisdom and customs. According to the Misak cosmology, their territory is a living, breathing entity, and it is important to exert close relationships to it through their traditions and customs. For example, as the Misak people originated from the highland lagoons of their territory, they consider these areas sacred and it is where their ancestors guard the "wisdom of all times". Consequently, they devote up to 70% of their territory to preserve the lagoons and *paramos* (cloud forests) in which the water of their territory is born (the other 30% they inhabit).

As a knowledge community attempting to bring back traditional customs and practices which are being lost due to a history of colonization and now modernity, the Misak University saw the possibility of forging an alliance with C.A.S.A. as a means of promoting territorial discussions, albeit within a context of deep suspicion by several students and teachers towards outsiders. (Based on a history of marginalization through government policies, academics writing about their traditions without their approval, and economic challenges of globalization, many Misak see the possibility of having their knowledge and resources exploited by outsiders who visit their territory.) Miscommunication caused a major rift already before *El Llamado* began. Unlike the hierarchical decision making process of the Hare Krishna community Varsana, the Misak university has a more horizontal (though multilevel) platform. The director of the University is the governor of the Indigenous territory Guambía, and decisions are made together with academic directors, coordinators, and to a degree, students. The C.A.S.A. organization committee followed protocols by asking permission to carry out *El Llamado* from the governor and the university coordinators. However, this permission was given before students were informed. This caused discontent among some students who stood up against *El Llamado* arguing that it represented an outsider agenda to "steal" their knowledge and impose western ways on the community. C.A.S.A. organizers attempted to mend the situation by explaining the philosophy of *El Llamado* directly to students, emphasizing that all activities would be respectful to their territory and traditions, and that they were invited to share only as much as they felt comfortable with.

Although some students decided not to attend, others embraced the event, eager to exert their knowledge by ensuring that the meaningful practices and rituals of the Misak people were carried out in the proper way. Protecting these relations, the desire of students to take charge and "protect" their territory led to several incidents. One of the students (a skeptical opponent to the event taking place) took the opportunity to practice his studies into Misak medicine and rituals, and letting people believe he was an expert, conducted a cleansing ritual to harmonize the energy of the C.A.S.A. organizers with that of the territory. Afterwards, organizers were told that because the ritual was carried out during the day (and not the night) it did not count, and that a Misak elder with more expertise would have to repeat it (this was the cleansing ritual described in the previous section).

Another important point within the Misak ontology is the relationship with *el abuelo fuego* (grandfather fire). It is around *el fogón* (the hearth fire) that from an early age knowledge is imparted by family and elders. In keeping with this tradition, a central hearth fire is continuously burning in the main University hall, around which many discussions are held. During the event, Misak organizers decided to appoint one of their teachers to maintain the central fire, which involved nurturing it by placing the logs in a special way, and periodically feeding the fire with tobacco and coca leafs.

It is not clear why the Misak organizers chose this person to take care of such important Misak tradition, as he himself was not a Misak person born in the territory but of urban origin. Nevertheless, he was a knowledgeable academic, who knew not only about the Misak traditions but also had contact with other Indigenous groups of Colombia. This task gave great authority to this person who meticulously controlled the fire and those helping him. However, as the days passed by, the situation became an increasing source of tension as the teacher became rude and impatient with those around him, reprimanding those who were not "following" the Misak ways. On the energetic level, non-Misak elders began talking of a strong negative energy pervading the event, and hence the need for neutralizing the situation at the energetic level. C.A.S.A. organizers and participants tried to remain open minded and respectful of the teacher's appointed role, but tensions became unbearable when his actions increasingly became offensive and divisive. Misak organizers were informed of the situation and all parties agreed to talk around another fire to understand what was happening. After a long session where the teacher kept verbally attacking the event and its participants, one of the highest ranked Misak organizers stood up and stamping the floor with his staff, stated: *"I am Misak, I am my territory, and I only accept the positive. What I have seen here [in this event] are only positive things for my territory."* Everyone stood up in support of this statement, and although asked to stay, the teacher left the event. With this episode, the role of elders as keepers of harmony at the energetic and spiritual levels was better understood by C.A.S.A. organizers, who were unaware of the struggles at this invisible level. Furthermore, during post-event evaluation and reflection, a learning point amongst organizers was the extent to which roles and positions appointed during *El Llamado* give power to people, thus they must be well understood and only given if strictly necessary, as power struggles can cause a disharmonized environment not conducive for learning.

5. Discussion

"One feels that nobody is right here, I mean, nobody owns the truth . . . each person is a link within the chain we are all part of." [50]

The above results sections have attempted to display the ontological politics that play out in enacting an "environmentalism of everyday life" and what it would mean to experience an intercultural gathering from the perspective of a pluriverse. By bringing the reader through a day in *El Llamado*, and highlighting two ontological "hotspots" of potential transgressive learning, as well as the experiential clashes of the first author in her personal journey through *El Llamado*, we have witnessed something which at least hints at the encounters between different worlds. What we want to bring to the discussion table now is the extent to which we can talk about a pluriverse in practice, and how engaging with these worlds can lead to transgressive learning towards more sustainable everyday living.

Reading through the results section it becomes clear that our modern day anthropocentric distinctions between humans and nature do not articulate well with what takes place in intercultural events such as *El Llamado*. Thus non-human actors such el *abuelo fuego* (grandfather fire) and the territory as a sentient being, we maintain, cannot be seen as entities belonging to the realm of "Nature" while human actors (Hare Krishnas, ecovillagers and so on) belong to the domain of "Culture". These divides, as Latour [51] has shown, are problematic because, in practice, different categories of the so-called "natural" and the so-called "cultural" are heavily entwined. In order to side-step the rather arbitrary division between "Nature" and "Culture" we resorted to the notion of "ontology". We find it important to note though that "ontology" is not just another word for "culture", as some anthropologists have recently suggested [52]. To us, taking the stance that ontology is another word for culture means taking on board unwarranted notions of multiculturalism and a host of accompanying dichotomies such as those between Nature/Culture, facts/beliefs or truth/superstition [26]. Instead, we have adopted a "multinaturalist" stance [22,26], that is, the understanding that there exist many kinds of Natures—possibly as many as there are cultures. This alternative frame, apparent in many Indigenous Amazonian philosophies [53–56], but also elsewhere [8,57–62] has consequences for the understanding of "Nature" and demands a renewed attention to ontological politics. In adopting such a "multinaturalist" or "pluriversal" stance we have taken the concerns of our research subjects seriously, providing an account of ontological politics which we believe the protagonists of our story would not disagree with. In other words, our account was not unduly shaped by our analytical and ontological concepts; rather, it was fashioned by "what we found"—what is perhaps the strongest point of the ontological turn [63]. This was clearly not an easy task: no matter how we talk about wanting to be open minded and inclusive to other ontologies, engaging with them in practice can be extremely difficult and, to put it in Helen Verran's [64] words, thoroughly disconcerting yet extremely valuable for critical reflection. Although a "cleansing" for the territory may be necessary for harmonizing energy between people and place, a four-hour long ritual with no information outside in the cold is cumbersome. The same goes, of course, for those who partake in *El Llamado*: not to be able to participate in most activities because one has "the moon", or not being able to cook with knives brought from home because they have been in contact with meat is challenging to say the least—and inevitably involves an ontological politics.

In the context of socio-ecological problems, ontological politics such as those described above tend to stay below the radar, or are misrepresented. At best, they would seem to be only indirectly related to environmental change; at worst, they could be presented as essentialist discourses or "New Age" ruminations. In order to more fully grasp what goes on during the different *Llamados*, we argue that it is expedient to attend to the ontological premises (i.e., different realities) upon which C.A.S.A.'s gatherings are based. Only then can we put in proper perspective that what participants are actually engaged in. This is not a plea to romanticize these gatherings, but rather an appeal to focus on crucial ontological politics. Attending to the participants´ alternative way of measuring and evidencing "environmental problems" is important for two reasons. First, because it feeds the way in which they imagine and shape their responses to it; second, and for practical reasons, because understanding how they do so may help shape transgressive learning in crucial transition processes.

It is important to note here that we as researchers actively participate in an ontological politics as well. The choices we make (that is: writing in the way we write) are not only of an epistemological kind but involve, at the same time, moral, ethical and political issues. In fact, and here we follow Jensen [65], one could state that the things we bring to the fore through our writing (such as sustainability and learning) "collapse into ontology" and that, in so doing, we are effectively performing our own ontology—thus intervening in an ongoing ontological politics around these "things".

Despite its unique characteristics, we believe that our case study helps to shed light on more general processes that shape alternative and non-modern ways of dealing with complex socio-ecological problems. As we have shown, in order to shape an "environmentalism of everyday life" it is expedient for those involved to exercise some form of ontological politesse. Communicative disjunctures—dialogues of the deaf—could, we think, be turned into constructive encounters if ontological differences were explicitly allowed to enter the negotiation room. The problem in this "learning to play with strangers" [66] is that one first needs to let go of the idea that debates about the environment are wars fed by epistemological politics (in the sense of settling the issue of what party can best measure and represent the environment), but rather attempts to arrive at agreement by way of an ontological politics [18] or a "cosmopolitics" that works as a cure for the "malady of tolerance" [67]. The proposal here is that decisions about how to deal with socio-ecological problems must, one way or another, take place in the presence of those who will bear their consequences. Like in our case, this requires acknowledging that "complex socio-ecological problems" carry different meanings for different people in different places. In the literature there are modest (yet sufficient) and workable examples of people negotiating and working across ontologies [62,68–71]; for two fascinating accounts of ontological dovetailing in scientific institutions see Cussins [72] and Mol [61].

In general, however, most individuals and communities (including indigenous communities) have rather singular ontologies, which in a globalizing world with shrinking ecological boundaries they are forced do compromise or even abandon. During *El Llamado* we argue, though, that to some extent transgressive learning is taking place through these ontological encounters, which address the complexity of socio-ecological challenges. Having managed to transgress her moral boundaries on equality and male/female relations, the participant on "her moon" managed to engage and reflect on her own femininity not only in the context of the event, but also in her own day-to-day life. This was made possible through a woman's circle where these issues were discussed and rituals took place. The situation with the teacher also brought up learning points on the power of traditions and unseen forces, and the necessity of valuing energetic levels in resolving these tensions. Compare these examples to the situation of the participant whose knives were denied access in what we can see as an "ontological impasse" where little positive learning has taken place. As one Hare Krishna devotee noted in retrospect, a ritual cleansing of the knives and a reflexive talk around the principles of *ahimsa* could have been carried out. If this had occurred together with the participant then a two-way conversation could have taken place, perhaps coming to some form of understanding and compromise. On a more collective scale, the ritualistic cleansing would have been an opportune time for the Misak to facilitate "affective flows" [42] of feelings and emotions into the non-human realm, explaining to participants the harmonization of energetic levels through the forces of nature, and really creating an "experience" which would cross boundaries of reason and meaning. Instead, at best, the cleansing maintained the mystical quality of people with a very different ontology, and, at worst, created or maintained a disconnection and a barrier to entering the realities of the Misak. It is important to emphasis here that although we speak of a "Misak ontology" it would be a delusion to think that the ontology of Others can be fully apprehended or described. For an interesting debate about how to approach Others' ontologies see Blaser [9] and Jensen [65].)

The role and power of those that mediate between different worlds brings us to the second point. As noted about the pluriverse in section two, there is no one logic that can mediate between the power-saturated realities of different worlds [73]. This means that issues of power arise when negotiating these different worlds as there is no one person who is "right". However, each respective ontology has representatives of power. In the hierarchical ontology of the Hare Krishna, the guru holds ultimate power through the interpretation of the ancient Vedic scripts, and it was to him devotees came with their concerns over the transgressions of non-Hare Krishna participants. It was a positive sign to the devotees, participants in *El Llamado*, and possibilities for greater sustainability processes that a leader of such a religious community had the capacity to be flexible in the rules of the community and was able to grasp the type of inclusive social tissue being created beyond a single "truth". In the community of the Misak University, however, power is held by University directors and students, and wielded through the elders and medicine men who are knowledgeable of the customs and traditions of the community and have the power to communicate with the territory. It is ultimately they who can communicate with other entities through rituals such as the *mambeo* and the cleansing, and decide what is allowed and what is not.

This points to the difficulties of learning-based interventions overcoming structural power, whether they come from the one truth of modernity, or from other realities which make up the pluriverse. As the example of the teacher also demonstrated, the capacity to understand (or at least represent) an ontology can give tremendous power, and his inflexibility and aggressive stance in upholding what were for him Misak traditions created tensions and disrupted relations between the organizers and participants. Consequently, when we talk about a decolonization of knowledge where *"nobody owns the truth"*, we must be prepared for a strong degree of inflexibility in ontologies, and ever-present power negotiations between those who represent each ontology. It is then interesting that unlike the top-down conflict resolution of the Hare Krishna community (without dialogue between the parties) Misak and C.A.S.A. organizers were able to sit down around the fire to discuss the conflict with the teacher. For one of the *Taitas* to stand up against the "negative energy" of the teacher demonstrates a negotiation of power between representatives, though unfortunately after this negotiation the teacher left and there was no opportunity to continue a reflection on the situation which could have resulted in deeper and perhaps more transgressive learning for all parties. Nevertheless, for C.A.S.A. organizers this became a source of learning in understanding the invisible power struggles that emerge when enacting the pluriverse.

So what does engaging with the pluriverse tell us about facing the sustainability challenges of our time? Well clearly a "world in which many worlds fit" is an exciting but rather utopian idea, at least at this point in time. Accustomed as most of us to the natural world being around us—instead of us being part of it—accepting different constellations of human–nature relations is a long and complicated process. Although there may well have been transgressive "moments" for participants and organizers, where structural barriers became visible through ontological encounters, for it to be called transgressive "learning" implies a long-term process in which structures are not just made visible, but also broken, and to an extent surpassed. It is perhaps better to think of a transition towards more inclusive understandings of other worlds and their sustainability practices. This does not necessarily mean a harmony between different ways of being sustainable, but at least a conversation as equal partners. *El Llamado* represents an attempt at such a transition through a tangible engagement with the pluriverse whereby, for example, participating in a cleansing, a sacred march and in a collective effort to "plant water" participants have the potential to gain insights into different enacted realities where harmonizing foreign energy and planting trees increases the territory's satisfaction and thus creates water in the mountain's lagoons through which the ancestors guard the territory's wisdom. At the same time, however, the initial inflexibility of the Hare Krishna community clearly created a lot stress for the devotees and divides between participants and the host community, as well as the divide, in the Misak case, between those who can communicate with "earth beings" and hence represent and control that ontology in its relation to others.

It is therefore worth ending this discussion with what we can view as both an inspirational as well as perhaps naive message from the Hare Krishna guru and the Misak *Taita*: only the positive is welcome—if actions are carried out based on love and good intentions then we should be flexible enough to allow for their manifestations and embrace the challenge that they bring to us as stimulants to improve our own ways. Although this philosophy contributed to the reduction of tensions in both the previous examples, it also raises the uneasy issue of the extent to which we should compromise what we believe in in the name of a tolerant and inclusive social tissue, especially if we should consider other ontologies inherently unsustainable and unfair. Vegetarianism is an example of an issue in which no compromise could be made at the level of writing a common manifesto. With the complexities of different ontologies, it could be useful to keep in mind the idea that our understandings with other ontologies will only ever be "partial connections" [25] in which different ontologies are entangled with one another, and which we will never really "get".

Hence, a question which is left for further research is how can we make a better use of these "partial connections" to engage the pluriverse in transitions towards a more sustainable and inclusive future? This, of course, begs the question of how to productively engage with ontological conflicts entailing "radical difference"; that is, conflicts in which interlocutors are unwilling to collaborate in bringing about transgressive learning processes. One way to go here would be, as Bonelli [74] proposes, to try to create a greater awareness (amongst conflicting parties) of their own particular ontological presuppositions so that room can be made for ontological diplomacy through "pragmatic encounters" [75].

6. Conclusions

Engaging with the pluriverse confronts us with differences in other people and in ourselves. Such confrontation or mirroring, and the frictions and dissonances it creates, has the potential to make us rethink our own norms, values and stubborn everyday routines and assumptions. This means leaving our comfort zones and experiencing other worlds where territories are literally alive, where food being prepared cannot be tasted out of respect for serving God, or where women "on their moon" have such strong and sacred energy that they cannot participate in communal activities. In a context of entrenched unsustainable practices which most of us partake in, such experiential learning can be transgressive to the extent that we really manage to cross the boundary of our own entrenched lifestyles and embrace not just the idea that we are different, but also what this means in practice.

It is important to note, however, that this process of boundary crossing and mirroring is not easy. As our example of *El Llamado de la Montaña* has hopefully demonstrated, organizing, facilitating and participating in intercultuscaled upral settings which are generative to transgressive learning is a challenging task. Diversity in itself is no panacea and often leads to misunderstanding and even conflicts, which must be addressed if reflexivity and learning to live together through difference and conviviality are to take place. Ontologies are not just very complicated; they may be more or less rigid and saturated with power, and negotiating them in the name of crossing boundaries towards different worlds will require new types of knowledge, skills and methodologies for the future.

It is also worth noting that not all people are willing to engage and experience the pluriverse as it means leaving the safety of a single "truth" and predictability of one's own comfort zone. This is a challenge for the pluriverse which implies participation by many actors. During *El Llamado* there is a pronounced absence of mainstream politicians, government officials and businessmen which gives the event a feeling of alternativeness, and a sense of disarticulation with wider society. This remains a great challenge for *El Llamado*, as well as society at large as the mainstream is also part of the pluriverse, and articulation with these groups is essential for transitioning towards truly deep changes in society.

Finally, we would like to point to the productive possibilities inherent to our case. Transgressive learning involves continuous negotiations between different worlds and realities. In these negotiations, worlds engage with one another—thus becoming sensitive to those ontological disjunctures that may lead to misunderstandings. Transgressive learning entails (as a minimum) an effort to translate different but oftentimes partially connected realities—a translation akin to the method of "controlled equivocation" proposed by Viveiros de Castro [26]. While difficult, transgressive learning demands an active awareness of other ways of being in the world. Insensitivity to this, as our case demonstrates, may result in a failure to learn. Openness to the possibility to be "moved over" [76] or, in our terms, being open to ontological politics thus has practical and political value as it allows for the (cosmo)political task of shaping the "environmentalism of everyday life" as a process of ontological dialogue. To be sure, the solution of complex socio-ecological problems requires the nurturing of diplomacy: the capacity to move in and relate ways of knowing and being that partly overflow one another, yet without a-priori assuming one to be superior. We therefore wish to stress ontological difference as a positive, productive capacity; a "useful complication" that stimulates thinking and reflection. Transgressive learning cherishes this difference, and renders it productive: it allows for trying out ways to find or create a "middle ground" [77] in the pluriverse.

Acknowledgments: Funding for this study has been provided by the Administrative Department of Science, Technology and Innovation of Colombia, COLCIENCIAS. Special acknowledgments to the C.A.S.A. network Colombia and its organizing team for collaborating in this research and for their commitment in building social laboratories for sustainability around Colombia through volunteer work.

Author Contributions: All authors helped conceive and design the research for this paper, while the first author Martha Chaves carried out the main fieldwork with the aid of the second author Thomas Macintyre. The first two authors also analyzed the data and wrote the paper, but with substantial contributions from the last two authors, Gerard Verschoor and Arjen Wals, in terms of contributing analysis tools, providing constructive feedback, and copyediting expertise.

Conflicts of Interest: The authors declare no conflict of interest.

Abbreviations

The following abbreviation is used in this manuscript:

C.A.S.A.   Consejo de Asentamientos Sustentables de las Américas (the Council of Human Settlements of Latin America)

References

1.   Council of Women Elders of Colombia. Available online: http://consejodeabuelas.wix.com/consejodeabuelas (accessed on 20 December 2016).
2.   Parr, A. *Hijacking Sustainability*; The MIT Press Cambridge: Cambridge, MA, USA, 2009.
3.   Wals, A.E.J. Beyond Unreasonable Doubt: Education and Learning for Socio-Ecological Sustainability in the Anthropocene. Available online: http://library.wur.nl/WebQuery/clc/redes/2103430 (accessed on 20 December 2016).
4.   Wals, A.E.J. Learning Our Way to Sustainability. *J. Educ. Sustain. Dev.* **2011**, *5*, 177–186. [CrossRef]
5.   Chaves, M. Answering the "Call of the Mountain": Co-Creating Sustainability through Networks of Change in Colombia. Ph.D. Thesis, Wageningen University, Wegeningen, The Netherlands, 2016.
6.   Mcmichael, P. Rethinking Land Grab Ontology. *Rural Sociol.* **2014**, *79*, 34–55. [CrossRef]

7.   Gudynas, E. Buen vivir: Germinando alternativas al desarrollo. *Am. Lat. Mov.* **2011**, *462*, 1–20. (In Spanish)

8.   De La Cadena, M. Indigenous cosmopolitics in the andes: Conceptual reflections beyond "politics". *Cult. Anthropol.* **2010**, *25*, 334–370. [CrossRef]

9.   Blaser, M. *Storytelling Globalization from the Chaco and beyond*; Duke University Press: Durham/London, UK, 2010.

10.   Escobar, A. Sustainability: Design for the pluriverse. *Development* **2011**, *54*, 137–140. [CrossRef]

11.   Martin, P.M.; Glesne, C. From the Global Village to the Pluriverse? "Other" Ethics for Cross-Cultural Qualitative Research. *Ethics Place Environ.* **2002**, *5*, 205–221. [CrossRef]

12.   Grosfoguel, R. Decolonizing Post-Colonial Studies and Paradigms of Political Economy: Transmodernity, Decolonial Thinking, and Global Coloniality. *Transmodernity* **2011**, *1*, 1–36.

13.   Escobar, A. *Encountering Development: The Making and Unmaking of the Third World*; Princeton University Press: Princeton, NJ, USA, 2011.

14.   Wals, A.E.J.; Jickling, B. "Sustainability" in higher education. *Int. J. Sustain. High. Educ.* **2002**, *3*, 221–232. [CrossRef]

15.   C.A.S.A. Colombia. Available online: http://www.casacontinental.org/ (accessed on 20 December 2016). (In Spanish)

16.   Schlosberg, D.; Coles, R. The new environmentalism of everyday life: Sustainability, material flows and movements. *Contemp. Polit. Theory* **2016**, *15*, 160–181. [CrossRef]

17.   Holbraad, M.; Pedersen, M.; de Castro, E. The Politics of Ontology: Anthropological Positions. Available online: https://culanth.org/fieldsights/462-the-politics-of-ontology-anthropological-positions (accessed on 18 May 2016).

18.   Mol, A. Ontological politics. A word and some questions. *Sociol. Rev.* **1999**, *47*, 74–89.

19.   Blaser, M. Ontology and indigeneity: On the political ontology of heterogeneous assemblages. *Cult. Geogr.* **2014**, *21*, 49–58. [CrossRef]

20.   Blaser, M. Political Ontology: Cultural Studies without "cultures"? *Cult. Stud.* **2009**, *23*, 873–896. [CrossRef]

21.   Palecek, M.; Risjord, M. Relativism and the Ontological Turn within Anthropology. *Philos. Soc. Sci.* **2012**, *43*, 3–23. [CrossRef]

22.   De Castro, E.V. Cosmological Deixis and Amerindian Perspectivism. *J. R. Anthropol. Inst.* **1998**, *4*, 469–488. [CrossRef]

23.   Bacchi, C.L. Strategic interventions and ontological politics: Research as political practice. In *Engaging with Carol Bacchi: Strategic Interventions and Exchanges*; University of Adelaide Press: Adelaide, Australia, 2012; pp. 141–156.

24.   Latour, B. Whose Cosmos, which Cosmopolitics? *Common Knowl.* **2004**, *10*, 450–462. [CrossRef]

25.   Strathern, M. *Partial Connections*; Rowman & Littlefield: Savage, MD, USA, 1991.

26.   De Castro, E.V. Perspectival anthropology and the method of controlled equivocation. *Tipiti J. Soc. Anthropol. Lowl. S. Am.* **2004**, *2*, 2–22.

27.   Lotz-Sisitka, H.; Wals, A.E.; Kronlid, D.; McGarry, D. Transformative, transgressive social learning: Rethinking higher education pedagogy in times of systemic global dysfunction. *Curr. Opin. Environ. Sustain.* **2015**, *16*, 73–80. [CrossRef]

28.   Peters, M.A.; Wals, A.E.J. Transgressive learning in times of global systemic dysfunction: Interview with Arjen Wals. *Open Rev. Educ. Res.* **2016**, *3*, 179–189. [CrossRef]

29.   Transgressive Learning. Available online: http://transgressivelearning.org/ (accessed on 20 December 2016).

30.   Mezirow, J. Transformative Learning: Theory to Practice. *New Dir. Adult Contin. Educ.* **1997**, *1997*, 5–12. [CrossRef]

31.   De Lissovoy, N. Decolonial pedagogy and the ethics of the global. *Discourse Stud. Cult. Political Educ.* **2010**, *31*, 279–293. [CrossRef]

32.   Freire, P. *Pedagogy of the Oppressed*; Continuum: New York, NY, USA, 1970.

33.   Geels, F.W. Ontologies, socio-technical transitions (to sustainability), and the multi-level perspective. *Res. Policy* **2010**, *39*, 495–510. [CrossRef]

34.   Wals, A.; Heymann, F. Learning on the edge: Exploring the change potential of conflict in social learning for sustainable living. In *Educating for a Culture of Social and Ecological Peace*; Wenden, A., Ed.; State University of New York Press: New York, NY, USA, 2004; pp. 123–145.

35. Sol, J.; Beers, P.J.; Wals, A.E.J. Social learning in regional innovation networks: Trust, commitment and reframing as emergent properties of interaction. *J. Clean. Prod.* **2013**, *49*, 35–43. [CrossRef]
36. Escobar, A. Beyond the Third World: Imperial globality, global coloniality and anti-globalisation social movements. *Third World Q.* **2004**, *25*, 207–230. [CrossRef]
37. Escobar, A. Más allá del desarrollo: Postdesarrollo y transiciones hacia el pluriverso. *Rev. Antropol. Soc.* **2012**, *21*, 23–62. (In Spanish) [CrossRef]
38. UNESCO (United Nations Educational Scientific and Cultural Orgaianization). *Education for People and Planet: Creating Sustainable Futures for All*; UNESCO: Paris, France, 2016; pp. 1–535.
39. Wijesooriya, C.; Heales, J.; McCoy, S. Multi-Dimensional Views for Sustainability: Ontological Approach. Available online: http://aisel.aisnet.org/cgi/viewcontent.cgi?article=1194&context=amcis2015 (accessed on 20 December 2016).
40. Global Footprint Network. Available online: http://www.footprintnetwork.org/en/index.php/GFN/page/trends/colombia/ (accessed on 15 September 2016).
41. Australian Public Service Commission. *Tackling Wicked Problems: A Public Policy Perspective*; Australian Government: Canberra, Australia, 2007.
42. Fox, N.J.; Alldred, P. New materialist social inquiry: Designs, methods and the research-assemblage. *Int. J. Soc. Res. Methodol.* **2015**, *18*, 399–414. [CrossRef]
43. Spry, T. Performing Autoethnography: An Embodied Methodological Praxis. *Qual. Inq.* **2001**, *7*, 706–732. [CrossRef]
44. De Castro, E.V. Who is afraid of the ontological wolf? Some comments on a recent anthropological debate. *Camb. J. Anthropol.* **2015**, *33*, 2–17.
45. Goodall, H.L. *Writing the New Ethnography*; AltaMira Press: Walnut Creek, CA, USA, 2000.
46. Anderson, L. Analytic Autoethnography. *J. Contemp. Ethnogr.* **2006**, *35*, 373–395. [CrossRef]
47. Law, J. Seeing like a survey. *Cult. Sociol.* **2009**, *3*, 239–256. [CrossRef]
48. Litfin, K. Reinventing the future: The global ecovillage movement as a holistic knowledge community. In *Environmental Governance: Power and Knowledge in a Local-Global World*; Kütting, G., Lipschutz, R.D., Eds.; The MIT Press: Cambridge, MA, USA, 2009.
49. Chaves, M.; Cauca Territory, Colombia. Personal communication, 2016.
50. Anonymous Participant; El Llamado de la Montaña, Cauca Territory, Colombia. Personal communication, 2014.
51. Latour, B. *We Have Never Been Modern*; Harvard University Press: Cambridge, MA, USA, 1993.
52. Carrithers, M.; Candea, M.; Sykes, K.; Holbraad, M.; Venkatesan, S. Ontology Is Just Another Word for Culture: Motion Tabled at the 2008 Meeting of the Group for Debates in Anthropological Theory, University of Manchester. *Crit. Anthropol.* **2010**, *30*, 152–200. [CrossRef]
53. Arhem, K. The cosmic food web. In *Nature and Society: Anthropological Perspectives*; Routledge: London, UK, 1996; pp. 185–204.
54. Descola, P. *In the Society of Nature: A Native Ecology in Amazonia*; Cambridge University Press: Cambridge, MA, USA, 1994.
55. Reichel-Dolmatoff, G. Cosmology as Ecological Analysis: A View from the Rain Forest. *Man* **1976**, *11*, 307–318. [CrossRef]
56. Rival, L. The Growth of Family Trees: Understanding Huaorani Perceptions of the Forest. *Man* **1993**, *28*, 635–652. [CrossRef]
57. Wagner, R. *The Invention of Culture*; University of Chicago Press: Chicago, IL, USA, 1981.
58. Verran, H. Re-imagining land ownership in Australia. *Postcolonial Stud.* **1998**, *1*, 237–254. [CrossRef]
59. Bonelli, C. Ontological disorders: Nightmares, psychotropic drugs and evil spirits in southern Chile. *Anthropol. Theory* **2013**, *12*, 407–426. [CrossRef]
60. Muller, S. Co-motion: Making space to care for country. *Geoforum* **2014**, *54*, 132–141. [CrossRef]
61. Mol, A. *The Body Multiple: Ontology in Medical Practice*; Duke University Press: Durham, NC, USA; London, UK, 2002.
62. Umans, L.; Arce, A. Fixing rural development cooperation? Not in situations involving blurring and fluidity. *J. Rural Stud.* **2014**, *34*, 337–344. [CrossRef]
63. Henare, A.; Holbraad, M.; Wastell, S. *Thinking through Things: Theorising Artefacts Ethnographically*; Routledge: London, UK; New York, NY, USA, 2007.
64. Verran, H. *Science and an African Logic*; University of Chicago Press: Chicago, IL, USA, 2001.

65. Jensen, C.B. Practical Ontologies. Available online: https://culanth.org/fieldsights/466-practical-ontologies (accessed on 20 December 2016).

66. Haraway, D. *When Species Meet*; University of Minnesota Press: Minneapolis, MN, USA, 2008.

67. Stengers, I. *Cosmopolitics I*; University of Minnesota Press: Minneapolis, MN, USA, 2010.

68. Castleden, H.; Garvin, T.; Nation, H.F. "Hishuk Tsawak" (Everything Is One/Connected): A Huu-ay-aht Worldview for Seeing Forestry in British Columbia, Canada. *Soc. Nat. Resour.* **2009**, *22*, 789–804. [CrossRef]

69. Echeverri, J.A. Territory as body and territory as nature: Intercultural dialogue? In *The Land Within: Indigenous Territory and the Perception of Environment*; Surrallés, A., Hierro, P.G., Eds.; Centraltrykkeriet Skive: Copenhagen, Denmark, 2005; pp. 230–247.

70. Verran, H. A Postcolonial Moment in Science Studies: Alternative Firing Regimes of Environmental Scientists and Aboriginal Landowners. *Soc. Stud. Sci.* **2002**, *32*, 729–762. [CrossRef]

71. Gombay, N. "Poaching"—What's in a name? Debates about law, property, and protection in the context of settler colonialism. *Geoforum* **2014**, *55*, 1–12.

72. Cussins, C. Ontological Choreography: Agency through Objectification in Infertility Clinics. *Soc. Stud. Sci.* **2016**, *26*, 575–610. [CrossRef]

73. Law, J. What's Wrong with a One-World World. Available online: http://www.heterogeneities.net/publications/Law2011WhatsWrongWithAOneWorldWorld.pdf (accessed on 10 April 2016).

74. Bonelli, C. To see that which cannot be seen: Ontological differences and public health policies in Southern Chile. *J. R. Anthropol. Inst.* **2015**, *21*, 872–891. [CrossRef]

75. De Almeida, M.B. Caipora e outros conflitos ontológicos. *Rev. Antropol. UFSCar* **2013**, *5*, 7–28.

76. Kohn, E. *How Forests Think: Towards an Anthropology beyond the Human*; University of California Press: Berkeley, CA, USA, 2013.

77. White, R. *The Middle Ground: Indians, Empires, and Republics in the Great Lakes Region, 1650–1815*; Cambridge University Press: New York, NY, USA, 1991.

9

The Retail Chain Design for Perishable Food: The Case of Price Strategy and Shelf Space Allocation

Yujie Xiao [1,2] and Shuai Yang [3,*]

[1] Jiangsu Key Laboratory of Modern Logistics, School of Marketing and Logistic Management,
    Nanjing University of Finance and Economics, Nanjing 210046, China; yujiexiao@njue.edu.cn
[2] Business School, Nanjing University, Nanjing 210093, China
[3] School of Economics and Management, Changshu Institute of Technology, Changshu 215500, China
* Correspondence: 1393655123@163.com

Academic Editor: Marc A. Rosen

Abstract: Managing perishable food in a retail store is quite difficult because of the product's short lifetime and deterioration. Many elements, such as price, shelf space allocation, and quality, which can affect the consumption rate, should be taken into account when the perishable food retail chain is designed. The modern tracking technologies provide good opportunities to improve the management of the perishable food retail chain. In this research, we develop a mathematical model for a single-item retail chain and determine the pricing strategy, shelf space allocation, and order quantity to maximize the retailer's total profit with the application of tracking technologies. Then the single-item retail chain is extended into a multi-item one with a shelf space capacity and a simple algorithm is developed to find the optimal allocation of shelf space among these items. Finally, numerical experiments and real-life examples are conducted to illustrate the proposed models.

Keywords: food retail chain; shelf space allocation; perishable food; pricing strategy; quality deterioration

1. Introduction

With the accelerated pace of modern life, more and more perishable food is sold in marts or retail groceries. The increase in demand for perishable food brings about more profit, while the increased demand also makes it more difficult to manage with more quantities and varieties. Because of the short life of perishable food, all the items should be sold out before their "sell-by date", or they have to be thrown away. Every year, the scale of spoilage is quite large in retail stores. According to Ferguson and Ketzenberg [1], the attrition rate reaches as high as 15% during the time perishable food is sold in retail stores. At the same time, the gross profit of perishable food is relatively high because of its mass spoilage and the difficulty in managing it. Therefore, a highly efficient operation and rational design of the food retail chain may provide great potential to reduce spoilage and increase profit [2,3]. Since the quality of perishable food is decaying all the time and the consumption rate is affected by many factors, such as price, shelf space, discount rate, and so on, we should comprehensively consider all these important factors in designing the perishable food retail chain. In this paper, we determine the pricing and order strategy and shelf space allocation to maximize the food retailer's profit.

The supply chain design for perishable products has attracted a great deal of attention in the industry recently. As the loss rate in retail stores is not quite satisfactory, a great number of models for managing perishable products have been developed. There are three main approaches to model the perishability of food. First, it is assumed that the perishable product has a fixed or random lifetime. For example, Zhou and Yang [4] determined the optimal replenishment policy for items with a stock-dependent demand rate and a fixed lifetime. Second, perishability is defined as a proportion of the product disappearing or becoming outdated, while the value of the remainder does not change. Related work by Mandal and Phaujdar [5], which presented an inventory model for deteriorating items, assumed that the demand rate is a linear function of the current stock level and the deterioration rate could be constant or time-dependent. Third, some models assume that the value of a perishable product deteriorates as over time. Zanoni and Zavanella [6] considered the economic aspects and energy efforts together with the assumption that the quality of perishable food decays exponentially. As the third assumption is realistic in most perishable food cases, it is also applied in this research. Since the value of a perishable product decreases over time, the demand for the product also goes down. Hence, the pricing and discount strategy plays an important role in the sales promotion, especially when the product approaches its expiration date. Hong and Lee [7] decided the price, lead time, and lateness penalty to maximize the total profit for a price- and time-sensitive market. Wang and Li [8] developed a pricing model to maximize the total profit for perishable food supply chains.

Besides the perishable product's quality and price, the shelf space allocated is also an important factor which may affect the consumption rate. Larger shelf space can attract more visibility and bring about more sales; further, larger shelf space may also imply the product is popular and in season. Lynch and Curhan [9] assumed a quadratic relationship between the shelf space allocated to a product and the consumption rate in supermarkets. Wang and Gerchak [10] developed a coordination mechanism where the manufacturer offers a holding cost subsidy to coordinate the channel when the retailer's demand is shelf space–dependent. Mohsen et al. [11] proposed an integrated vendor-buyer inventory model to maximize the channel profit, with the assumption that the buyer has a two-stage inventory, a warehouse and a display shelf, and the demand is dependent on the amount of items displayed. Since few papers have considered shelf space allocation and pricing strategies simultaneously in the management of the perishable food retail chain, this research deals with the problem and provides an efficient food retail chain design for food retailers.

2. Model Formulation for the Single-Item Food Retail Chain

In this section, we model the whole course of the perishable food sold in a retail store, from the arrival of the product, selling it on the shelf and discount rack, to the disposal of the expired goods. The following notations are used through the whole paper. Additional notations are introduced when required.

q	quality of the perishable food
$q(t)$	quality of the perishable food at time t
q_0	initial quality of the perishable food
λ	deterioration rate of the perishable food
p	selling price of the perishable food
n_g	shelf space allocated to the perishable food
a	parameter of market scale for the perishable food
b	parameter of price elasticity for the perishable food
c	parameter of demand sensitivity to shelf space allocated to the food
d	parameter of demand sensitivity to quality of the food
m	parameter of opportunity cost for the current shelf space allocation
θ	ratio of discount price to original price of the perishable food ($1 - \theta$ is the discount rate)
f	discount attraction rate of the perishable food
t_0	food's sales time on shelf
T	shelf lifetime of the perishable food
$D_1(t)$	demand of the perishable food sold on shelf without discount at time t
$D_2(t)$	demand of the perishable food sold on discount rack at time t
$D(t)$	demand of the perishable food during its shelf life
Q	order quantity of the perishable food

C_0	purchasing cost per unit of the perishable food
α	no discount price for purchasing the perishable food
β	discount sensitivity to order quantity Q for purchasing the perishable food
π	total profit of the perishable food retail chain
L	shelf space capacity

2.1. Quality Deterioration

Quality deterioration is a complex course for perishable food. Tracking and predicting the quality of perishable food was quite a difficult task before modern technologies were developed, such as radio frequency identification (RFID), wireless sensor technology and the humidity-temperature sensor. Nowadays, with these technologies and quality prediction models, we can predict the remaining shelf life of a perishable product as its quality, which is the main interest for retailers and customers [12–15]. According to Labuza [12], the quality degradation of perishable food is affected by several factors: the storage time, the ambient temperature, and the ambient atmosphere condition. In more detail, the quality degradation can be expressed by the following equation:

$$dq/dt = -k_0 e^{-(E_a/RT_0)} q^n \tag{1}$$

In Equation (1), n is the chemical order of the reaction. In this equation, n can be equal 0 or 1, depending on the degradation course of the product. When $n = 0$, the quality decays at a constant rate. When $n = 1$, the quality decays exponentially, which is more realistic and hence used in this research. k_0 is a constant rate, E_a is the activation energy, which is an empirical parameter reflecting the exponential temperature, R is the gas constant, and T_0 is the absolute temperature. By solving Equation (1), the quality of a perishable product at time t can be described as:

$$q(t) = q_0 e^{-k_0 t e^{-(E_a/RT_0)}} \tag{2}$$

In a retail grocery store, the temperature and atmosphere condition are usually stable. Therefore, we introduce λ as the deterioration rate to reduce the complexity. Let $\lambda = k_0 e^{-(E_a/RT_0)}$, and hence the quality at time t becomes:

$$q(t) = q_0 e^{-\lambda t} \tag{3}$$

2.2. Demand Model

In a mart or retail grocery store, the consumption rate of perishable food is affected by many elements, such as the arrival rate of customers, price, quality, discount rate, and so on. In this research, we assume the consumption rate depends on three factors at first: p, $q(t)$, and n_g. First, customers are assumed to be able to distinguish the quality of the food. Then, larger shelf space can attract more visibility and imply that the product is popular, as mentioned in the literature. Consequently, the demand function can be described as:

$$D_1(t) = a - bp + cn_g + dq(t) \tag{4}$$

In Equation (4), as we allocate n_g shelves to the product, there is a corresponding cost for these shelves, mn_g^2 per unit time. The assumption has been used in Swami and Shah [16].

While after a certain period, the quality of perishable food decreases and its shelf life is about to end, the consumption rate will therefore go down. In a mart or retail grocery, the perishable products near their shelf life are usually taken from the shelves and sold on a discount rack. In this case, the shelf space does not affect the product's sale and the discount rack can attract much attention especially when the discount policy is stable and normalized. Therefore, we assume the demand function after discount is:

$$D_2(t) = a - b\theta p + dq(t) + f \tag{5}$$

Then, the demand function over the shelf life time can be described as:

$$D(t) = \begin{cases} a - bp + cn_g + dq(t) & 0 \leq t \leq t_0 \\ a - b\theta p + dq(t) + f & t_0 < t \leq T \end{cases} \tag{6}$$

2.3. Modeling the Single-Item Food Retail Chain

In the above part, we have modeled the quality and demand of perishable food in a retail store. Now the total profit model for the whole food retail chain is developed. There are several elements in the retailer's profit: the sales revenue, shelf space cost, purchasing cost, and disposal cost.

First, we calculate the profit earned when the product is sold on the shelves by:

$$p \int_0^{t_0} D_1(t)dt = p(a - bp + cn_g)t_0 + pdq_0(1 - e^{-\lambda t_0})/\lambda \tag{7}$$

Second, since n_g shelves are allocated to the product from the beginning to t_0, the corresponding cost is $t_0 mn_g^2$.

Third, the profit earned during the time the product is sold on the discount rack (this profit may be negative) is:

$$\theta p \int_{t_0}^{T} D_2(t)dt = \theta p(a - b\theta p + f)(T - t_0) + \theta pdq_0(e^{-\lambda t_0} - e^{-\lambda T})/\lambda \tag{8}$$

Fourth, let Q and C_0 denote the order quantity and the purchasing cost per unit, respectively. For small retailers, C_0 is usually a constant. For big retailers, the supplier may offer a wholesale discount to promote the product. Here, we assume the supplier offers a continuous discount function $C_0 = \alpha - \beta Q$, where α is the no-discount price and β is the discount sensitivity to Q. Then, the total purchasing cost is $C_0Q = (\alpha - \beta Q)Q$.

Finally, all the perishable products may not be sold out by the end of their shelf life. The remainder needs disposing, which may cause carbon emissions, environmental pollution, and corresponding cost. Let C_d denote the disposal cost per unit. Then, the total disposal cost is:

$$C_d \left[Q - \int_0^{t_0} D_1(t)dt - \int_{t_0}^{T} D_2(t)dt \right] \\ = C_d[Q - (a - bp + cn_g)t_0 - dq_0(1 - e^{-\lambda t_0})/\lambda - (a - b\theta p + f)(T - t_0) - dq_0(e^{-\lambda t_0} - e^{-\lambda T})/\lambda] \tag{9}$$

Consequently, the total profit of the food retail chain is:

$$Max\ \pi(p, n_g, Q) = p\int_0^{t_0} D_1(t)dt + \theta p\int_{t_0}^{T} D_2(t)dt - t_0 mn_g^2 - C_0Q - C_d\left[Q - \int_0^{t_0} D_1(t)dt - \int_{t_0}^{T} D_2(t)dt\right] \\ = p(a - bp + cn_g)t_0 + pdq_0(1 - e^{-\lambda t_0})/\lambda + \theta p(a - b\theta p + f)(T - t_0) + \theta pdq_0(e^{-\lambda t_0} - e^{-\lambda T})/\lambda - mn_g^2 - (\alpha - \beta Q)Q \\ -C_d[Q - (a - bp + cn_g)t_0 - dq_0(1 - e^{-\lambda t_0})/\lambda - (a - b\theta p + f)(T - t_0) - dq_0(e^{-\lambda t_0} - e^{-\lambda T})/\lambda] \\ subject\ to\ Q \geq \int_0^{t_0} D_1(t)dt + \int_{t_0}^{T} D_2(t)dt \tag{10}$$

In Equation (10), p, n_g and Q are determined to maximize the total profit. The other parameters are assumed to be given.

3. Optimal Solution for the Single-Item Food Retail Chain

In this section, we prove there is an optimal solution to maximize the total profit. First, we can see that π is a polynomial of Q once p and n_g are given. Taking the first and second derivative of π with respect to Q, we get:

$$\theta p \int_{t_0}^{T} D_2(t)dt = \theta p(a - b\theta p + f)(T - t_0) + \theta pdq_0(e^{-\lambda t_0} - e^{-\lambda T})/\lambda \tag{11}$$

$$\theta p \int_{t_0}^{T} D_2(t)dt = \theta p(a - b\theta p + f)(T - t_0) + \theta p d q_0 (e^{-\lambda t_0} - e^{-\lambda T})/\lambda \tag{12}$$

From $\partial \pi / \partial Q = 0$, we can obtain $Q = (\alpha + C_d)/2\beta$. Since the only constraint is $Q \geq \int_0^{t_0} D_1(t)dt + \int_{t_0}^{T} D_2(t)dt$, if $\int_0^{t_0} D_1(t)dt + \int_{t_0}^{T} D_2(t)dt \geq (\alpha + C_d)/2\beta$, then $\lim_{Q \to +\infty} \pi = +\infty$. It is obvious that the parameter values are not reasonable. The value of β may be too large. Therefore, $\int_0^{t_0} D_1(t)dt + \int_{t_0}^{T} D_2(t)dt < (\alpha + C_d)/2\beta$. For π decreasing on $[0, (\alpha + C_d)/2\beta]$, $Q = \int_0^{t_0} D_1(t)dt + \int_{t_0}^{T} D_2(t)dt$ is the optimal solution, which means all the products should be sold by the end of their shelf life. Therefore, there is no disposal cost. Substituting $Q = \int_0^{t_0} D_1(t)dt + \int_{t_0}^{T} D_2(t)dt$ into π, the total profit becomes:

$$Max \ \pi(p, n_g) = p \int_0^{t_0} D_1(t)dt + \theta p \int_{t_0}^{T} D_2(t)dt - t_0 m n_g^2 - C_0 \left[\int_0^{t_0} D_1(t)dt + \int_{t_0}^{T} D_2(t)dt \right] \tag{13}$$

The Hessian matrix of π is:

$$H = \begin{vmatrix} \partial^2 \pi/\partial p^2 & \partial^2 \pi/\partial p \partial n_g \\ \partial^2 \pi/\partial n_g \partial p & \partial^2 \pi/\partial n_g^2 \end{vmatrix} = \partial^2 \pi/\partial p^2 \cdot \partial^2 \pi/\partial n_g^2 - (\partial^2 \pi/\partial p \partial n_g)^2 \tag{14}$$

$$\partial^2 \pi/\partial p^2 = -2bt_0 - 2b\theta^2(T - t_0) + 2\beta[bt_0 + b\theta(T - t_0)]^2 \tag{15}$$

$$\partial^2 \pi/\partial n_g^2 = -2t_0 m + 2\beta t_0^2 c^2 \tag{16}$$

$$\partial^2 \pi/\partial p \partial n_g = \partial^2 \pi/\partial n_g \partial p = ct_0 - 2\beta[bt_0 + b\theta(T - t_0)] \tag{17}$$

From Equations (15)–(17), we can see that $\partial^2 \pi/\partial p^2$, $\partial^2 \pi/\partial n_g^2$, $\partial^2 \pi/\partial p \partial n_g$ and $\partial^2 \pi/\partial n_g \partial p$ are all constants. Since β is quite small, $\partial^2 \pi/\partial p^2$ and $\partial^2 \pi/\partial n_g^2$ are usually less than 0. In this situation, if $H > 0$, the optimal point (p^*, n_g^*) obtained from $\partial \pi/\partial p = \partial \pi/\partial n_g = 0$ is the maximum point; if $H < 0$, (p^*, n_g^*) is a saddle point, then the maximum point is at the boundary. Obviously, it is not realistic in practice; if $H = 0$, it cannot be judged what the point (p^*, n_g^*) is. This situation is almost impossible because $\partial^2 \pi/\partial p^2$, $\partial^2 \pi/\partial n_g^2$, $\partial^2 \pi/\partial p \partial n_g$ and $\partial^2 \pi/\partial n_g \partial p$ are all constants. When $\partial^2 \pi/\partial p^2 > 0$ (or $\partial^2 \pi/\partial n_g^2 > 0$), which means when p (or n_g) tends to infinity, the profit is infinity; this result is obviously not realistic. In this situation, the parameter values are not reasonable.

4. The Multi-Item Food Retail Chain

In this section, we extend the single-item retail chain model into a multi-item one. The following assumptions are needed to simplify the problem. First, the consumption of each item is assumed to be independent from that of the others. Second, the capacity of the shelf space is limited. Third, since the deterioration rate and replenishment cycle of each item are different from those of the others, all the items are replenished separately. Fourth, the objective of the multi-item retail chain is to maximize the summation of each item's profit per unit time rather than the total profit per unit cycle, because the replenishment cycle of each item is different. Let superscript i denote the index of each item and L denote the shelf space capacity. Therefore, the objective function of the multi-item retail chain is:

$$Max \sum_i \pi^i(p^i, n_g^i, m) = \frac{1}{T^i}[p^i \int_0^{t_0^i} D_1^i(t)dt + \theta^i p^i \int_{t_0^i}^{T^i} D_2^i(t)dt - t_0^i m n_g^{i\,2} - C_0^i(\int_0^{t_0^i} D_1^i(t)dta + \int_{t_0^i}^{T^i} D_2^i(t)dt)]$$

$$= 1/T^i \left\{ p^i(a^i - b^i p^i + c^i n_g^i)t_0^i + p^i d^i q_0^i(1 - e^{-\lambda^i t_0^i})/\lambda^i + \theta^i p^i(a^i - b^i \theta^i p^i + f^i)(T^i - t_0^i) + \theta^i p^i d^i q_0^i(e^{-\lambda^i t_0^i} - e^{-\lambda^i T^i})/\lambda^i \right.$$

$$- t_0^i m n_g^{i\,2} - \alpha^i[(a^i - b^i p^i + c^i n_g^i)t_0^i + d^i q_0^i(1 - e^{-\lambda^i t_0^i})/\lambda^i + (a^i - b^i \theta^i p^i + f^i)(T^i - t_0^i) + d^i q_0^i(e^{-\lambda^i t_0^i} - e^{-\lambda^i T^i})/\lambda^i] \tag{18}$$

$$\left. + \beta^i[(a^i - b^i p^i + c^i n_g^i)t_0^i + d^i q_0^i(1 - e^{-\lambda^i t_0^i})/\lambda^i + (a^i - b^i \theta^i p^i + f^i)(T^i - t_0^i) + d^i q_0^i(e^{-\lambda^i t_0^i} - e^{-\lambda^i T^i})/\lambda^i]^2 \right\}$$

$$subject \ to \ \sum_i n_g^i \leq L$$

The shelf space cost in the above objective function is useful when we determine each item's shelf space. However, when the total profit per unit time is calculated, the retailer does not take the shelf space cost into account. The shelf space is considered as a limited resource. Therefore, the total profit per unit time without the shelf space cost is:

$$\prod = \sum_i \frac{1}{T^i}[p^i \int_0^{t_0^i} D_1^i(t)dt + \theta^i p^i \int_{t_0^i}^{T^i} D_2^i(t)dt - C_0^i(\int_0^{t_0^i} D_1^i(t)dt + \int_{t_0^i}^{T^i} D_2^i(t)dt)] \qquad (19)$$

The summation of each item's shelf space may be larger or smaller than the shelf space capacity when the opportunity cost of the shelf space (m) is given. From the objective function we can see that each item's shelf space allocated and the total profit per unit time are negatively correlated with the shelf space opportunity cost, so we could adjust the value of m to make the summation of each item's shelf space equal to the shelf space capacity.

In the multi-item retail chain, the objective is to maximize the total profit per unit time with the shelf space cost. Since the replenishment cycle of each item is constant, the optimal solution obtained in Section 3 will not change when the value of m is given. Ji et al. [17] developed a mathematical model for a multi-commodity, two-stage transportation and inventory problem, which was solved by CPLEX Optimizer. CPLEX Optimizer is a high-performance mathematical programming solver for linear programming, mixed integer programming and quadratic programming. In this part, we develop a simple algorithm to find the optimal solution for the multi-item retail chain.

Algorithm

Step 1: Input all the values of the parameters and find the upper and lower bound of m.

Step 2: Let m_k ($k = 1, 2 \ldots , 10$) be equally distributed between the upper and lower bound. Set an acceptable error e. Calculate each item's n_g^i, p^i, Q^i and π^i with each m_k and $\prod$.

Step 3: If there is an m_k that makes $0 \leq L - \sum_i n_g^i \leq e$, the current result is the optimal solution. Otherwise, the m_k that makes $\sum_i n_g^i \geq L$ and the m_{k+1} that makes $\sum_i n_g^i \leq L$ become the new upper and lower bound of m. Go to Step 2.

5. Numerical Experiments and Real-Life Examples

In this section, we first evaluate the effects of some key parameters and then use some real-life data from physical retail stores to validate the model.

5.1. Parameters Evaluation

We input different values of the key parameters, the discount rate and the opportunity cost of shelf space, and show the effects on the total profit and optimal solutions in the single-item retail chain. Table 1 shows the values of the input parameters.

Table 1. Input parameters.

Parameter	Value	Parameter	Value	Parameter	Value	Parameter	Value	Parameter	Value
a	30	d	1.5	t_0	10	α	10	θ	0.65
b	1.8	f	2	T	12	β	0.005	q_0	0.9
c	2	λ	0.01	m	0.9				

Since the values of the parameters are given, we can calculate the value of the Hessian matrix, where $H = 147.20 > 0$. (p^*, ng^*) is the maximum point. In the model, we assume that the discount rate $1 - \theta$ is a constant, while it is also an important decision variable in the retail store. Therefore, we attempt to find an approximation of the optimal discount rate by numerical analysis.

Table 2 shows that as the discount rate ($1 - \theta$) decreases, the shelf space allocated, price, and order quantity also decrease. That is because the decrease in the discount rate reduces the demand on the discount rack. The retailer has to lower the price to promote the market, and hence lower the shelf space to save some shelf space cost. From Figure 1, we can see that the total profit seems concave in θ and reaches its maximum around $\theta = 0.65$, while in the reality, θ is usually correlated with the discount rack attraction rate, which needs further research in the retail store to investigate their relationship.

Table 2. The optimal solution with different values of θ.

θ	n_g (Decimeter)	p (Dollar)	Q (Pound)
0.5	15.28	21.08	267.55
0.55	14.98	20.86	262.05
0.6	14.64	20.61	256.56
0.65	14.27	20.33	251.11
0.7	13.87	20.02	245.73
0.75	13.44	19.69	240.46

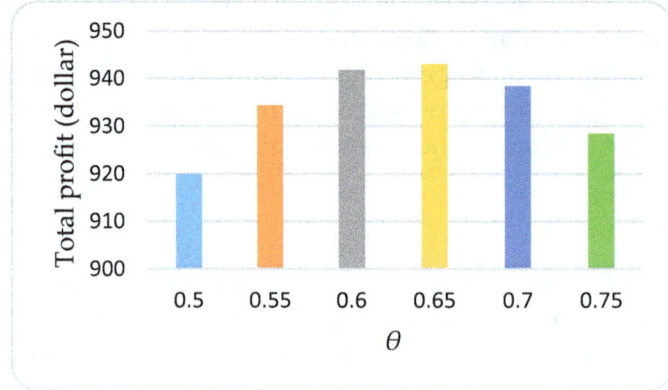

Figure 1. Total profit of the retail chain with different θ.

The opportunity cost of the shelf space is also an important parameter which may affect the space allocation and total profit significantly. Table 3 and Figure 2 show the effects of m on the total profit and decision variables.

Table 3. The optimal solution with different m.

m	n_g (Decimeter)	p (Dollar)	Q (Pound)
0.7	37.03	30.96	495.6
0.8	20.98	23.54	324.28
0.9	14.64	20.61	256.56
1	11.24	19.4	220.27
1.1	9.12	18.06	197.66

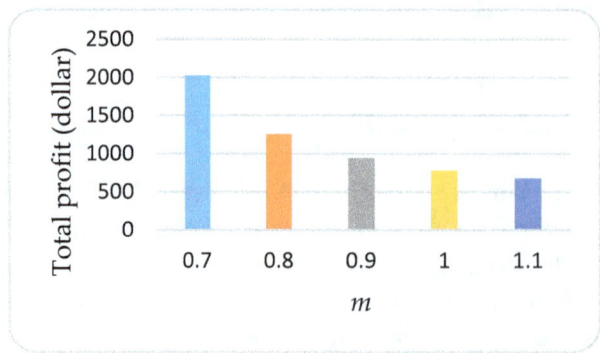

Figure 2. Total profit of the retail chain with different m.

Table 3 and Figure 2 show that the total profit and all the variables decrease when m goes up. That's due to the fact that as the opportunity cost of the shelf space increases, the retailer reduces the shelf space and lowers the price to improve the utilization rate of the shelves. That can explain why the product on the shelves with the best position in a mart always changes. Usually, the best or largest shelves are only allocated to the product in season or on discount.

In the second case, we show the effects of the shelf space capacity on the total profit per unit time in the multi-item retail chain. It is assumed that there are five items in the retail chain. Table 4 shows the value of the input parameters.

Table 4. Input parameters.

Parameter	Value	Parameter	Value
a^i	27, 30, 25, 30, 22	q_0^i	0.9, 0.85, 1, 0.75, 0.8
b^i	1.4, 1.8, 2, 1.5, 1.2	θ^i	0.65, 0.75, 0.6, 0.85, 0.9
c^i	1.8, 2, 1.5, 1.3, 2.1	λ^i	0.005, 0.01, 0.012, 0.12, 0.001
d^i	1.2, 1.5, 1.6, 1, 0.9	T^i	24, 12, 12, 10, 48
f^i	3, 2, 2.5, 2, 1.8	t_0^i	20, 10, 10, 8, 44
β^i	0.005, 0.002, 0.001, 0.007, 0.0002	α^i	15, 10, 5, 17, 10
e	0.1	L	40, 60, 80, 100

In the multi-item retail chain, the shelf space capacity is the main constraint which is a fundamental resource for the retailer. Usually, customers are willing to go to large marts, which can provide comfortable shopping conditions, a variety of goods, and low prices.

From Figure 3, we can find that larger shelf space could reduce the opportunity cost of the shelf space. Figure 4 shows that the total profit grows as the shelf space capacity increases. In reality, besides the increase of the shelf space capacity, the retailers also try to expand the types of goods, which may attract more customers and sales to increase the total profit rapidly.

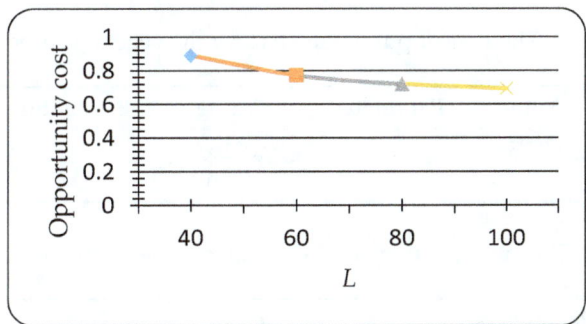

Figure 3. Opportunity cost of shelf space with different L.

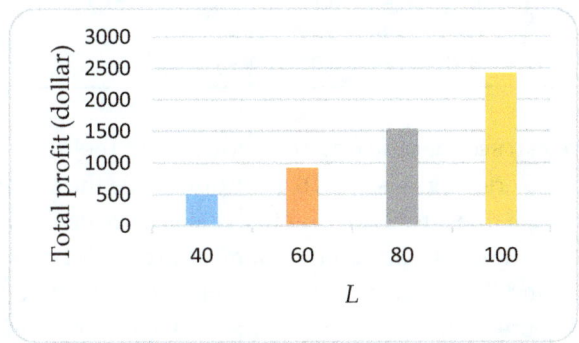

Figure 4. Total profit per unit time with different L.

5.2. Real-Life Examples

In this part, we use some real-life data to show the model's application in a practical case. The data we use are obtained from physical retail stores, which have been applied in Wang and Li [8] and Desmet and Renaudin [18]. To satisfy the assumptions well in Section 4, two kinds of perishable products are considered in the case, meat and vegetables. Meat and vegetables are the two kinds of most common perishable foods in the market, and they are quite representative in real life. Additionally, empirical data can be obtained easily from published reference paper. The common parameters of a meat product sold in different stores are shown in Table 5. Table 6 shows the different parameters and results. Then, we consider a vegetable product in different stores. The parameters and results are shown in Tables 7 and 8. Due to the fact that there is no prior empirical data of the effect of product quality on the consumption rate, we assumed that the effect of quality equals the price sensitivity, as Wang and Li [8] did.

Table 5. The same input parameters for the meat product.

Parameter	Value	Parameter	Value	Parameter	Value
θ	0.75	λ	0.0067/h	t_0	168 h
C_0	3.5	T	240 h		

Table 6. The input parameters and results for the meat product.

Store	Parameters							Price (p)	Shelf Space (n_g)	Order Quantity (Q)	Total Profit (π)
	a	b	d	f	q_0	c	m				
1	9.79	1.83	1.83	0.7	0.95	0.33	0.2	5.12	1.34	601.80	554.12
2	8.32	1.56	1.56	1.1	0.97	0.39	0.15	5.30	2.33	575.49	551.75
3	12.73	2.38	2.38	1.5	0.88	0.46	0.18	5.20	2.18	831.82	760.82
4	7.83	1.46	1.46	0.5	0.90	0.33	0.21	5.13	1.28	481.21	443.32

Table 7. The same input parameters for the vegetable product.

Parameter	Value	Parameter	Value	Parameter	Value
θ	0.75	λ	0.0216/h	t_0	48 h
C_0	1	T	72 h		

Table 8. The input parameters and results for the vegetable product.

Store	Parameters							Price (p)	Shelf Space (n_g)	Order Quantity (Q)	Total Profit (π)
	a	b	d	f	q_0	c	m				
1	7.92	4.86	4.86	2.1	0.90	0.58	0.1	1.84	2.45	257.23	138.30
2	6.73	4.13	4.13	2.0	0.92	0.57	0.12	1.85	2.02	221.97	120.50
3	10.30	6.32	6.32	2.5	0.83	0.56	0.08	1.80	2.80	317.24	162.22
4	6.34	3.89	3.89	1.6	0.85	0.59	0.09	1.88	2.90	214.05	114.94

From Tables 5 and 7 we can see that the deterioration rate of the meat product is 0.0067 per hour and the vegetable product's deterioration is 0.0216 per hour. Since the vegetable perishes much more quickly than the meat, the vegetable's replenishment cycle is shorter than the meat's. From Desmet and Renaudin [18], we get that the consumption rate of meat products is not quite sensitive to the shelf space allocated (around 0.35) and that of the vegetable is relatively sensitive to the shelf space allocated (around 0.57). Furthermore, the shelves for meat products contain a refrigerating system, which reduces the meat's deterioration rate but makes the shelf space for meat quite expensive. Therefore, the retailer prefers to allocate smaller shelf space to each meat product than the vegetable product.

As there is little empirical data on perishable products, we assume the meat product in different stores as different products in a retail store and validate the multi-item retail chain model. The input parameters are shown in Table 5. The results for the multi-item retail chain are shown in Table 9.

From Table 9 we can see that each item's price goes up as the shelf space capacity increases. That is because larger shelf space decreases the opportunity cost of the shelf space, and larger shelf space attracts more sales, which could offset the effect of the increased price. This may not conform to the actual situation. The reason is that as the shelf space capacity increases, more customers are attracted and hence the market scale is expanded. In our case, the market scale remains the same. Another reason is that as the shelf space capacity increases, the retailer usually adds more varieties of items. When one item is added, the total profit per unit time will increase. If not, the retailer will not choose to add the item. The increased profit means the opportunity cost of the shelf space goes up, and the retailer will decrease the price to improve the utilization rate of the shelves. However, the total profit may not always increase with the number of items, for too much variety of items may increase the operation cost and also dazzle customers. Meanwhile, the items on a shelf may be replenished too frequently with small shelf space, which may increase the labor cost.

Table 9. The input parameters and results for the multi-item retail chain.

Capacity (L)	Total Profit Per Unit Time (Π)	Shelf Space Cost (m)	Item	Price (p^i)	Shelf Space ($n_g{}^i$)	Order Quantity (Q^i)	Item Profit Per Unit Time (π^i)
8	11.64	0.1613	1	5.14	1.68	610.77	2.69
			2	5.28	2.15	569.78	2.78
			3	5.23	2.46	842.08	3.92
			4	5.17	1.70	492.24	2.25
12	13.56	0.1130	1	5.20	2.49	631.63	3.00
			2	5.39	3.26	603.84	3.34
			3	5.32	3.71	887.26	4.64
			4	5.24	2.54	514.01	2.58
16	15.57	0.0889	1	5.26	3.27	651.82	3.31
			2	5.50	4.39	638.59	3.94
			3	5.42	4.97	932.80	5.40
			4	5.32	3.37	535.53	2.92

6. Conclusions

This research attempts to improve the retail chain design for perishable food in a retail store by comprehensively evaluating the pricing strategy, shelf space allocation, and replenishment policy with the help of modern tracking technologies. This paper may have great potential in spoilage reduction and profit improvement for the food retail chain management. First, we developed a mathematical model for perishable food with a single item. The results showed that the discount rate is positively related to the decision variables (price, shelf space, and order quantity) and its optimal value can be obtained by numerical analysis. Larger shelf space could reduce the opportunity cost of the shelf space and bring about more profit. The results also showed that the discount and pricing strategy and shelf space can greatly affect the performance of the retail chain, so the design of the food retail chain should evaluate all the relevant parameters carefully. Then, the single-item retail chain was extended into a multi-item one with a shelf space capacity. As shown in the results, when larger shelf space is available, the retailer should allocate more space to the products whose demands are sensitive to the shelf space and expand the types of goods, which may attract more customers and sales.

Acknowledgments: The authors are grateful to the editors and anonymous reviewers for providing the valuable comments. The research was supported by the National Natural Science Foundation of China (Grant No. 71501090), the Natural Science Foundation of Jiangsu Higher Education Institutions of China (Grant No. 14KJD410001) and the Jiangsu Philosophical-social Science Program (2016SJB630113).

Author Contributions: Yujie Xiao conceived and designed the experiments; Shuai Yang performed the experiments; Yujie Xiao and Shuai Yang analyzed the data; Yujie Xiao contributed analysis tools; Yujie Xiao and Shuai Yang wrote the paper.

Conflicts of Interest: The authors declare no conflict of interest. The founding sponsors had no role in the design of the study; in the collection, analyses, or interpretation of data; in the writing of the manuscript, and in the decision to publish the results.

References

1.   Ferguson, M.E.; Ketzenberg, M.E. Information Sharing to Improve Retail Product Freshness of Perishables. *Prod. Oper. Manag.* **2006**, *15*, 57–73.

2.   Mouron, P.; Willersinn, C.; Möbius, S.; Lansche, J. Environmental Profile of the Swiss Supply Chain for French Fries: Effects of Food Loss Reduction, Loss Treatments and Process Modifications. *Sustainability* **2016**, *8*, 1214–1223. [CrossRef]

3.   Aiello, G.; la Scalia, G.; Micale, R. Simulation analysis of cold chain performance based on time–temperature data. *Prod. Plan. Control* **2012**, *23*, 468–476. [CrossRef]

4.   Zhou, Y.W.; Yang, S.L. An Optimal Replenishment Policy for Items with Inventory-level-dependent Demand and Fixed Lifetime under the LIFO Policy. *J. Oper. Res. Soc.* **2003**, *54*, 585–593. [CrossRef]

5.   Mandal, B.N.; Phaujdar, S. An Inventory Model for Deteriorating Items and Stock-dependent Consumption Rate. *J. Oper. Res. Soc.* **1989**, *40*, 483–488. [CrossRef]

6.   Zanoni, S.; Zavanella, L. Chilled or Frozen? Decision Strategies for Sustainable Food Supply Chains. *Int. J. Prod. Econ.* **2012**, *140*, 731–736. [CrossRef]

7.   Hong, K.S.; Lee, C. Optimal Pricing and Guaranteed Lead Time with Lateness Penalties. *Int. J. Ind. Eng. Theory* **2013**, *20*, 153–162.

8.   Wang, X.; Li, D. A Dynamic Product Quality Evaluation Based Pricing Model for Perishable Food Supply Chains. *Omega* **2012**, *40*, 906–917. [CrossRef]

9.   Lynch, M.; Curhan, R.C. A Comment on Curhan's "The Relationship between Shelf Space and Unit Sales in Supermarkets". *J. Mark.* **1974**, *11*, 218–220. [CrossRef]

10. Wang, Y.; Gerchak, Y. Supply Chain Coordination When Demand is Shelf-space Dependent. *Manuf. Serv. Oper.* **2001**, *3*, 82–87. [CrossRef]

11. Mohsen, S.S.; Anders, T.; Mohammad, R.A.J. An Integrated Vendor–buyer Model with Stock-dependent Demand. *Transport. Res. E-Log.* **2010**, *46*, 963–974.

12. Labuza, T.P. Application of Chemical-kinetics to Deterioration of Foods. *J. Chem. Educ.* **1984**, *61*, 348–358. [CrossRef]

13. Hertog, M.L.; Uysal, I.; McCarthy, U.; Verlinden, B.M.; Nicolaï, B.M. Shelf Life Modelling for First-expired-first-out Warehouse Management. *Philos. Trans. R. Soc. A Math. Phys. Eng. Sci.* **2014**, *372*, 1–15. [CrossRef] [PubMed]

14. Sciortino, R.; Micale, R.; Enea, M.; la Scalia, G. A WebGIS-based System for Real Time Shelf Life Prediction. *Comput. Electr. Agric.* **2016**, *127*, 451–459. [CrossRef]

15. La Scalia, G.; Settanni, L.; Corona, O.; Nasca, A.; Micale, R. An Innovative Shelf Life Model based on Smart Logistic Unit for an Efficient Management of the Perishable Food Supply Chain. *J. Food Process Eng.* **2015**. [CrossRef]

16. Swami, S.; Shah, J. Channel Coordination in Green Supply Chain Management: The Case of Package Size and Shelf-space Allocation. *Technol. Oper. Manag.* **2011**, *2*, 50–59. [CrossRef]

17. Ji, P.; Chen, K.J.; Yan, Q.P. A Mathematical Model for a Multi-Commodity, Two-Stage Transportation and Inventory Problem. *Int. J. Ind. Eng. Theory* **2008**, *15*, 278–285.

18. Desmet, P.; Renaudin, V. Estimation of Product Category Sales Responsiveness to Allocated Shelf Space. *Int. J. Res. Mark.* **1998**, *15*, 443–457. [CrossRef]

10

Cheat Electricity? The Political Economy of Green Electricity Delivery on the Dutch Market for Households and Small Business

J. A. M. Hufen

QA+ Research and Consultancy, Boddens Hosangweg 83, 2481KX Woubrugge, The Netherlands;
hh@qaplus.info

Academic Editors: Michiel Heldeweg, Ellen van Bueren, Anna Butenko, Thomas Hoppe, Séverine Saintier and Victoria Daskalova

Abstract: The European Commission's renewable energy directive introduced a market-based Guarantees of Origin (GO)-trade system that gives consumers the choice of buying "real" green energy. This has been successful, as the market share of Dutch households that buy green energy grew to 64% in 2015. However, societal organizations are dissatisfied with the green energy offered, categorizing it as "cheat" electricity. This article aims to solve this riddle of a successful product created under the GO-trade system but also heavily criticized. Research reveals a lively marketplace with buyers eager to buy green energy and energy producers offering a wide range of labels. Marketplace mechanisms are strongly influenced by political choices, and financial support for energy suppliers makes green energy a credible option. Societal groups, however, argue that the information provided is incomplete and misleading, that buying green energy does not impact positively on greenhouse gas reduction, and that better information and structural reform are required. The GO-trade system is strongly influenced by member states' national energy politics. Societal organizations have helped to optimize the implementation of the GO-trade system in the Netherlands, but they are not expected to be able to support the creation of a level playing field in which an optimal GO-trade system will flourish.

Keywords: renewable energy directive; directive 2009/28; green electricity; cheat electricity; Guarantees of Origin; market-based system

1. Introduction

1.1. Background

The liberalization process in energy markets over the last two decades offered the European Commission an opportunity to stimulate the introduction of a market-based system to buy green energy. Guarantees of Origin (GOs) are the keystone in the market-based system that should ensure the origin of "real" green energy. After the idea was launched in 2001, Directive 2009/28 elaborated the new system both for issuing, transferring, and cancelling GOs and for organizing issuing bodies and supervision. The idea of the renewable energy directive was that consumer preferences would influence the switch to reduce CO_2 emissions as scheduled [1].

The market for households and small business that is the focus in this article opened in the Netherlands July 2001 before the market for all electricity products started. Since then, the consumption of green energy has grown steadily to a share of 64% of all households and small companies in 2015 [2]. Consumers were apparently excited by the idea of buying green energy and contributing to the mitigation of negative climate effects. Dutch households' enthusiasm was not, however, matched in any other European country. This is remarkable, considering the low share of renewable energy in overall Dutch consumption in comparison to that in the EU-27 countries [3].

Despite the success of green electricity in terms of increased Dutch market share, societal organizations' criticism of green electricity for households and small business made green electricity a contaminated market label in the Netherlands. Their argument was firstly that the market for green energy was not transparent because the information provided by companies offering green energy was misleading for consumers [4,5]. Furthermore, Dutch households buying green electricity would not in any way contribute to diminishing CO_2 emissions. National media reported regularly that green energy as sold in the Netherlands did not contribute to diminishing CO_2 emissions [6,7].

1.2. What Is the Market-Based GO-Trade System in Directive 2009/28?

In the Netherlands, a green energy certificate system started in the 1990s as a burden-sharing system used by energy companies to account for their part in renewable energy production. This private system, set up by the energy sector, was transformed into a public system under national legislation of the Ministry of Economic Affairs [8–10]. The new system was no longer used for the energy sector itself. It was operational from July 2001 as the first part of an energy market for households and small business was liberalized. Because of tax exemptions, the market for green electricity was popular amongst consumers but criticized for excessive stimulation and lack of supervision [11].

The Commission's idea was to give energy consumers the choice of buying green energy and thereby enhance consumption and production of renewable energy [1,12]. A European Commission directive proposed the introduction of a market-based system for issuing and trading green energy certificates; such trading fits well in the process of liberalizing energy markets. The system as proposed by the Commission included definitions of "green energy", rules about the process of issuing green certificates, trade in green energy, surveillance of the process, and the role of different participants [1,12]. In the Netherlands, the European system replaced the national trade system that had been in use for several years.

The Commission asked member states to take responsibility for the arrangements to enable renewable energy producers' requests for a GO to be accommodated. A GO is defined in article 1 as "an electronic document which has the sole function of providing proof to a final customer that a given share or quantity of energy was produced from renewable sources" [1]. The Commission obliged member states to set up an electronic system with high quality standards [1] (pp. 19–20). A GO is the equivalent of one megawatt of energy and is valid for a period of one year, after which it expires [1]. The GO administration should contain: the energy source, type of energy (electricity, cooling, heating), description of the production installation, whether the production of renewable energy was financially supported, unique identification number including date and country of issue [1,12].

The highlights of the GO-trade system as proposed in Directive 2009/28 are visualized in Figure 1. The Commission requires member states to set up a system of GO trade with an issuing body in a central position. Member states can designate organizations to act on their behalf to put in place appropriate mechanisms to ensure that GOs are issued, transferred, and cancelled [1]. The issuing body is supposed to be an organization that is independent of energy production, transfer, or supply.

The Commission obliges member states to create a system that confirms the origin of electricity produced from renewable energy sources. The GO market as set up by member states should meet high quality standards such as objectivity, transparency, and non-discrimination [1,12]. Member states should construct their own system to supervise the issuance, transfer, and cancellation of GOs. The Commission chose to be itself responsible for the supervision of the international GO-trade system.

Energy market liberalization makes each member state responsible for ensuring that its country's energy suppliers communicate essential information to customers [1,13]. Energy suppliers should explain every year: the contribution of each energy source to the overall fuel mix and information on the environmental impact (at least CO_2 emissions and radioactive waste). A model of the label was developed in order to structure the communication between energy suppliers and consumers [14].

Since 2015, an element has been added in the Dutch electricity label: the origin of renewable energy in terms of the location of the production in or outside the Netherlands.

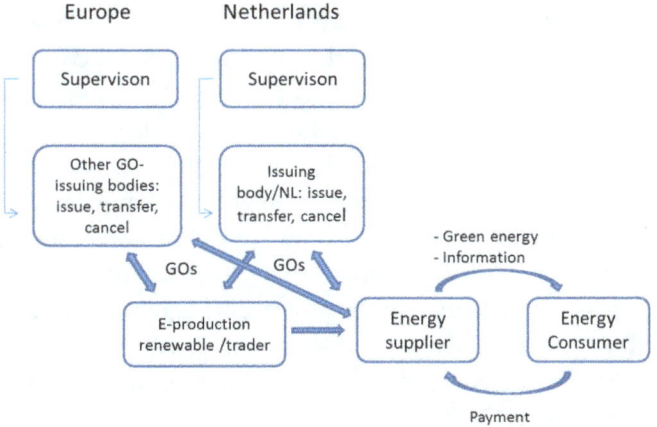

Figure 1. Design of the GO-trade system according to Directive 2009/28.

A consequence of the member states' high quality market-based GO-trade system is that GOs should be mutually recognized by participants. A member state has the right to refuse GOs from other countries according to the Commission. The refusal should in that case be explained to the Commission in terms of violations of the high quality standards (2001/77; 2009/28). The renewable directive explicitly makes clear that GOs do not play a role in the monitoring of a country's performance as far as its goal attainment is concerned [1]. A country's performance is monitored by measuring its own production of renewable energy. The consumption of renewable energy does not influence a member state's performance.

1.3. Theory and Policy Theory Regarding GO Trade

The introduction of a European trade system for GOs is an interesting subject because of the connection to current debates and discussions about the liberalization of European energy markets, market-based trade systems, the role of citizens and consumers in the market system, and the relation between public actors and the energy production and supply sector. Below, we connect the GO-trade system with some ongoing discussions, theories, and conceptual models.

The design of a market-based GO-trade system can be interpreted as a development that is part of the liberalization of the European energy market. The realization of a competitive market and the protection of consumer interests was an important step in this process, but energy market liberalization is proving to be a difficult process. The old mechanisms in the centralized power of large companies and national preoccupations with the energy sector are persistent [15]. Governments had problems adapting to the realities of a liberalized market and chose their own interpretation during implementation [16]. The question here is whether the GO-trade system overcame these problems, as green energy trade seems to be a great success in the Netherlands. Were national interests not an obstacle in this case?

Dutch experiences with the predecessor of the GO-trade system showed that the design of a good trade system is demanding. An evaluation of the system revealed some weak spots: the quality of the administration, the administrator's independence of the energy sector, adequate supervision [11]. The Commission similarly found that the design of government-mandated market-based systems is not easy. The Emission Trading System (ETS) is a cornerstone in the Commission's climate policy, yet, according to some, design flaws limit its effectiveness [17]. In several periods, the CO_2 prices were much lower than expected due to a lack of confidence in the system as well as uncertainty about political responses [18,19]. ETS implementation requires an active role on the part of the governments of major emitting countries to limit emissions [17].

In principle, green energy consumers have a key role in a liberalized energy system. The idea is that their preferences determine the supply of energy or renewable energy to be delivered. The Commission is well aware that the interests of consumers or other societal groups are subordinate in European decision-making processes [20]. The same goes for policy processes in the field of energy in the Netherlands [15]. The Commission points out that an active role for consumers and other groups is a major challenge that should be addressed. According to some, the active participation of these groups in decision-making must be translated into new criteria for legislation processes [21]. The failure to involve consumers' representatives or other societal groups in the design of the market-based GO-trade system could perhaps be a reason for their dissatisfaction with it. Additionally, it should be questioned why the success of the GO trade was not countered by the lack of consumer participation.

In policy sciences, there is growing awareness that government action is limited because of the networks in which private as well as semi-public and public actors participate [22]. This means that governmental activities like the design of a new system such as GO trade can only be understood when the actors directly and indirectly involved are considered. Jan Rotmans points out that, in Dutch energy transition, special attention needs to be given to the large energy companies in the market [23]. In his opinion, many private, semi-public, and public actors are involved in the energy market; the large energy companies in particular influence the outcomes of decision-making. Others see the combination of market-driven forces and public policy as the key to understanding energy consumption and production [24]. Energy market liberalization did not make government involvement obsolete [25].

2. Methodology

2.1. Domain

The domain of research is the implementation structure realized to implement Directive 2009/28. This structure is visualized in Figure 2.

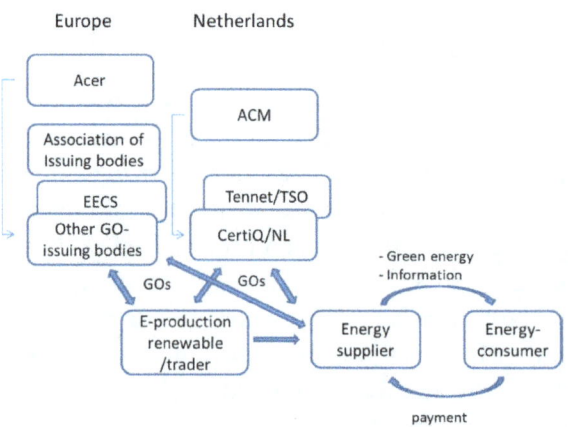

Figure 2. GO-implementation structure realized in the Netherlands as implemented on the basis of Directive 2009/28.

The starting point in the implementation structure is the marketplace in which energy suppliers and consumers meet. The energy supplier provides information about the offered energy labels as well as energy. Consumers are free to buy the energy; this means that they choose a contract type (e.g., annual contract), the price, and the origin of the energy (e.g., renewable energy). In the Dutch marketplace, the energy market is dominated by three energy suppliers that satisfy around 80% of demand [2]. In sum, around 7.7 million households use energy in the Netherlands [26]. This group accounts for around 20% of national energy consumption in the Netherlands [26].

CertiQ was established in July 2003 and is a 100% subsidiary of the national Transport System Organization (TenneT). CertiQ has been appointed by the Ministry of Economic Affairs as the national implementing organization. CertiQ certifies renewable energy and issues GOs as evidence that the energy is truly green and describes the source: sun, wind, water, and biomass. When a GO is issued, CertiQ credits it to the account of a trader or a producer. Certificates issued by CertiQ can be traded within Europe. CertiQ is a member of the Association of Issuing Bodies (AIB), an international partnership of European Guarantee of Origin organizations.

AIB is an international platform of European organizations responsible for GO standards and measurement. The energy certificate system administrators across Europe are members of AIB, mostly transmission system operators, electricity regulators, and energy market operators. The AIB's purpose is to develop, use, and promote a standardized system: the European Energy Certificate System—EECS. EECS offers a framework for creating and transferring electronic documents. The harmonized standards enable the owners of EECS certificates to transfer them to other domestic and international accountholders. EECS is based on structures and procedures that ensure the reliable operation of an international certificate scheme.

The Autoriteit Consument en Markt (ACM) is charged with competition oversight, sector-specific regulation of several sectors, and enforcement of consumer protection laws. ACM aims to create a level playing field, where all businesses play by the rules and where well-informed consumers exercise their rights. ACM holds in its tool box: investigations, monitoring, publishing (sanction) decisions. ACM is open to contact with consumers and communicates actively through Consuwijzer. Consumers can file complaints against companies if they feel that economic law has been violated.

2.2. Concepts, Goal, Research Questions, Method

The main concepts in this article are defined in Sections 1.2 and 2.1.

Green electricity is electricity whose green credentials can be proved by a GO issued by a body like CertiQ. Suppliers buy GOs and cancel them to cover the green energy demand of their clientele. The issuing body is responsible for checking whether the renewable energy is actually produced, and a supervisor controls the issuing, transfer, and cancellation of GOs.

Green energy in the GO-market system as defined in the renewable energy directive is a product that is supposed to give an impetus to an energy transition in the electricity market. The market-based GO system is objective, transparent, and non-discriminative. Energy suppliers are supposed to inform consumers about the offered product.

"Cheat" green electricity is either electricity whose origin as renewable energy cannot be proved by a GO or electricity that does not comply with the standards and norms of the renewable energy directive.

The goal of this research is to describe and explain the success of Dutch households' consumption of renewable energy as well as to describe and explain the criticism regarding the lack of transparency and lack of additionality of the market-based GO-trade system.

Research questions:

Q1. How successful has the consumption of green energy (in terms of the size of the market share) been in the Netherlands, and how can we explain this in terms of the implementation mechanisms of the renewable energy directive and its broader context?

Q2. How have representatives of consumers and other societal groups criticized the GO-trade system, and how can we understand this criticism from their motives and their position in the implementation structure?

Q3. What actions were taken by societal groups as a follow-up to their criticism? Do their activities affect the market-based GO-trade system?

Our data collection strategy was to describe and analyze the green energy delivery process. We conceptualized this delivery process as an interaction process between actors in the implementation structure of Directive 2009/28 and actors in the environment such as national governments, consumer organizations, and other societal groups. We started with the renewable energy directive and organizations directly involved in the GO system and in addition addressed the broader implementation context.

The method included three steps. As government designed the GO system, we started with the description of the GO system as designed in the EC directive and national legislation (Step 1). Texts were gathered and analyzed. In several interviews, questions were asked about the directive and specific subjects therein.

We continued (Step 2) by describing and analyzing the marketplace in which the green energy was traded. During this step, statistical information about GO trade in the Netherlands and imports to and exports from the Netherlands were analyzed using information from CertiQ, AIB, and RECS. We used a combination of interviews and document research to describe and analyze the differences in the implementation structure. An interview was held with one of the three energy suppliers and representatives of the energy sector (Energie Nederland, RECS International, Vereniging voor Energie, Milieu en Water (VEMW). Because consumers as such cannot be observed, we used secondary material, especially questionnaires, to reconstruct consumers' purchases and understand their motives.

In addition, we reconstructed the broader context of the actors involved (Step 3) as the marketplace is a network that is open to influences by other actors:

- National government (Ministry of Economic Affairs);
- Consumers' representatives (Consumentenbond (CB), and Eigen Huis (EH)), nature and environmental groups (Greenpeace, Natuur en Milieuorganisatie, Wereld Natuur Fonds, and Wise), and other organizations: Hivos, Hier Opgewekt;
- Energy companies' representatives (Energie Nederland VEMV); and
- Representatives of the energy sector and consumers.

Finally, two experts were consulted: one specializing in green energy trade (ECN) and one specializing in market design (TUD).

In sum, 21 interviews were held, of which 13 face-to-face and 8 interviews by phone. Several interviewees were contacted several times. In many cases, interviewees provided additional material that was analysed. Document research included statistics from CertiQ, AIB, RECS, and the Central Bureau of Statistics.

The answer to research question one is based on the research during Steps 1 and 2, although information from Step 3 was also used. The answer to research question one is presented in Section 3.1. The answers to research questions two and three are especially based on Step 3, but the results from Steps 1 and 2 were also used. The answer to research question two is presented in Section 3.2. The answer to research question three can be found in Section 3.3.

3. Results

3.1. Successful Implementation of EC 2009/28, Large Demand and Supply

Directive 2009/28 introduced a GO-trade system that set in motion an active trade in green energy. In the period 2003–2015, the green energy market share rose to 64% of every household as shown in Table 1. The development of the consumer market for green energy was obviously successful. Energy suppliers were able to accommodate Dutch consumers' large and rising demand. Consumers were very interested in green energy from the start, and eventually two out of every three consumers switched to it.

Table 1. Market share of green energy on the retail market in the Netherlands 2003–2015 (%).

Year	2003	2004	2005	2006	2007	2008	2009	2010	2011	2012	2013	2014	2015
Market-share green energy	29	38	36	31	31	35	39	44	53	61	63	64	64

Sources: [2,27–31].

What is remarkable is green energy consumption's high market share in 2003. This is to some degree explained by the quick rise in demand for green energy when it was introduced under the regime of the Dutch predecessor of Directive 2009/28 [11]. After the first three years (2003) of green energy trade, one of every three consumers bought green energy.

Consumers' interest in green energy is not unconditional however, as research shows. The price of energy is by far the most important motive for consumers' energy choice [32]. Market research reveals that only a fraction of all consumers (20%–25%) are interested in green energy if prices are high [33]. The most important motive for consumers to switch to another energy contract is a better price. Around 10% of consumers mention the greenness of renewable energy as a motive for choosing another energy supplier [31]. Apparently, the large market share cannot be interpreted one dimensionally as resulting from a deliberate choice to reduce CO_2 emissions.

A relevant question is how to explain why consumers buy renewable energy if prices are the most influential motive. Part of the answer is that the preference for renewable energy on the Dutch market was not obstructed by higher prices. Prices for renewable energy were competitive and sometimes even in the low price range because of successive governments' financial support to energy suppliers between 2001 and 2015. At first, tax exemptions (2001–2004) kept down the price of green energy [11]. Later on, production cost differences were mitigated by the MEP subsidy (2005–2006) and the SDE$^+$ (2008–2015). The Dutch issuing body found that demand for GOs depended strongly on subsidies [34]. The tax exemptions were replaced by the MEP subsidy, which was abolished soon, and so the market dropped 20%.

The rise in green energy's market share shows consumers' interest in green energy as well as the ability of energy suppliers to accommodate this demand. Each of the energy companies tried to secure its share of the green energy market with active marketing of their labels [35–37]. The Dutch energy suppliers grasped the opportunity offered by liberalization to provide green energy. Each of the larger energy companies (Nuon/Vattenval, Essent/RWE and Eneco) offered its green energy labels to the market. Some new energy suppliers entered the market offering renewable energy [35].

Statistics demonstrate that Dutch production covers around 28% of Dutch renewable energy demand. Around 75% of the remaining demand (72% of Dutch renewable demand) is covered by GO imports, especially water energy [34,38]. This strategy was a necessary condition for energy suppliers to accommodate consumer demand. In 2015, 70% of the GO water energy imports were covered by water energy from the Scandinavian countries (Norway, Sweden, and Finland) and Iceland. Although the precise number of water energy imports and each country's share vary, the dominance of water energy in the imports from these countries is evident [39].

ACM interprets green energy as an essential product characteristic for competition for a homogeneous good such as electricity. To some extent, green energy was a lever to enter the market, although the market structure did not change [35]. The number of green energy labels was from the start more than sufficient to create choice [34]. Recently, energy companies have been using green energy labels to differentiate their prices strongly [2]. No societal organization filed complaints against GO trade as developed under Directive 2009/28. ACM's attitude as a supervisor regarding green energy was benevolent, and this gave green energy a competitive edge.

Although the green energy trade was criticized by national media and consumer organizations, CertiQ's GO administration developed into an undisputed institution. GOs were accepted in the market as proof that renewable energy was produced. CertiQ provided the independent institution that was needed after the energy sector itself had previously been completely responsible [11].

As explained above, national government provided different national schemes to support the demand for green energy. An additional support was a consequence of the match between GOs issued and national support schemes. GOs are an obligatory part of the submissions for support of energy production companies. This match between the two resulted in an accurate GO administration because almost all investments in renewable energy are covered by submissions. In comparison to other countries, the Dutch GO-trade system covers more than 90% of the production of renewable energy; this is the highest in Europe [39].

In sum, the answer to research question one is that green energy consumption in the Netherlands was successful because green energy became the most popular label bought by households and small business. Economic and political forces in the marketplace under the rule of the freshly introduced market-based GO system explain the success of green energy labels in the Netherland. A large proportion of Dutch households chose to buy green energy at competitive prices. Energy suppliers accommodated the demand by importing Scandinavian green energy in particular. The fact that energy suppliers could prove that their energy was green stimulated the market. National support schemes were important because they enabled energy companies to offer competitive prices. As price is the most important factor in consumers' choice of an energy label, this political action was crucial. The issuing body and the supervisor were satisfied about the GO system and the development of the market because in their opinion it operated within the legal rules.

3.2. Criticism of the GO-Trade System: Transparency, Additionality

3.2.1. Lack of Transparency

Directive 2009/28 presupposes that GO trade takes place in an objective and transparent market (see Section 2). In order to substantiate this transparency, GOs must contain descriptive information about, amongst other things, the origin of the renewable energy. Furthermore, energy suppliers are obliged to produce an electricity label for consumers every year with information about the sustainability of their energy supply as well as the share of renewable energy in the fuel mix.

In order to provide clarity about the meaning of the legislation, ACM developed a method to calculate the sustainability of energy supplied by suppliers active on the Dutch market. In addition, a model of companies' electricity label was developed in Dutch legislation. Since 2004, ACM checks the content of electricity labels and the calculation methods as the law prescribes. In their opinion, the transparency of the green electricity market is guaranteed by issuing well-defined certificates and the check on the energy suppliers' electricity labels. Their appraisal of the market as "objective" and "transparent" is based on the idea the Dutch energy suppliers operate within the boundaries prescribed by European and Dutch law.

Since liberalization, societal organizations like Consumentenbond and Greenpeace have monitored the energy market. After joint research by several societal organizations, their conclusion was that the green electricity market was not transparent. Energy suppliers did not provide the information necessary for consumers to choose the energy label that fitted their preferences. In their opinion, energy suppliers were inaccurate about the nature of the green energy delivered, especially about the background of the green energy. Figure 3 shows two pathways by which green energy can be delivered but that cannot be discerned in energy suppliers' offers.

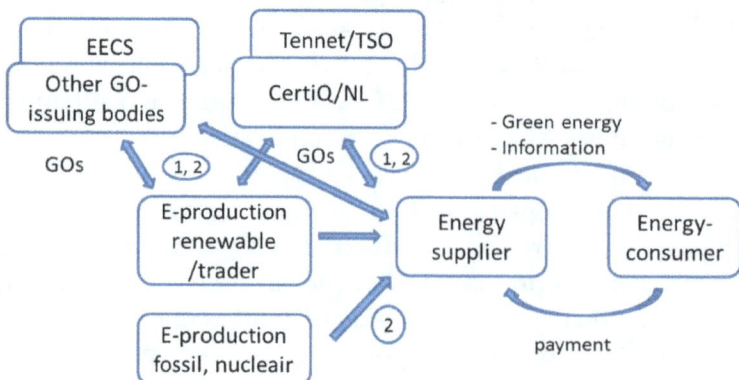

Figure 3. Green pathway (1) and red pathway (2) by which energy suppliers can deliver green electricity to consumers in the Netherlands.

Green pathway (1): GOs are bought by a trader, energy producer, or energy supplier as legal proof that renewable energy was produced. An energy supplier buys sufficient renewable energy to cover its customers' green energy needs.

Red pathway (2): GOs are bought by a trader, energy producer, or energy supplier as legal proof that renewable energy was produced. An energy supplier buys a combination of fossil, nuclear, or renewable energy from an energy producer to cover its customers' energy needs.

The existence of two different pathways is a consequence of the existence of two separate markets: the electricity market and the GO market. Consequently, it is possible to sell to all consumers a 100% renewable energy label although the fuel mix is 100% energy from coals. The coalition of societal organizations claims that the distinction between the green and the red pathway to deliver green energy should be made explicitly clear in communications with consumers.

The information as communicated through the electricity label is, in their opinion, not sufficient to adequately inform consumers. Consumers are therefore not aware of the true nature of products and are therefore not able to choose the product that fits their needs. The coalition partners try to provide additional information to support the decision-making process.

Energy suppliers did not all respond in the same way to the criticism from the societal organizations. One of the big energy suppliers chose an active sustainability strategy in 2008, announced a number of measures, and cooperated with the societal organizations. The two other energy suppliers did not respond to the criticism, chose a strategy that defended the red pathway, and did not cooperate with the societal organizations.

Difficulties in understanding the products provided by energy suppliers are manifest in consumers' opinions, given that half of consumers say that the offer on an energy label is unclear [31]. In addition, a substantial proportion of consumers (around 40%) think that buying green energy results in additional investment in renewable energy [32]. Furthermore, the nature of green energy is a difficult subject as one in every eight consumers thinks that green energy does not exist [40]. A group of consumers indicates that it has lost confidence in energy companies and sees them as an obstacle in the switch to a zero-emission society [41].

Market transparency is a matter of national interest, yet outdated from a European perspective. In 2006, soon after the market liberalization started, the Commission recognized the insufficient indication of the origin of electricity as a main deficiency of the European legal framework [42]. In an elaborate research project on the disclosure of market information, different European countries proved to be aware of the deficiencies of information disclosure about renewable energy. Not only were these deficiencies discovered, but also they have been the subjects of deliberation of possible improvements [43].

This subsection provides part of the answer to research question two. Representatives of consumers and other societal groups criticize the lack of transparency of the green energy market. In the opinion of societal groups, energy consumers are not adequately informed about the nature of green energy. Especially the red pathway for buying GOs to account for green energy delivery and buying a combination of fossil, nuclear, and renewable energy is not acceptable. From their perspective, it is not acceptable in a market-based system that consumers are not informed properly.

3.2.2. GO-Trade System

The idea of GO trade as introduced by the Commission is to enable consumers to buy green energy, thereby stimulating the consumption of renewable energy and supporting an energy transition. In the interviews, societal organizations expressed criticism of the effects of GO trade. In most interviews, the absence of any renewable energy consumption effect was mentioned. The existence of a second pathway to supply green energy without actually buying or producing renewable energy was seen as a problem of the GO-trade system.

The second pathway (red way) to supply renewable energy results in GO trade without a contribution to a switch or energy transition (see Figure 4). Several interviewees hinted at the absence of any effects of water energy imports, which is the most traded form of renewable energy through the GO-trade system [44]. Societal organizations referred especially to counties like Iceland, Norway, Sweden, and Finland. These groups obviously claim that the GO-trade system is not an effective instrument to realize countries' national overall goals regarding CO_2-emission reduction.

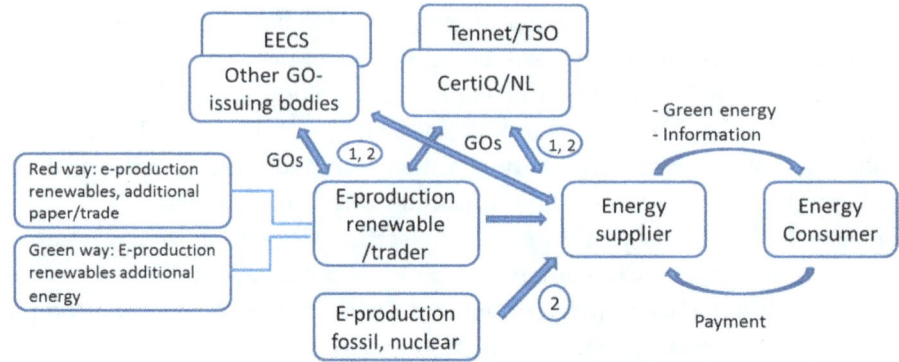

Figure 4. Estimated effects of GO trade according to societal groups in a green pathway (1) and a red pathway (2).

Closer inspection of statistics about GO trade can clarify the interviewees' opinion. Iceland produces all its energy needs by renewable energy and operates as an autonomous country. Energy imports to Iceland or exports from the island are not possible because there is no pipeline infrastructure to other countries. Iceland participates in GO trade as designed by the European countries and sells almost all GOs to interested energy companies or traders [38,44,45]. The inhabitants of Iceland are not interested in buying renewable energy, as indicated by the less than 1% of GOs that are redeemed [46]. The GO export provides additional revenues for the energy suppliers. Any stimulus on additional production or consumption is physical impossible because there is no pipeline infrastructure to the mainland.

The same lack of impetus on additional water energy production seems to hold for GO imports from Norway, Sweden, and Finland.

- The level of production of water energy in Norway, Sweden, and Finland after the introduction of Directive 2009/28 (2008–2014) did not change, as statistics of the International Energy Agency show [47]. The amount of water energy produced is different every year but does not rise over the course of years. Statistics of the Norwegian Ministry of Petroleum and Energy show the same pattern in Norway, being the country that exports the largest amount of water GOs [48].

- Prices for GOs are low (in 2015: around 22 eurocent) and therefore do not provide a stimulus for investment in water energy. Norway received an estimated €2 million for its GO exports, Sweden €1.25 million, and Finland €0.5 million [43,49]. GO trade gives energy companies or traders an additional source of income but is at this moment not an influential parameter in the behavior of private organizations in the energy sector.

- GO trade creates a reality on paper that differs from the usual trade in goods and services. For example: Norway produces 98% of its energy needs by means of water energy [46]. This is a convenient technology, using water and height differences provided by the country's natural terrain. Around 12% of its water energy is actually exported through the existing pipeline infrastructure to other countries, amongst which the Netherlands [50]. GO trade resulted for Norway in an export of 88% of Norwegian renewable energy [46]. Renewable energy is not a popular product, as around 22% of Nordic citizens buy renewable energy to cover their electricity needs. The Norwegian energy companies' GO-trade system is not of much interest to consumers or citizens.

In interviews with the Ministry of Economic Affairs, CertiQ, and ACM, the existence of the two pathways was recognized. The imports and exports of Scandinavian GOs were interpreted by these organizations as in line with the European legislation in the renewable energy directive. The failure of the GO-trade system to stimulate additional production of renewable energy was not seen as a problem by the aforementioned organizations. The point of view of the Ministry of Economic Affairs matches with the procurement procedures of national and local governments. A substantial proportion of local and national governments' energy needs is covered by energy from the red pathway [51,52].

Societal groups' criticism regarding the effectiveness of GO trade refers to the conduct of specific countries not only in selling their GOs but also in keeping their energy. The idea of several societal groups is that the mechanisms of the market-based system do not contribute to realizing goals regarding CO_2-emission reductions. In the interviews, several shortcomings in the GO-trade system were mentioned. The comments in these interviews are in line with the information in statistical data mentioned earlier in this section in relation to the lack of impetus for GO exports in Scandinavian countries.

In the first place, low prices for GOs were mentioned as an obstacle in the GO-trade system. Interviewees in the energy sector stated that GO demand is smaller than GO supply, with resulting low prices, as confirmed by RECS statistics [53]. Research reveals that Scandinavian GO prices varied in the period 2008–2015 between 0.10 eurocent and 0.25 eurocent [49]. Low prices are a structural problem in a market-based system because they do not result in an acceptable return on suppliers' investments.

Secondly, green energy producers in the two largest EU countries, that is, Germany and France, do not participate in GO submissions [39]. When countries other than France and Germany are added to the calculations, in sum 32% of energy production is not eligible for GOs because of national supports. These countries use the renewable energy directive to discourage GO submissions. According to article 15 of Directive 2009/28, national governments can chose to let energy producers lose the right to financial support for their renewable energy when the renewable energy is submitted for GOs. The revenues from GO trade are much lower than national financial support, and therefore energy producers do not participate in the GO-trade system. Another 20% is eligible for GOs but is not submitted. GOs that are issued cover 49% of the renewable energy produced [53].

Finally, GO trade as a market-based system aims to stimulate demand, whereas many national policies aim at energy production. Therefore, a dominant position for consumers in the market-based system cannot be realized. Several countries such as Germany and France stick to their reflex for the active involvement with energy supply experienced at the start of the market liberalization. The Netherlands is no exception, because Dutch tax exemptions and financial support for energy producers made the prices of renewable energy competitive. Directive 2009/28 sets the goals for renewable energy in terms of performance on energy supply, although while the policy for the directive was being developed a market-based system was discussed.

This subsection provides the other part of the answer to research question two. Representatives of consumer organizations and societal organizations criticize the GO-trade system because in their opinion it does not result in additional investments in renewable energy and does not contribute to an energy transition. The market-based GO-trade system results in very low prices that do not stimulate additional investments. Furthermore, more than half of renewable energy is not included in GO submissions. The largest economies in Europe choose active national involvement in energy policy instead of full cooperation in the market-based GO system.

3.3. System of GO Trade under Pressure: Activities and Effects of Societal Groups' Criticism?

The implementation of Directive 2009/28 in the Netherlands resulted in an intriguing situation. On the one hand, the demand for renewable energy rose to 64% in 2015 and became a very popular product amongst households and small business. On the other hand, green energy labels became a contaminated product, and green energy trade was criticized by consumer representatives as well as by nature and environmental groups and other groups.

The question raised related to what societal groups did and the effect of their activities. The interviews and secondary research gave a perspective on this question. The societal groups most active in the field of green energy trade are characterized in Table 2.

Table 2. Characteristics of societal organizations active in the field of green energy trade.

	Type	Focus	Constituency	Related to e-Sector
1. Consumentenbond (CB)	Association of consumers	National	Consumers 2015: 486, 403	No
2. Eigen Huis (EH)	Association of homeowners	National	House owners: 2015: 712, 954	No
3. WNF	Association of nature friends	National International	Nature friends	Eneco
4. Greenpeace	Environmental pressure group	National International	Citizens	No
5. Natuur-en Milieufederaties (NME)	Environmental activity group	National	Nature friends	No
6. Wise	Anti-nuclear action group	National	Anti-nuclear antagonists	No
7. HIVOS	Sustainability	National International	Citizens	No
8. Hier Opgewekt (HO)	Cooperation NME and Hier Klimaatbureau	National	See NME, TSO, VNG	RWE/Essent

Wise = World Information Service on Energy; Hivos = Humanisch Instituut voor Ontwikkelingssamenwerking; TSO = Transfer System Organization; VNG = Vereniging Nederlandse Gemeenten (Association of Dutch local government).

Societal groups' activities aimed especially at improving the mechanisms of the market-based GO-trade system. The activities can be categorized as: (1) improving transparency; and (2) stimulating directly and indirectly the consumption of real green energy.

(a)   Improving transparency

Because information on energy labels did not disclose enough information, Greenpeace and the Consumentenbond invested in research on the sustainability profile of energy suppliers [54,55]. The research focused on the following areas: (1) investments and disinvestments; (2) energy production; (3) energy sales; and (4) energy delivery. The outcomes of the research were firstly meant to inform societal groups about the sustainability of Dutch energy suppliers [37,56].

In addition, individual societal organizations invested in research on subjects like the number of green energy clients of the Dutch energy companies, price development of GO trade, among others. According to the societal organizations, this information is a more accurate description of the energy market. It supports the objective of green energy and the transparency of the market. The research created a solid knowledge base for societal organizations that gives additional information about the green energy market.

The knowledge resulted in a small improvement in transparency as the aforementioned research was translated into a useful measurement. The result of the sustainability profile, expressed on a scale of 1 to 10 (1 = not sustainable, 10 = very sustainable), was very brief and clear. The list of energy suppliers was classified into: frontrunners, peloton, and laggards [4,5]. This classification was used by all of the societal groups to inform their constituency and others interested. In addition, the Consumentenbond developed an electronic tool to support consumers' decision-making process.

Furthermore, the societal organizations collaborated with members of parliament to change rules about the content of electricity labels. This resulted in a change in the rules concerning energy suppliers' communication with their clients. The origin of renewable energy now has to be communicated (in terms of produced in the Netherlands or not produced in the Netherlands).

The societal organizations draw attention to the need for further steps to improve transparency. A new method to improve information disclosure about the energy market is backed by the societal groups, the so-called ReDiss-method. It was developed by AIB members in cooperation with the issuing bodies of several countries [39]. The national supervisor, ACM, was also involved in the ReDiss-project. A coalition of societal groups, representatives of the GO-trade system, and the energy sector support the new idea.

(b)   Stimulating green energy trade

Furthermore, societal groups were directly or indirectly involved in stimulating the purchase of "real" green energy:

In 2012, Wise introduced the concept of "cheat" electricity as a categorization for green energy bought especially in Scandinavia and Iceland (water energy). Since then, "cheat" electricity has been integrated in the idiom of Dutch consumers because newspapers and national media adopted the concept. Wise, Hier Opgewekt, and Greenpeace are very explicit in their communications about blaming energy companies that sell this "cheat" energy.

Wise and Hier Opgewekt are strict in their advice to consumers. If consumers choose green energy, the green energy labels that are not really green are excluded. "Not green" is interpreted by Wise as 75% produced in the Netherlands and covered by Dutch GOs. Hier Opgewekt demands at least 75% production in the Netherlands. These organizations plead for consumers to choose Dutch green electricity.

Two societal organizations arrange collective procurement procedures for their members, using green energy criteria that are in line with those for "real" green energy. Large groups of consumers participate in these green energy procurement procedures. Thus far, 489,000 consumers have chosen another energy supplier as a result of Eigen Huis' collective procurement procedure.

The effect of these activities is not so much a destabilization of the market-based GO-trade system as a demand stimulus within the boundaries of the current rules. It is difficult to assess the effect of societal organizations on the green energy trade. However, a demand stimulus for Dutch green electricity seems to have occurred after the activities of societal organizations since 2012 [2]. Energy suppliers offer substantially more Dutch green energy labels than before [2]. In the period 2012 to 2016, the price of GOs for Dutch wind energy and Dutch sun energy rose considerably [57]. It seems that the societal organizations have influenced the green energy market and the GO market effectively.

The different organizations in the GO-implementation system do not agree with the categorization of green energy as "cheat" energy. According to CertiQ, this categorization jeopardizes the GO system and puts pressure on it. According to CertiQ, Greenpeace and Wise's plea to buy Dutch GOs is not necessary because foreign GOs are a valid proof of the authenticity of the renewable energy. Energy suppliers complained to the ACM about categorizations like "cheat" electricity. Their grievances were not substantiated by official complaints that ACM would have to investigate and assess. Neither is ACM itself willing to investigate the categorization as a supervisor. The societal groups' communication is welcomed as an acceptable additional bit of information for consumers.

Societal groups have a pragmatic approach to improving the market-based GO-trade system. Although some of these groups operate in different countries, the international aspects are not under scrutiny. GO trade is seen as an imperfect system, but so far there is no lobbying about it in European decision-making processes.

The Ministry of Economic Affairs acts as loyal member state that implements the Commission's renewable energy directive as it is. The ministry is well aware of the difficulties regarding imports of green energy, as it was responsible for such imports under the green certificate system that preceded

GO trade. Approval was not initially given for the import of green water energy, but, soon after, it was allowed as long as the energy was fed in on energy transport lines [8–10].

This subsection answers the third research question concerning the action of societal groups. Consequent to societal groups' criticisms, their main resulting activities are attempts to improve the transparency of the market and stimulate "real" green energy. Their activities contributed to an adaptation of the rules concerning the provision of information about green energy by energy companies. Furthermore, the demand for Dutch green energy seems to have been raised as a result of communication, advice, and collective procurement procedures for households. Prices for this type of GOs have risen lately. Thus far, no attention has been paid by societal groups to mitigating the flaws of the GO-trade system.

4. Discussion

Contradictory opinions regarding the Dutch green energy market for households and small business provided the starting point of our research. After more than a decade, green energy has become the most popular type of electricity bought by Dutch households. At the same time, green energy labels have become a contaminated product in the Netherlands. Societal organizations criticize the lack of transparency and the failure of the GO-trade system to contribute to CO_2 reductions.

We have tried to understand the contradictory aspects of the GO-trade system by looking at the actors involved in the process of green energy delivery. The interaction between public, semi-public, and private organizations in the market-based GO-trade system elucidates the underlying mechanisms. Our research shows that market forces and political factors in the GO-trade system determine the dynamics of the market. Both the success and the criticism of the GO-based system are better understood in terms of the economics–politics dichotomy.

The idea of liberalizing the energy market was to support the economics in the system. The renewable energy directive (Directive 2009/28) opens up the possibilities of GO trade between energy suppliers, energy producers, and households. Economic forces incentivized energy suppliers to offer green energy labels from the start. The big three energy suppliers in the Netherlands responded quickly to the strong demand for the new products. Green energy labels were attractive to households, as demonstrated by the growing market share. The implementation of the directive by CertiQ supervised by ACM became an undisputed institution.

However, the success of the GO-trade system was not solely a market adaptation to demand and supply. The price of green energy was influenced by the Dutch government, which provided three different financial support measures in the period 2001 to 2015. These measures offered support to energy suppliers that compensated the additional production costs of renewable energy. It was therefore possible for energy suppliers to offer green energy at a competitive and sometimes even low price. As price is by far the most important factor in households' decision-making process, this was a critical factor. The impact of the governmental support proved to be critical, as demand dropped 20% when one of the three support measures became less attractive. Energy consumers are sympathetic towards green energy but are only to a limited extent willing to pay more for it. In the absence of governmental support, the Dutch share of green energy consumption would have been lower and in line with other European countries. Obviously, the dynamics in the GO market and its success result from the economic forces of the GO-trade system and the governmental support.

Economic and political forces are also determining factors in the failure of the GO-trade system to stimulate additional green energy production.

The GO-trade system stimulates energy imports and exports by producers and traders in different EU countries. For the Netherlands, GO trade was unavoidable because of the need to accumulate enough certificates to accommodate the national demand for green energy. Countries like Norway, with an abundance of renewable energy and low domestic demand, export GOs, and this provides around €20 million in additional revenues. The red pathway that many energy suppliers in the Netherlands use is legitimized by the trade system but does not stimulate additional production.

Thus far, the accuracy of the competent bodies for issuing, transferring, and cancelling GOs has been criticized, as also the role of national supervisors.

The research in this article demonstrates some structural shortcomings in the GO-trade system that are in line with experiences with the liberalization of European energy markets as well as market-based systems like the ETS. The GO-trade system cannot contribute to additional renewable energy supply in Norway or any other country because the GO prices are too low, as was also clear in the ETS [18,19]. Low GO prices result in low revenues and do not stimulate investment in renewable energy.

Different political factors prohibit a harmonized implementation of Directive 2009/28. In the first place, national governments prefer to implement their own energy policy rather than actively participate in the market-based GO-trade system. Of the renewable energy produced, 32% is ineligible for GO trade because countries prioritize their own national support schemes [17]. National governments withdraw the right to financial support if energy suppliers submit a request for GOs. An additional 20% of renewable energy is not submitted for GOs; this seems to reflect a lack of interest in GOs. As less than 50% of energy production is covered by GOs, the market is obviously not perfect yet.

In the second place, national governments choose to stimulate renewable energy in their own style. The old reflexes of preoccupation with the energy sector exist in the GO-trade system as they did before energy market liberalization [15]. Lack of harmonization is a well-known obstacle in the liberalization process, and the GO-trade system is no exception to the rule. Neither the Dutch government, nor the German and French governments, have implemented the idea of market liberalization or active support of GO trade. Even the European Commission itself is not an exception. Directive 2009/28 explicitly does not include the outcomes of the GO system as a performance indicator for progress in renewable energy policy. Renewable energy production is used as a performance indicator—a fact that does not encourage the development of a liberalized market.

The Commission is aware that, in particular, large companies and governments play a major role in European decision-making, such as Directive 2009/28. Interestingly, societal organizations were important in the implementation of Directive 2009/28 in the Netherlands. Their role was both critical and constructive, because their involvement resulted in an improvement in the transparency of the green energy market. Because consumer organizations and other societal groups played an active part on the market, the GO prices for Dutch renewable energy rose. These groups improved the GO-trade system mechanism within the limits of the current rules. The case of the market-based GO-trade system supports the idea of an active involvement of societal groups [21,58].

The constructive criticism of societal groups regarding the GO-trade mechanisms was translated into a probably successful attempt to add a more national color. However, their severe criticism of the market-based system deserves a greater audience than their constituency and Dutch consumers. The shortcomings of the GO-trade system are in line with structural problems encountered in the energy market liberalization process and in other market-based systems such as the ETS. The Dutch government is well aware of the problems because problems concerning green certificates were already manifest in the national trade system that preceded the GO-trade system. However, national energy policy persists in an approach of passive implementation of Directive 2009/28. Hopefully, the evaluation and amendment of the renewable energy directive will give some momentum to solving deficiencies in the GO-trade system.

In this article, we have studied the political economy of the Dutch green energy market for households and small business. We have identified political and economic factors that determine the introduction of the market-based GO system. Although our research focuses on the Dutch green energy market, we have also dealt with the international aspects of the GO market. The Netherlands can only account for their green energy sales because of the red pathway in which Scandinavian water energy (GOs) is imported. We would like to recommend an international study of the market-based system in all participating countries. To use the full potential of this market-based system, it is necessary to understand the dynamics of the GO-trade system and correct its flaws. The Dutch experiences are an interesting but small part of the puzzle.

The subject of this article lies at the crossroads of different debates, theories, and conceptual models. The outcomes of our research provide several ideas for theoretical elaborations. In the first place, analysis of the introduction of the market-based GO-trade system in the Netherlands clarifies that political factors such as national interests and limited harmonized implementation of Directive 2009/28 are important prohibiting factors. An international comparative study of the European GO system could be interpreted in terms of these experiences in European decision-making. Furthermore, the critical and very constructive role of consumer representatives deserves further attention. The Commission is well aware of its limits concerning the involvement of societal groups and citizens in European decision-making. This article demonstrates the added value of an active involvement of societal groups and should be followed up by an investigation of the potential of their involvement in other decision-making processes.

Acknowledgments: The author would like to thank QA$^+$ Research and Consultancy in Woubrugge for supporting this research project.

Conflicts of Interest: The author declares no conflict of interest.

References

1. European Parliament; Council of the European Union. Directive 2009/28/EC of the European Parliament and of the Council of 23 April 2009. *Off. J. Eur. Union* **2009**, *L140*, 16–62.
2. Autoriteit Consument en Markt. *Annual Report 2015*; ACM: Den Haag, The Netherlands, 2016. (In Dutch)
3. European Environment Agency (EEA). Share of Renewable Energy in Gross Final Energy Consumption. Available online: http://www.eea.europa.eu/data-and-maps/indicators/renewable-gross-final-energy-consumption-4/assessment (accessed on 21 December 2016).
4. Consumentenbond; Greenpeace; Hivos; Natuur & Milieu; Vereniging Eigen Huis; World Wildlife Fund (WWF); WISE. Onderzoek Duurzaamheid Elektriciteitsleveranciers. 2014. Available online: http://www.consumentenbond.nl/tests/bestanden/woning-huishouden/onderzoek-duurzaamheid-elektriciteitsleveranciers2014.pdf (accessed on 21 December 2016). (In Dutch)
5. Consumentenbond; Greenpeace; Hivos; Natuur & Milieu; World Wildlife Fund (WWF); WISE. Onderzoek Duurzaamheid Nederlandse Stroom-Leveranciers. 2015. Available online: https://www.natuurenmilieu.nl/wp-content/uploads/2015/10/Duurzaamheid_Leveranciers_rapport_2015.pdf (accessed on 14 October 2016).
6. Nieuwe Rotterdamse Courant Handelsblad (NRC). Groene Stroom Is nu Vooral Een Imago Kwestie. Available online: http://vorige.nrc.nl/opinie/article1880794.exe/Groene_stroom_is__nu_vooral_imagokwestie (accessed on 14 October 2016). (In Dutch)
7. Volkskrant. Groene Stroom of Sjoemelstroom? 2015. Available online: http://www.volkskrant.nl/economie/groene-stroom-of-sjoemelstroom~a3821247/ (accessed on 21 December 2016). (In Dutch)
8. Ministerie van Economische Zaken (MEZ). Regeling Groencertificaten Elektriciteitswet 1998, Nr. WJZ 01022598. *Staatscourant*, 7 May 2011. (In Dutch)
9. Ministerie van Economische Zaken (MEZ). *Brief van de Minister van Economische Zaken, Vergaderjaar 2000–2001, Kamerstuk 25097*; No. 53; Staatsuitgeverij: The Hague, The Netherlands, 2001. (In Dutch)
10. Ministerie van Economische Zaken (MEZ); EZ. Wijziging Regeling Groencertificaten Electriciteitswet 1998, Nr. WJZ 01053730. *Staatscourant*, 26 October 2001. (In Dutch)
11. Algemene Rekenkamer. *Groene Stroom*; Algemene Rekenkamer: Den Haag, The Netherlands, 2004. (In Dutch)
12. European Parliament; Council of the European Union. Directive 2001/77/EC of the European Parliament and of the Council of 27 September 2001. On the Promotion of Electricity Produced from Renewable Energy Sources in the Internal Electricity Market. *Off. J. Eur. Union* **2001**, *L283*, 33–40.
13. European Parliament; Council of the European Union. Directive 2003/54/EC of the European Parliament and of the Council of 26 June 2003 Concerning Common Rules for the Internal Market in Electricity and Repealing Directive 96/92/EC. *Off. J. Eur. Union* **2003**, *L176*, 37–56.
14. Ministerie van Economische Zaken (MEZ). Regeling Afnemers en Monitoring Elektriciteitswet 1998 en Gaswet. *Staatscourant*, 14 July 2004. (In Dutch)

15. Pront-van Bommel, S. Het Derde Energiepakket. *Tijdschrift voor Europees en Economisch Recht* **2010**, *58*, 455–467. (In Dutch)
16. Pront-van Bommel, S. (Ed.) *De Consument en de Andere Kant van de Electriciteitsmarkt*; Universiteit van Amsterdam, Centrum voor Energievraagstukken: Amsterdam, The Netherlands, 2010. (In Dutch)
17. Jones, B.; Keen, M.; Norregaard, J.; Strand, J. Global prospects and policy issues. In *World Economic Outlook, World Economic and Financial Surveys, Globalization and Inequality*; International Monetary Fund (IMF): Washington, DC, USA, 2007.
18. Grubb, M.; Brewer, T.L.; Sato, M.; Heilmary, R.; Fazekas, D. *Climate Policy and Industrial Competitiveness: Ten Insights from Europe on the EU Emissions Trading System*; Climate Strategies, The German Marshall Fund of the United States: Washington, DC, USA, 2009.
19. Energy Centre Netherlands (ECN). *Nationale Energieverkenning 2015*; ECN: Petten, The Netherlands, 2015. (In Dutch)
20. European Commission. *Communication from the Commission Launching the Public Consultation Process on a New Energy Market Design, COM (2015) 340 Final*; European Commission: Brussels, Belgium, 2015.
21. Alemanno, A. Unpacking the Principle of Openness in EU Law: Transparency, Participation and Democracy. *Eur. Law Rev.* **2014**, *39*, 72–90.
22. Klijn, E.H.; Koppenjan, J. *Governance Networks in the Public Sector*; Routledge: Abingdon, UK, 2016.
23. Rotmans, J. *In Het oog van de Orkaan*; Aeneus: Boxtel, The Netherlands, 2012. (In Dutch)
24. Arentsen, M.J.; Fuchs, D. Green electricity in the market place revisited. In *Im Hürdenlauf zur Energiewende: Von Transformationen, Reformen und Innovationen*; Brunnengräber, A., di Nucci, M.R., Eds.; Springer: Wiesbaden, Germany, 2014; pp. 201–214.
25. Pront-van Bommel, S. *Een Redelijke Energieprijs, de Mythe van de Marktwerking*; University of Amsterdam Press: Amsterdam, The Netherlands, 2012. (In Dutch)
26. Centraal Bureau voor de Statistiek (CBS); Rijksdienst voor Ondernemend Nederland (RVO). *Protocol Monitoring Hernieuwbare Energie*; CBS: The Hague, The Netherlands, 2015. (In Dutch)
27. Planbureau voor de Leefomgeving. *Milieubalans 2012*; Planbureau voor de Leefomgeving: Den Haag, The Netherlands, 2013. (In Dutch)
28. Centraal Bureau voor de Statistiek (CBS). Gebruik van Groene Stroom Neemt Toe. Available online: https://www.cbs.nl/nl-nl/nieuws/2015/27/sterke-groei-aandeel-hernieuwbare-energie (accessed on 21 December 2016). (In Dutch)
29. Algemeen Nederlands Persbureau (ANP). Populariteit groene stroom gedaald. *Boerderij*, 13 December 2006.
30. Energeia. Hoeveelheid Groene Stroom Klanten Stijgt Opnieuw. 2008. Available online: http://energeia.nl/nieuws/2008/11/19/hoeveelheid-groene-stroom-klanten-stijgt-opnieuw-met-12-in-de-lift (accessed on 14 October 2016). (In Dutch)
31. Autoriteit Consument en Markt (ACM). *Trendrapportage Marktwerking en Consumentenvertrouwen in de Energiemarkt (Tweede Halfjaar 2013)*; ACM: Den Haag, The Netherlands, 2014. (In Dutch)
32. Marketresponse Stand van Zaken op de Energiemarkt. Available via Autoriteit Consument en Markt. 2014. Available online: https://www.acm.nl/nl/download/publicatie/?id=12909 (accessed on 14 October 2016). (In Dutch)
33. Autoriteit Consument en Markt (ACM). *Trendrapportage Marktwerking en Consumentenvertrouwen in de Energiemarkt (Tweede Halfjaar 2012)*; ACM: Den Haag, The Netherlands, 2013. (In Dutch)
34. CertiQ. *Annual Report 2008*; CertiQ: Arnhem, The Netherlands, 2009.
35. Nederlandse Markt Autoriteit. *Een Markt Zonder Spanning*; Marktmonitor, Ontwikkeling van de Nederlandse Kleinverbruikersmarkt voor Elektriciteit; Directie Toezicht Energie: Den Haag, The Netherlands, 2006. (In Dutch)
36. Nederlandse Mededigings Autoriteit. *Monitor Kleinverbruikersmarkten Gas en Elektriciteit 2009*; Energiekamer, Nederlandse Mededingings Autoriteit: Den Haag, The Netherlands, 2009. (In Dutch)
37. Afman, M.R.; Bles, M.; Schepers, B.L.; Wielders, L.M.L. *Electriciteitsleveranciers in Kaart, Update 2015*; Centrum voor Energiebesparing (CE) Delft: Delft, The Netherlands, 2015. (In Dutch)
38. CertiQ. *Annual Report 2015*; CertiQ: Arnhem, The Netherlands, 2016.
39. Association of Issuing Bodies. Annual Report 2015. 2016. Available online: http://www.aib-net.org/portal/page/portal/AIB_HOME/NEWSEVENTS/Annual_reports (accessed on 14 October 2016).

40. Hier. Een op de Acht Nederlanders Denkt dat Groene Stroom Niet Bestaat. 2016. Available online: https://hier.nu/klimaatbureau/nieuws/persbericht-een-op-de-acht-nederlanders-denkt-dat-groene-stroom-niet-bestaat (accessed on 14 October 2016). (In Dutch)

41. Ministerie van Economische Zaken (MEZ). *Motivaction, Energievoorziening 2015–2050: Publieksonderzoek Naar Draagvlak voor Verduurzaming van Energie*; Rijksoverheid: Den Haag, The Netherlands, 2016. (In Dutch)

42. European Commission. *Rules Concerning Communication from the Commission on Prospects for the Internal Gas and Electricity Market COM(2006) 841 Final*; European Commission: Brussels, Belgium, 2007.

43. Reliable Disclosure Systems for Europe. Country Reports: Norway, Iceland, Sweden, Finland. 2015. Available online: http://www.reliable-disclosure.org/documents/ (accessed on 14 October 2016).

44. Association of Issuing Bodies. Newsletter 25. 2016. Available online: http://www.aib-net.org/portal/page/portal/AIB_HOME/NEWSEVENTS/AIB_Newsletter%2025.pdf (accessed on 14 October 2016).

45. Stichting Onderzoek Multinationale Ondernemingen. *Kortsluiting op de Groene Electriciteitsmarkt*; SOMO (Stichting Onderzoek Multinationale Ondernemingen): Amsterdam, The Netherlands, 2016. (In Dutch)

46. Reliable Disclosure Systems for Europe. Country Report: Iceland. 2015. Available online: http://www.reliable-disclosure.org/upload/180-RE-DISSII_Country_Profile_IS_2015-06-26_Final.pdf (accessed on 14 October 2016).

47. Statistics of the International Energy Agency. Available online: https://www.iea.org/statistics/statisticssearch/ (accessed on 19 November 2016).

48. Norwegian Ministry of Petroleum and Energy. *Facts, Energy and Water Resources in Norway*; Norwegian Ministry of Petroleum and Energy: Oslo, Norway, 2015.

49. Afman, M.; Wielders, L. *Factsheet: Ontwikkeling Prijzen Garanties van Oorsprong*; Centrum voor Energiebesparing (CE) Delft: Delft, The Netherlands, 2016. (In Dutch)

50. Statnet. Key Figures 1974–2012, Market and Operations. Available online: http://www.statnett.no/en/Market-and-operations/Data-from-the-power-system/Key-figures-1974-20121/ (accessed on 14 October 2016).

51. Hartlief, I.; Kiezebrink, V. *Kortsluiting op de Groene Energiemarkt*; SOMO (Stichting Onderzoek Multinationale Ondernemingen): Amsterdam, The Netherlands, 2016. (In Dutch)

52. Schmid, M.; Sledsens, T. *Rijksoverheid & Sjoemelstroom*; World Information Service Energy: Amsterdam, The Netherlands, 2016. (In Dutch)

53. Renewable Energy Certificates (RECS). Annual Report 2015. 2016. Available online: http://www.recs.org/documents/annual-report-2015 (accessed on 14 October 2016).

54. Centre for Research on Multinational Corporations. Sustainability in the Dutch Power Sector. Fact Sheets Series. 2009 Update. Available online: http://www.greenpeace.nl/Global/nederland/report/2010/5/sustainability-in-the-dutch-po-2.pdf (accessed on 14 October 2016).

55. Centre for Research on Multinational Corporations. Sustainability in the Dutch Electricity Sector. 2012. Available online: https://www.somo.nl/nl/duurzaamheid-in-de-nederlandse-elektriciteitssector/ (accessed on 14 October 2016).

56. Afman, M.R.; Bles, M.; Schepers, B.L.; Wielders, L.M.L. *Electriciteitsleveranciers in Kaart, Updated 2014*; Centrum voor Energiebesparing (CE) Delft: Delft, The Netherlands, 2014. (In Dutch)

57. Wielders, L.; Cherif, S.; Schepers, B.L. *Groene Stroom per Product voor de Consumentenmarkt*; Centrum voor Energiebesparing, Centrum voor Energiebesparing (CE) Delft: Delft, The Netherlands, 2014. (In Dutch)

58. Lavrijssen, S. *Waarborgen voor de Energieconsument in de Energietransitie*; Tilburg University: Tilburg, The Netherlands, 2016. (In Dutch)

11

Environmental Performance of Miscanthus, Switchgrass and Maize: Can C4 Perennials Increase the Sustainability of Biogas Production?

Andreas Kiesel *, Moritz Wagner and Iris Lewandowski

Department Biobased Products and Energy Crops, Institute of Crop Science, University of Hohenheim, Fruwirthstrasse 23, 70599 Stuttgart, Germany; moritz.wagner@uni-hohenheim.de (M.W.); Iris_Lewandowski@uni-hohenheim.de (I.L.)
* Correspondence: a.kiesel@uni-hohenheim.de

Academic Editor: Michael Wachendorf

Abstract: Biogas is considered a promising option for complementing the fluctuating energy supply from other renewable sources. Maize is currently the dominant biogas crop, but its environmental performance is questionable. Through its replacement with high-yielding and nutrient-efficient perennial C4 grasses, the environmental impact of biogas could be considerably improved. The objective of this paper is to assess and compare the environmental performance of the biogas production and utilization of perennial miscanthus and switchgrass and annual maize. An LCA was performed using data from field trials, assessing the impact in the five categories: climate change (CC), fossil fuel depletion (FFD), terrestrial acidification (TA), freshwater eutrophication (FE) and marine eutrophication (ME). A system expansion approach was adopted to include a fossil reference. All three crops showed significantly lower CC and FFD potentials than the fossil reference, but higher TA and FE potentials, with nitrogen fertilizer production and fertilizer-induced emissions identified as hot spots. Miscanthus performed best and changing the input substrate from maize to miscanthus led to average reductions of -66% CC; -74% FFD; -63% FE; -60% ME and -21% TA. These results show that perennial C4 grasses and miscanthus in particular have the potential to improve the sustainability of the biogas sector.

Keywords: anaerobic digestion; *Miscanthus x giganteus*; *Panicum virgatum*; *Zea mays*; LCA; GWP; carbon mitigation; fossil fuel depletion; acidification; eutrophication

---

1. Introduction

Biogas is a renewable energy carrier produced by anaerobic digestion of biomass. Various kinds of biomass can be utilized for biogas production, such as sewage sludge, agricultural residues (e.g., manure), biogenic waste and energy crops [1]. Power production based on biogas is more reliable than other renewable energy sources, e.g., wind and solar, and can be used to cover power demand peaks or fluctuations in production due to unfavorable weather conditions. Biogas can be utilized directly in combined heat and power units (CHP) or can be upgraded to biomethane and transported to large gas power stations via the gas grid.

The Renewable Energy Act (EEG) and its amendments have led to a rapid increase in biogas exploration in Germany [1]. Here, approximately 8075 biogas plants with a total installed capacity of 4.1 GW were in operation in 2016 [2]. The latest amendments promote the restructuring of biogas plants to flexible operation, and approximately 31% of the installed capacity [2] have already been modernized. This allows power production to be adapted more to demand. Currently, 182 biogas plants upgrade biogas to biomethane and inject it into the gas grid [2]. These numbers show that, in Germany, there is a significant biogas infrastructure in place and the process of adapting it to the needs of a future renewable power supply has already begun. However, to allow an economically and environmentally viable operation, this infrastructure needs a reliable, affordable and sustainable supply of biomass. In 2014, substrate input (based on mass) was composed of 52% energy crops (of which 73% was maize) and 43% manure [2]. However, the proportion of biogas produced from energy crops is considerably higher than their proportion by mass, because they have a higher specific biogas and methane yield than other biogas substrates, e.g., manure. In Germany, about 1.4 million ha energy crops are grown for biogas production, of which 0.9 million ha are biogas maize [2]. This reveals the great importance of energy crops—and in particular energy maize—in Germany. The high economic viability of maize [3] for biogas production is given by its high methane yield, easy digestibility, and well-established, optimized crop production and harvest logistics, including storage as silage.

However, the strong reliance of the biogas sector on maize as substrate crop can lead to environmental problems and a low acceptance in public opinion. The environmental profile of maize cultivation is characterized by a high nitrogen fertilizer input, high risk of erosion and leaching, and negative impact on biodiversity [4–6]. In particular, the regional concentration in areas with high biogas plant densities can lead to environmental problems, such as surface and groundwater pollution through erosion and leaching, and losses in biodiversity and soil organic matter due to the high proportion of maize in crop rotations [7]. Other aspects are also criticized, such as the high concentration of maize in the landscape and the use of good agricultural land for growing energy instead of food crops. For these reasons, the sustainability of the biogas sector is often questioned not only by environmentalists but also by the general public.

The replacement of maize (*Zea mays*) by crops with a more benign environmental profile is seen as one route towards more sustainable biogas production. These crops, however, should have an equally high yield and biomass supply potential as maize. The high-yielding and nutrient-efficient perennial C4 grasses miscanthus and switchgrass are considered promising options.

The miscanthus genotype, *Miscanthus x giganteus*, was introduced into Europe in 1935 and is today still the only commercial genotype available on the market [8]. However, promising breeding efforts have begun in recent years and latest results show the suitability of novel genotypes for marginal lands and the potential contribution of miscanthus to greenhouse gas (GHG) mitigation [9]. Progress in upscaling miscanthus cultivation and crop production has also raised interest in the industrial sector [10]. Miscanthus' beneficial environmental profile is mainly due to its perennial nature and because soil organic carbon tends to increase when arable land is converted to its cultivation [11]. It is a very resource- and land-use efficient crop with efficient nutrient-recycling mechanisms and high net energy yields per unit area [12,13]. For this reason, the global warming potential (GWP) and the resource depletion potential of miscanthus cultivation is low [14,15]. Miscanthus is suitable for biogas production and has a high methane yield potential per unit area [16–18]. For anaerobic digestion, the biomass is harvested before winter, which increases the yield and digestibility [18]. Whittaker et al. [19] proved storage of green miscanthus via ensilaging to be feasible with losses in a similar range as for maize. These losses were significantly reduced by the addition of silage additives [19]. Compared to the conventional harvest of dry biomass in early spring, a green harvest in late autumn prevents leaf fall over winter, which leads to a higher nutrient removal than at spring harvest [13,18]. However, the recycling of fermentation residues is assumed to at least partially compensate for this removal and contribute to the formation of soil organic carbon. Nevertheless, the effects of a green cut on the development of soil carbon and fertility needs to be further investigated and is for this reason not considered in this study.

The crop production and environmental profile of switchgrass (*Panicum virgatum*) is comparable to that of miscanthus, except establishment via seeds and not rhizomes. Switchgrass is native to the US and Canada, where it has been developed as a promising energy grass [20]. It is also suitable for biogas production as harvest of green biomass and even double-cutting is possible [21]. Although yields are generally lower than with *Miscanthus x giganteus* [22], switchgrass can perform equally well under abiotic stress, such as cold and drought [23]. Its major advantage over the miscanthus genotypes presently available (mainly propagated clonally via rhizomes) is its low-cost establishment via seeds. Currently, switchgrass is not commercially cultivated in Germany and miscanthus is grown on an estimated area of 4000 hectares, mainly for combustion purposes [9]. Extending the utilization to anaerobic digestion could contribute to the sustainability and crop diversity (important for biodiversity) of the biogas sector.

The objective of this paper is to assess and compare the environmental performance in biogas production of the perennial C4 grasses miscanthus and switchgrass and the annual C4 crop maize. This was done in a Life Cycle Assessment (LCA) according to ISO standards 14040 and 14044 [24,25], using data from a field trial and laboratory measurements. Wagner and Lewandowski [26] showed that, when analyzing the environmental performance of biobased value chains, it is crucial to consider more impact categories than just global warming potential (GWP). Therefore, the following impact categories were assessed to estimate the environmental performance of the crops and their subsequent utilization: climate change (CC)—which corresponds to the GWP, freshwater eutrophication (FE), marine eutrophication (ME), terrestrial acidification (TA) and fossil fuel depletion (FFD). The impact categories FE, ME and TA were chosen as eutrophication and acidification have been identified as important impact categories for agricultural systems. The category marine eutrophication represents the impact of nitrogen on biomass growth in aquatic ecosystems. Freshwater eutrophication represents the same impact, but caused by phosphorus [27,28].

The data for the LCA were collected from a randomized split-block field trial, where miscanthus, switchgrass and maize were grown under *ceteris paribus* conditions. The field trial was started in 2002 and allows a comparison of annual and perennial crops. Samples and yield measurements for this LCA were taken in 2012 and 2013 and laboratory analyses were performed to estimate biogas and methane yield and biomass quality.

2. Material and Methods

2.1. Scope and Boundaries

The scope of the present study is an assessment of the environmental performance of the cultivation of three dedicated energy crops ((i) miscanthus (*Miscanthus x giganteus*); (ii) switchgrass (*Panicum virgatum* L.) "Kanlow"; and (iii) silage maize (*Zea mays*) "Mikado") and their subsequent fermentation in a biogas plant. The biogas produced is utilized in a CHP unit (Combined Heat and Power) to produce electricity and heat. The cultivation as well as the utilization of the biomass takes place in Germany. One kilowatt hour of electricity ($kWh_{el.}$) was chosen as the functional unit (FU). The environmental impacts of these biobased value chains were compared with the German electricity mix as a fossil reference. In order to do this, a system expansion approach was applied which enables the inclusion of fossil reference system hot spots.

The systems are described in Figure 1. On the right side the maize cultivation is shown, on the left side the cultivation of the perennial crops miscanthus and switchgrass. The system boundaries include the production of the mineral fertilizers and the herbicides used, the production of the propagation material (miscanthus rhizomes as well as switchgrass and maize seeds), and the agricultural management (soil preparation, planting, mulching, fertilizing, spraying of herbicides, harvesting, recultivation resp. stubble cultivation) over the whole cultivation period which is for maize 1 year, for switchgrass 15 years and for miscanthus 20 years. Miscanthus and switchgrass are mulched in the first year and harvested from the second year onwards. All crops are harvested with a self-propelled forage harvester. The biomass is then transported to the biomass plant where it is fermented to biogas which is combusted in a CHP unit to produce electricity and heat. The fermentation residues are rich in nutrients and are used as fertilizer.

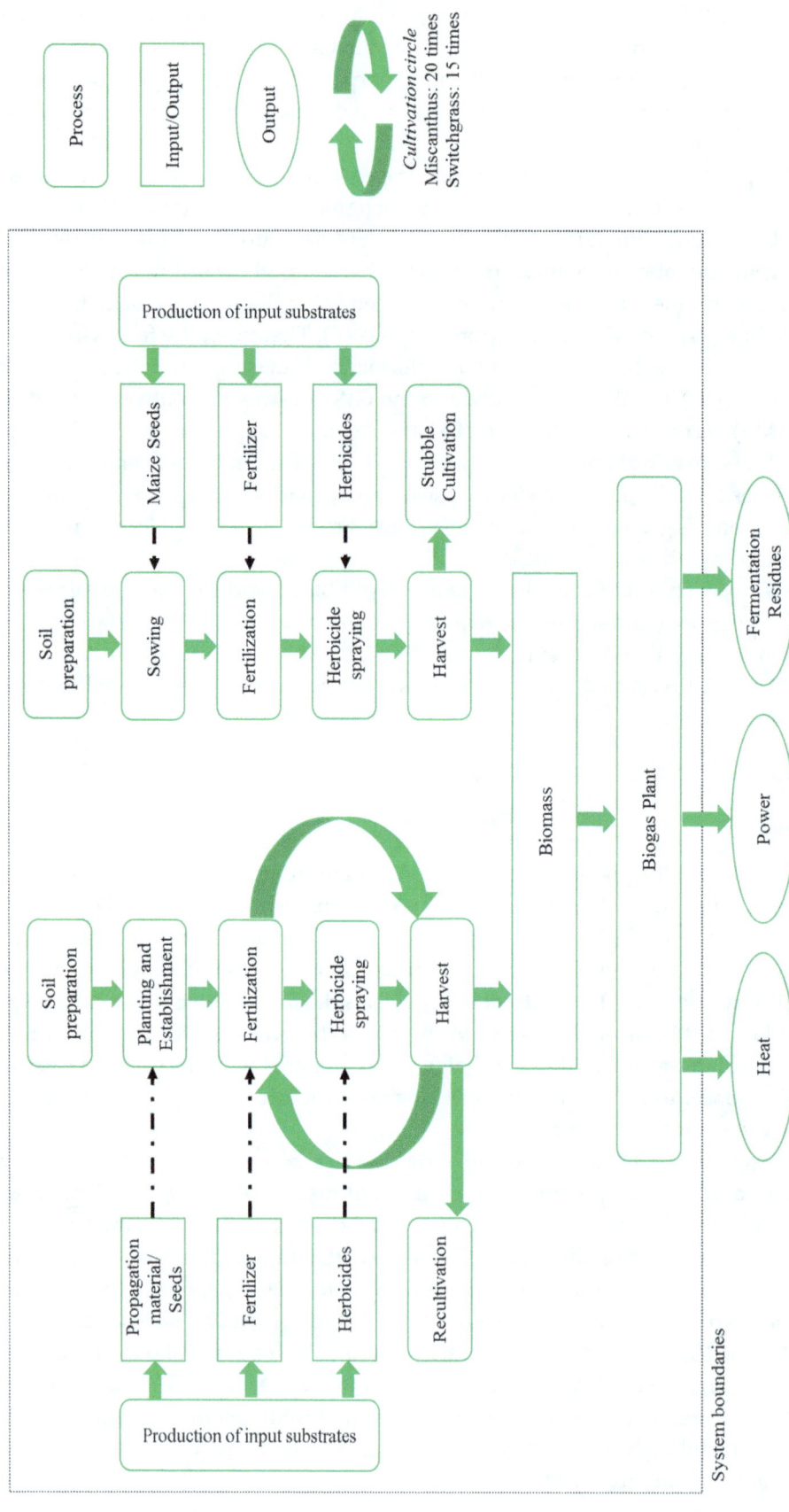

Figure 1. System description and boundaries for miscanthus, switchgrass (**left**) and maize (**right**) biomass cultivation, the fermentation to biogas and the subsequent utilization in a CHP unit.

2.2. Life Cycle Inventory

The data for the cultivation process used in this LCA were obtained from a multiannual field trial at Ihinger Hof. The Ihinger Hof is a research station of the University of Hohenheim and is located in southwest Germany (48.75°N and 8.92°E). The soil belongs to the soil class Haplic Luvisol. The long-term average annual air temperature and precipitation at the research station are 8.3 °C and 689 mm, respectively. The experimental design of the trial is described in Boehmel et al. [29].

Data on cultivation practices such as fertilizer and herbicide inputs were available for an 11-year period from 2002 to 2013. Miscanthus and switchgrass were established in spring 2002 by rhizome planting and sowing, respectively. Maize was sown on 27 April 2012 and 21 May 2013 at a density of 9.5 seeds m^{-2}. Nitrogen was applied as calcium ammonium nitrate (CAN), K_2O as potassium chloride and P_2O_5 as triplesuperphosphat (TSP). The use of herbicides during the miscanthus and switchgrass cultivation is described in Iqbal et al. [30]. For maize cultivation chemical weeding was performed using two conventional herbicides mixtures following good agricultural practice. The first application was a mixture of three herbicides (2.0 $L·ha^{-1}$ Stomp Aqua, BASF SE, active ingredient 455 $g·L^{-1}$ Pendimethalin; 1.0 $L·ha^{-1}$ Spektrum, BASF SE, active ingredient 720 g L^{-1} Dimethenamid-P; and 1.0 $L·ha^{-1}$ MaisTer power, Bayer, active ingredient 31.5 $g·L^{-1}$ Foramsulfuron + 1.0 $g·L^{-1}$ Iodosulfuron + 10.0 $g·L^{-1}$ Thiencarbazone + 15.0 $g·L^{-1}$ Cyprosulfamide). The second application was a mixture of two herbicides (1.7 $L·ha^{-1}$ Laudis, Bayer, active ingredient 44 $g·L^{-1}$ Tembotrione + 22 $g·L^{-1}$ Isoxadifen-ethyl; and 0.35 $L·ha^{-1}$ Buctril, Bayer, active ingredient 225 $g·L^{-1}$ Bromoxynil).

The principle data for the cultivation of miscanthus, switchgrass and silage maize used in this analysis are summarized in Table 1. The data are shown for the years 2012 and 2013. In the year 2013 the weather conditions were not ideal for silage maize cultivation in Germany which is an important reason for the significantly lower yield of silage maize in the year 2013 compared to 2012. After a serious frost period in February 2012, the weather conditions in 2012 where quite usual, spring was rather dry, but followed by plenty of rain in June (Figure 2). Weather conditions in 2013 were completely contrary and very challenging for agriculture. The spring and especially May was unusually cool and wet. Due to this challenging weather conditions, maize sowing was delayed to late May. In July, the temperatures were unusually high and the crops faced a serious drought followed by few days of rain from 24 to 29 July. In this period, 168.5 mm of rainfall occurred in 4 major events, which represents 97% of the rain of the complete month.

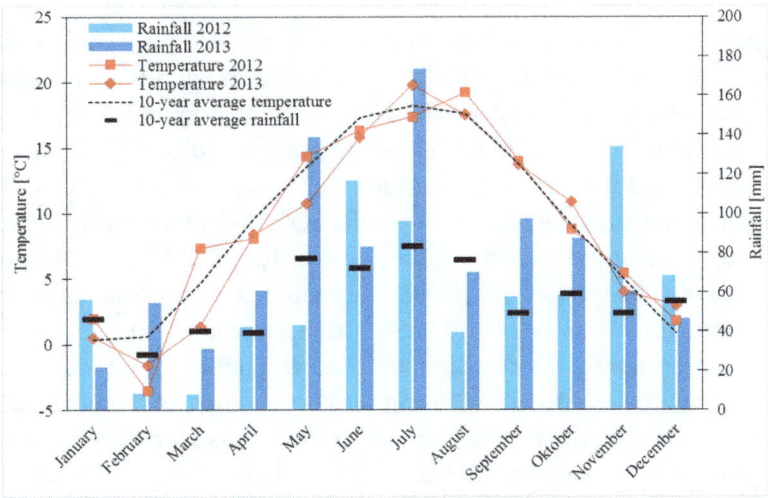

Figure 2. Temperature and rainfall in 2012 and 2013 at the field site on the research station "Ihinger Hof". For comparison the 10-year average temperature and rainfall from 2003 to 2012 is shown.

Maize was harvested at milk-ripe stage (end of September in 2012; late October in 2013) and miscanthus and switchgrass in late October in both years. The years 2012 and 2013 were selected to compare the environmental performance of perennial crops as an alternative to maize under different conditions for silage maize cultivation. The yield of maize, miscanthus and switchgrass is shown for the favorable year 2012 and non-favorable year 2013 in Table 1. However, the yield of the two perennial crops is the average yield over the whole cultivation period (20 years for miscanthus, and 15 for switchgrass) including the establishment phase based on the measured yield of the respective year. In the first year, miscanthus and switchgrass are mulched and not harvested. Full yields are only reached from the third year on. This calculation is exemplarily shown for the yield in 2012 for miscanthus in Equation (1) and for switchgrass in Equation (2) and was performed in the same way for the lower yields in 2013. The variable *yield_year2* describes the yield in the second cultivation year, which, for both crops, is slightly lower than the mean yield achieved in the following years.

$$Mean\ yield\ miscanthus\ [t\ DM\ ha^{-1}{\cdot}yr^{-1}] = \frac{yield\_year2\ +\ yield\_year\_2012 \times 18}{20} \tag{1}$$

$$Mean\ yield\ switchgrass\ [t\ DM\ ha^{-1}{\cdot}yr^{-1}] = \frac{yield\_year2\ +\ yield\_year\_2012 \times 13}{15} \tag{2}$$

The methane yield was measured as described in Kiesel and Lewandowski [18]. A biogas batch test was performed for 35 days at mesophilic conditions (39 °C) according to VDI guideline 4630. The approach of the biogas batch test was certified by the KTBL and VDLUFA interlaboratory comparison test 2014 and 2015. Each sample was assessed in four technical replicates.

Table 1. Summary of the in- and outputs of the three energy crops.

Input/Output	Unit	Maize		Switchgrass		Miscanthus	
		2012	2013	2012	2013	2012	2013
N	$Kg{\cdot}yr^{-1}{\cdot}ha^{-1}$	240	240	80	80	80	80
K_2O	$Kg{\cdot}yr^{-1}{\cdot}ha^{-1}$	304	204	137	137	128	128
P_2O_5	$Kg{\cdot}yr^{-1}{\cdot}ha^{-1}$	100	100	37	37	32	32
Herbicides	$Kg{\cdot}yr^{-1}{\cdot}ha^{-1}$	6.05	6.05	1.32	1.32	1.375	1.375
Dry matter yield	$Kg{\cdot}yr^{-1}{\cdot}ha^{-1}$	18915	12616	14227	8369	22760	18929
Dry matter content	%	25.4	21.1	38.9	36.2	43.4	41.2
Methane yield	$m^3\ CH_4\ yr^{-1}{\cdot}ha^{-1}$	5594	3635	3328	2095	5006	4542
Agricultural land required for biogas plant	$ha{\cdot}yr^{-1}$	173	266	291	461	194	213

The background data for the environmental impacts associated with the production of the input substrates (seeds, propagation material, herbicides and fertilizers) and the cultivation processes were taken from the GaBi database [31]. Direct N_2O and NO emissions from the mineral fertilizers used were calculated according to Bouwman et al. [32]. The estimations of indirect N_2O emissions from mineral fertilizers and N_2O emissions from harvest residues were done in accordance to IPCC [33]. Nitrate leaching to groundwater was calculated according to the SQCB—NO_3 model [34]. Ammonia emissions were calculated using emission factors from the Joint EMEP/CORINAIR Atmospheric Emission Inventory Guidebook [35]. Phosphate emissions were estimated according to van der Werf et al. [36].

In this study a transport distance of 100 km by truck for the input material such as herbicides or fertilizer and of 5 km by tractor for the biomass from the field to the biogas plant was assumed. This assumption is align with literature [37–39] and was done, since no data for the transport distance of the input substrates to the farmer and the biomass to the biogas plant were available. The emission stage for the truck used was assumed to be EUR5. The data for the transportation processes of the input material and the biomass were taken from the GaBi database [31].

After the harvest, the biomass of the different crops is ensiled. During the ensilage process dry matter losses of 12% were assumed [40]. The silage is subsequently fermented in a biogas plant. The methane hectare yield of the different crops is shown in Table 1. In the biogas plant methane losses of 1% were assumed [41]. The biogas is then combusted in a CHP with an electrical capacity of 500 kW to produce heat and power. The technical characteristics of the CHP used in this analysis are shown in Table 2. The inherent power consumption for miscanthus and switchgrass was assumed to be 12% and thus significantly higher than for maize. This is due to the more energy intensive pre-treatment of lignocellulosic biomass before the fermentation process. The emissions associated with the combustions of the biogas were taken from the ecoinvent database [42]. The electricity generated is fed into the grid. Twenty percent of the heat produced is used internally for the heating of the fermenter. In practice the remaining heat is partially used for heating nearby buildings thereby substituting heat produced by fossil sources. In this study, it was assumed that of the remaining heat 50% is used for this purpose.

Table 2. CHP unit—technical characteristics.

Technical Characteristics		Unit
Full load hours	7800	h
Plant output electrical	500	$kWh_{el.}$
Plant output total	1219	kWh
Electrical efficiency	41	% of plant total output
Thermal efficiency	41	% of plant total output
Inherent heat demand	20	% of total heat production
Inherent power consumption—perennial crops	12	% of total power production
Inherent power consumption—silage maize	6.6	% of total power production

The residues of the fermentation process are rich in nutrients. Table 3 shows the plant available nutrients, which can be recycled through the use of fermentation residues as fertilizers (related to the generation of the functional unit of 1 $kWh_{el.}$). The nutrient content is the average of the measured values of year 2012 and 2013. The phosphorus and the potassium content of the biomass fermented remains fully in the fermentation residues. Only 70% of the nitrogen compounds in the fermentation residues are available for the plants. That is why the nitrogen content can therefore not be taken fully into account. The nitrogen (N) content was analyzed according to the DUMAS principle (method EN ISO 16634/1 and VDLUFA Method Book III, method 4.1.2) using a Vario Macro Cube (Elementar Analysensysteme GmbH, Hanau, Germany) element analyzer. The phosphor (P) and potassium (K) contents were analyzed according to DIN EN ISO 15510 and VDLUFA Method Book III, method 10.8.2 [43] using ICP-OES and a ETHOS.lab microwave (MLS GmbH, Leutkirch, Germany).

Table 3. Nutrients in the biomass of the analyzed energy crops and plant available nutrients which can be recycled through the use of fermentation residues per FU.

Nutrient	Miscanthus		Switchgrass		Maize	
	in % of Biomass (d.b.)	in kg/FU	in % of Biomass (d.b.)	in kg/FU	in % of Biomass (d.b.)	in kg/FU
N	0.47	0.0036	0.50	0.0035	1.29	0.0058
P	0.09	0.0010	0.10	0.0010	0.18	0.0011
K	1.11	0.0119	1.03	0.0105	1.29	0.0083

2.3. Choice of Impact Categories

In this LCA study the life cycle impact assessment method ReCiPe was used [44]. The following impact categories were considered: climate change (CC), which corresponds to global warming potential (GWP); terrestrial acidification (TA); freshwater eutrophication (FE); marine eutrophication (ME); and fossil fuel depletion (FFD). Characterization factors were taken from Goedkoop et al. [44]. These impact categories were chosen according to their relevance for perennial biomass production, which was analyzed in the study by Wagner and Lewandowski [26].

3. Results

For each impact category analyzed, data are shown for the two climatically different production years 2012 and 2013 (2012 favorable and 2013 non-favorable for silage maize cultivation) and for two scenarios, one with and one without heat utilization. These are presented both in figures and in tables, depicting the results with (figures) and without (tables) a system expansion approach. The results are presented per functional unit (FU), which is kWh electricity. In the supplementary material (S1–S5), the same results are presented per kg dry biomass.

The value in each impact category shows the net impact or benefit of the substitution of the fossil reference through a biobased alternative. In this study, the German electricity mix was substituted by power generated through the fermentation of dedicated energy crops and the subsequent combustion of the biogas in a CHP unit. A negative value in this case is thus a net benefit while a positive value is a negative impact on the environment.

In contrast, the table shows the environmental impact of the generation of 1 $kWh_{el.}$ in each impact category without this substitution, separated into the main emission sources. In this context, the *recycling of nutrients* represents the emission savings associated with the reduction in fertilizer in other crops through the use of the fermentation residues. The *agricultural management* summarizes all operation steps from soil preparation, planting, mulching, fertilizing, and spraying of herbicides to recultivation. The *fertilizer-induced emissions* are emissions associated with the use of fertilizers, such as N_2O emissions, which occur after the application of nitrogen fertilizer. *Credits heat utilization* are credits given for the substitution of heat produced via a fossil reference (in the present study natural gas) by heat generated via the combustion of biogas in the CHP unit. In the heat utilization scenario, 20% of the heat produced is used internally in the biogas plant. Of the remaining 80%, one half (40% of total heat produced) is used to heat nearby buildings, thus substituting heat from conventional sources.

3.1. Climate Change and Fossil Fuel Depletion

The production and use of the analyzed C4 crops, both perennial and annual, leads to a net GHG emission reduction up to 0.66 $kg \cdot CO_2$-eqv. $(kWh_{el.})^{-1}$ through the substitution of a fossil reference (Figure 3). Furthermore, all scenarios show a net decrease of the fossil fuel depletion of up to 0.18 $kg \cdot oil$-eqv. $(kWh_{el.})^{-1}$ (Figure 4). As expected, the scenarios with heat utilization lead to both higher GHG emission and fossil fuel saving (Figures 3 and 4). On average, miscanthus shows the highest GHG emission and fossil fuels saving potentials. Both perennial grasses perform better than maize (Figures 3 and 4). The advantage of miscanthus over switchgrass is larger than the advantage of switchgrass over maize.

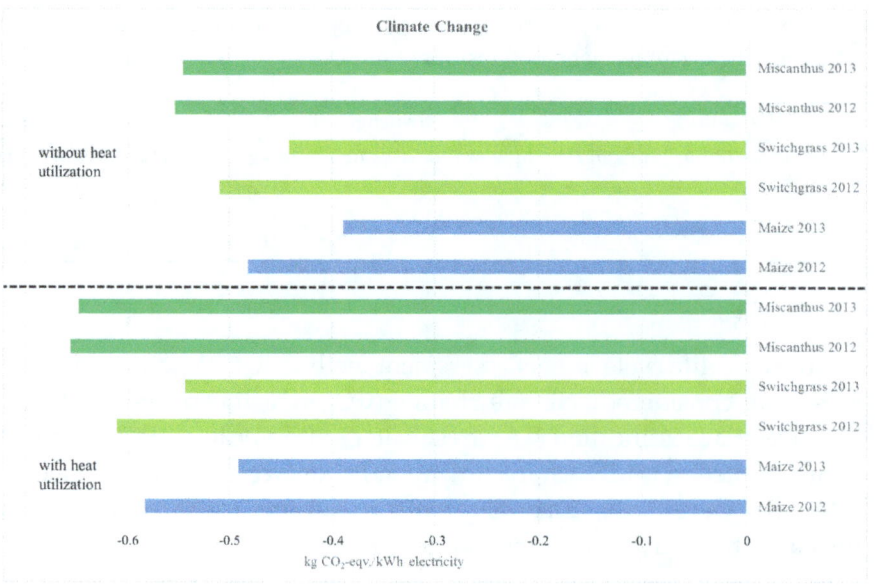

Figure 3. Assessment of the net benefits in $kg \cdot CO_2$-eqv. of substituting 1 $kWh_{el.}$ of the German electricity mix by power generated via combustion of the biogas in a CHP.

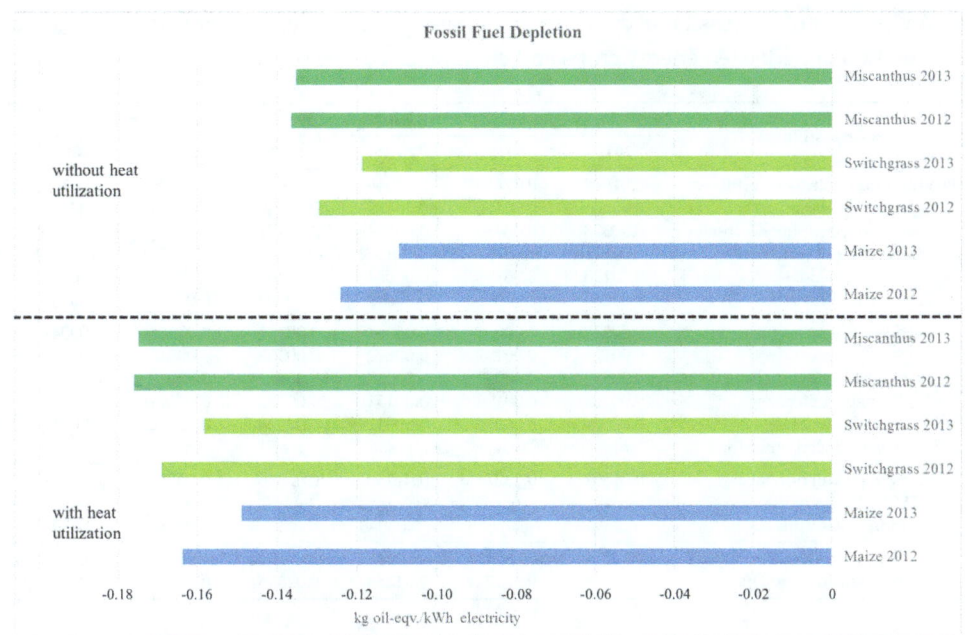

Figure 4. Assessment of the net benefits in kg·oil-eqv. of substituting 1 kWh$_{el}$. of the German electricity mix by power generated via combustion of the biogas in a CHP.

Table 4 shows the contribution of different processes to the GHG emissions and Table 5 the use of fossil fuels in these processes. The production of nitrogen fertilizer is responsible for the largest impact in both impact categories and for all crops. This is also the reason for the high credit—in terms of fossil energy savings—given for the recycling of nutrients from the fermentation residues (Table 5). Other processes with high impacts on GHG emissions and fossil energy consumption are harvest operation and biomass transport to the biogas plant (Tables 4 and 5).

Table 4. Assessment of the climate change in kg·CO$_2$-eqv. of 1 kWh$_{el}$. generated via the production and fermentation of dedicated energy crops and combustion of the biogas in CHP.

	Processes/Flows	Maize per FU		Switchgrass per FU		Miscanthus per FU		Unit
		2012	2013	2012	2013	2012	2013	
Input substrates	Production of nitrogen fertilizer	0.077	0.1185	0.0504	0.0800	0.0335	0.0369	kg·CO$_2$-eqv.
	Production of potassium fertilizer	0.0048	0.0075	0.0043	0.0068	0.0027	0.0029	kg·CO$_2$-eqv.
	Production of phosphate fertilizer	0.0064	0.0099	0.0047	0.0074	0.0027	0.0030	kg·CO$_2$-eqv.
	Recycling of nutrients	−0.0415	−0.0415	−0.0279	−0.0279	−0.0288	−0.0288	kg·CO$_2$-eqv.
	Herbicides	0.0028	0.0044	0.0012	0.0019	0.0008	0.0009	kg·CO$_2$-eqv.
	Seeds/Rhizomes	0.0002	0.0003	0.0001	0.0002	0.0003	0.0003	kg·CO$_2$-eqv.
Agricultural operations	Agricultural management	0.0075	0.0115	0.002	0.0032	0.0012	0.0013	kg·CO$_2$-eqv.
	Harvest	0.0038	0.0058	0.007	0.0111	0.0045	0.0049	kg·CO$_2$-eqv.
	Transport input substrates	0.0012	0.0018	0.0008	0.0013	0.0006	0.0006	kg·CO$_2$-eqv.
	Transport biomass	0.0049	0.0061	0.0047	0.0047	0.0045	0.0044	kg·CO$_2$-eqv.
	Ensilage	0.0003	0.0004	0.0005	0.0009	0.0004	0.0004	kg·CO$_2$-eqv.
	Fertilizer-induced emissions	0.0549	0.0906	0.0472	0.0725	0.0281	0.0311	kg·CO$_2$-eqv.
CHP	Biomass production system	0.1223	0.2154	0.0950	0.1622	0.0504	0.0580	kg·CO$_2$-eqv.
	CHP—Direct emissions	0	0	0	0	0	0	kg·CO$_2$-eqv.
	Credits heat utilization	−0.1021	−0.1021	−0.1021	−0.1021	−0.1021	−0.1021	kg·CO$_2$-eqv.
Total	**Total with credits**	**0.0202**	**0.1132**	**−0.0071**	**0.0600**	**−0.0518**	**−0.0441**	**kg·CO$_2$-eqv.**
	Total without credits	**0.1223**	**0.2154**	**0.0950**	**0.1622**	**0.0504**	**0.0580**	**kg·CO$_2$-eqv.**

Table 5. Assessment of the fossil fuel depletion in kg·oil-eqv. of 1 kWh$_{el}$. generated via the production and fermentation of dedicated energy crops and combustion of the biogas in CHP.

	Processes/Flows	Maize per FU		Switchgrass per FU		Miscanthus per FU		Unit
		2012	2013	2012	2013	2012	2013	
Input substrates	Production of nitrogen fertilizer	0.01598	0.02460	0.01046	0.01661	0.00695	0.00766	kg·oil-eqv.
	Production of potassium fertilizer	0.00206	0.00317	0.00182	0.00289	0.00113	0.00125	kg·oil-eqv.
	Production of phosphate fertilizer	0.00323	0.00497	0.00234	0.00372	0.00135	0.00148	kg·oil-eqv.
	Recycling of nutrients	−0.01020	−0.01020	−0.00742	−0.00742	−0.00774	−0.00774	kg·oil-eqv.
	Herbicides	0.00128	0.00196	0.00054	0.00087	0.00038	0.00042	kg·oil-eqv.
	Seeds/Rhizomes	0.00004	0.00005	0.00002	0.00003	0.00007	0.00008	kg·oil-eqv.
Agricultural operations	Agricultural management	0.00238	0.00367	0.00064	0.00101	0.00038	0.00042	kg·oil-eqv.
	Harvest	0.00121	0.00187	0.00222	0.00353	0.00143	0.00157	kg·oil-eqv.
	Transport input substrates	0.00037	0.00057	0.00027	0.00043	0.00019	0.00021	kg·oil-eqv.
	Transport biomass	0.00157	0.00194	0.00151	0.00152	0.00144	0.00139	kg·oil-eqv.
	Ensilage	0.00009	0.00014	0.00017	0.00028	0.00012	0.00013	kg·oil-eqv.
	Fertilizer-induced emissions	n.a.	n.a.	n.a.	n.a.	n.a.	n.a.	kg·oil-eqv.
CHP	Biomass production system	0.01801	0.03274	0.01258	0.02346	0.00569	0.00687	kg·oil-eqv.
	CHP—Direct emissions	n.a.	n.a.	n.a.	n.a.	n.a.	n.a.	kg·oil-eqv.
	Credits heat utilization	−0.03948	−0.03948	−0.03948	−0.03948	−0.03948	−0.03948	kg·oil-eqv.
Total	**Total with credits**	**−0.02147**	**−0.00674**	**−0.02691**	**−0.01602**	**−0.03379**	**−0.03262**	**kg·oil-eqv.**
	Total without credits	**0.01801**	**0.03274**	**0.01258**	**0.02346**	**0.00569**	**0.00687**	**kg·oil-eqv.**

3.2. Freshwater Eutrophication and Marine Eutrophication

The substitution of the fossil reference lead to a net increase in freshwater eutrophication of up to 3.5×10^{-5} kg·P-eqv. $(kWh_{el.})^{-1}$ in all scenarios (Figure 5). On average, the freshwater eutrophication potentials are lowest for miscanthus, followed by switchgrass and then maize (Figure 5).

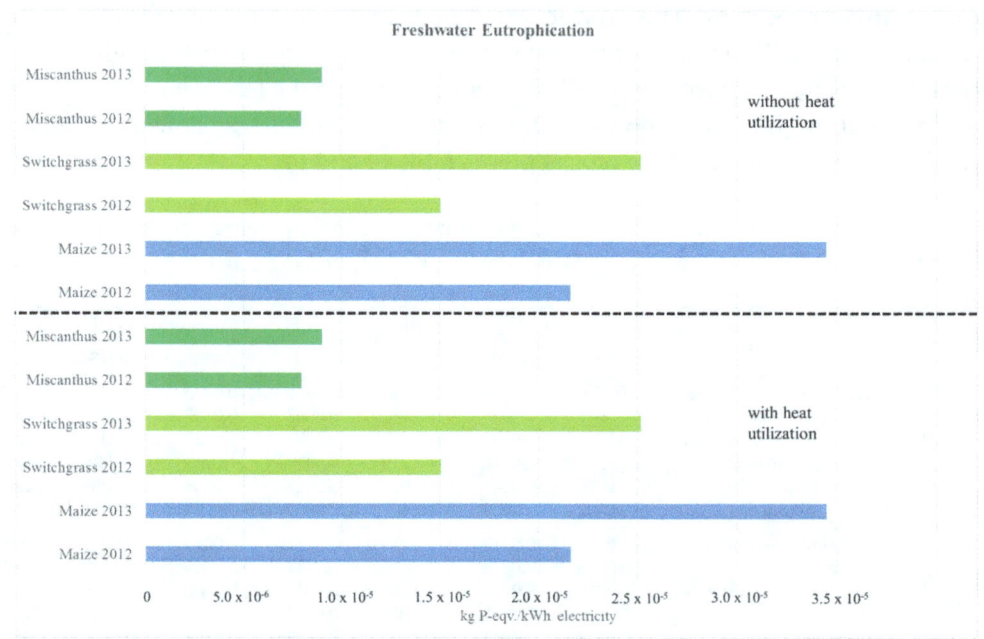

Figure 5. Assessment of the net impacts in kg·P-eqv. of substituting 1 kWh$_{el.}$ of the German electricity mix by power generated via combustion of the biogas in a CHP.

The recycling of nutrients leads to a high credit, which has a positive impact on the freshwater eutrophication (Table 6). In all scenarios, fertilizer-induced emissions account for the largest share of freshwater eutrophication. These are phosphate emissions associated with the use of phosphorus fertilizer, which are highest in maize and lowest in miscanthus (Table 6). The second-largest share comes from nitrogen fertilizer production, followed by the production of phosphate fertilizers (Table 6).

Table 6. Assessment of the freshwater eutrophication in kg·P-eqv. of 1 kWh$_{el.}$ generated via the production and fermentation of dedicated energy crops and combustion of the biogas in CHP.

	Processes/Flows	Maize per FU		Switchgrass per FU		Miscanthus per FU		Unit
		2012	2013	2012	2013	2012	2013	
Input substrates	Production of nitrogen fertilizer	1.18×10^{-7}	1.82×10^{-7}	7.74×10^{-8}	1.23×10^{-7}	5.14×10^{-8}	5.67×10^{-8}	kg·P-eqv.
	Production of potassium fertilizer	7.21×10^{-9}	1.11×10^{-8}	6.38×10^{-9}	1.01×10^{-8}	3.96×10^{-9}	4.36×10^{-9}	kg·P-eqv.
	Production of phosphate fertilizer	7.56×10^{-8}	1.16×10^{-7}	5.49×10^{-8}	8.72×10^{-8}	3.16×10^{-8}	3.48×10^{-8}	kg·P-eqv.
	Recycling of nutrients	-9.63×10^{-8}	-9.63×10^{-8}	-7.09×10^{-8}	-7.09×10^{-8}	-7.30×10^{-8}	-7.30×10^{-8}	kg·P-eqv.
	Herbicides	1.47×10^{-8}	2.26×10^{-8}	6.28×10^{-9}	9.97×10^{-9}	4.36×10^{-9}	4.80×10^{-9}	kg·P-eqv.
	Seeds/Rhizomes	1.34×10^{-7}	2.07×10^{-7}	2.88×10^{-8}	4.58×10^{-8}	2.76×10^{-7}	3.04×10^{-7}	kg·P-eqv.
Agricultural operations	Agricultural management	4.91×10^{-8}	7.56×10^{-8}	1.31×10^{-8}	2.09×10^{-8}	7.91×10^{-9}	8.72×10^{-9}	kg·P-eqv.
	Harvest	2.50×10^{-8}	3.85×10^{-8}	4.58×10^{-8}	7.28×10^{-8}	2.94×10^{-8}	3.24×10^{-8}	kg·P-eqv.
	Transport input substrates	7.63×10^{-9}	1.17×10^{-8}	5.55×10^{-9}	8.82×10^{-9}	3.87×10^{-9}	4.27×10^{-9}	kg·P-eqv.
	Transport biomass	3.23×10^{-8}	4.00×10^{-8}	3.12×10^{-8}	3.13×10^{-8}	2.97×10^{-8}	2.87×10^{-8}	kg·P-eqv.
	Ensilage	2.80×10^{-9}	2.80×10^{-9}	5.67×10^{-9}	5.67×10^{-9}	2.62×10^{-9}	2.62×10^{-9}	kg·P-eqv.
	Fertilizer-induced emissions	2.34×10^{-5}	3.60×10^{-5}	1.70×10^{-5}	2.70×10^{-5}	9.78×10^{-6}	1.08×10^{-5}	kg·P-eqv.
CHP	Biomass production system	2.38×10^{-5}	3.67×10^{-5}	1.72×10^{-5}	2.74×10^{-5}	1.01×10^{-5}	1.12×10^{-5}	kg·P-eqv.
	CHP—Direct emissions	0	0	0	0	0	0	kg·P-eqv.
	Credits heat utilization	-4.46×10^{-9}	-4.46×10^{-9}	-4.46×10^{-9}	-4.46×10^{-9}	-4.46×10^{-9}	-4.46×10^{-9}	kg·P-eqv.
Total	**Total with credits**	$\mathbf{2.38 \times 10^{-5}}$	$\mathbf{3.66 \times 10^{-5}}$	$\mathbf{1.72 \times 10^{-5}}$	$\mathbf{2.73 \times 10^{-5}}$	$\mathbf{1.01 \times 10^{-5}}$	$\mathbf{1.12 \times 10^{-5}}$	**kg·P-eqv.**
	Total without credits	$\mathbf{2.38 \times 10^{-5}}$	$\mathbf{3.67 \times 10^{-5}}$	$\mathbf{1.72 \times 10^{-5}}$	$\mathbf{2.74 \times 10^{-5}}$	$\mathbf{1.01 \times 10^{-5}}$	$\mathbf{1.12 \times 10^{-5}}$	**kg·P-eqv.**

A net benefit in the impact category marine eutrophication was achieved for the utilization of switchgrass and maize only in the year 2012—where the yield was significantly higher than in 2013—and when the heat utilization was accounted for (Figure 6). Miscanthus was the only crop that led to a reduction of marine eutrophication in comparison to the fossil reference in all years and scenarios. The maximum reduction was—4.6×10^{-5} kg·N-eqv. (kWh$_{el.}$)$^{-1}$ (Figure 6).

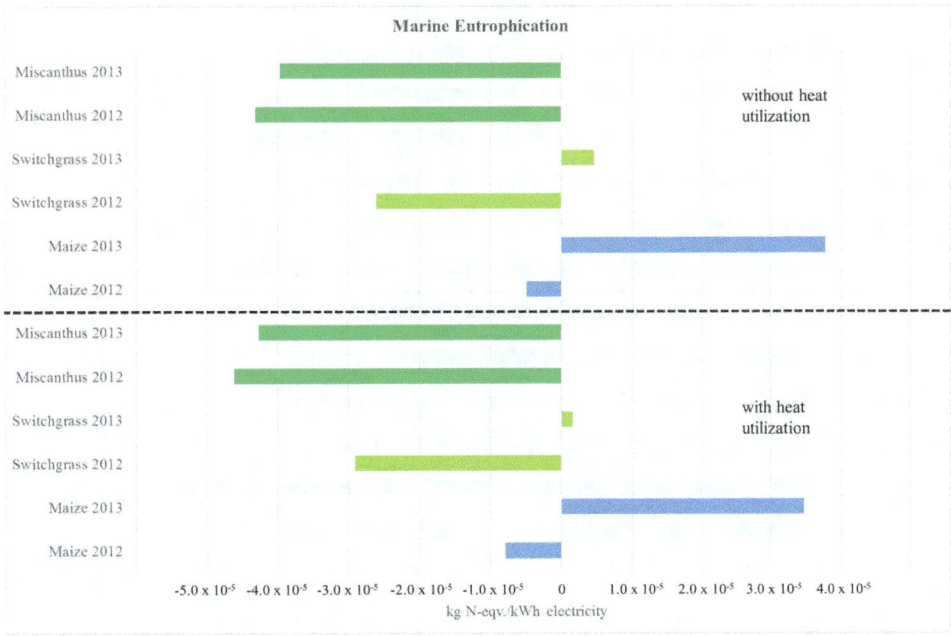

Figure 6. Assessment of the net benefits and impacts in kg·N-eqv. of substituting 1 kWh$_{el.}$ of the German electricity mix by power generated via combustion of the biogas in a CHP.

The production of nitrogen fertilizer had the strongest impact on marine eutrophication for all crops, followed by fertilizer-induced emissions. Ammonia emissions and nitrate leaching due to the use of nitrogen fertilizer play a particularly important role here. Both impacts were highest for maize and lowest for miscanthus (Table 7). The recycling of nutrients results in a significant credit (Table 7).

Table 7. Assessment of the marine eutrophication in kg·N-eqv. of 1 kWh$_{el}$. generated via the production and fermentation of dedicated energy crops and combustion of the biogas in CHP.

	Processes/Flows	Maize per FU		Switchgrass per FU		Miscanthus per FU		Unit
		2012	2013	2012	2013	2012	2013	
Input substrates	Production of nitrogen fertilizer	2.60×10^{-5}	4.01×10^{-5}	1.70×10^{-5}	2.70×10^{-5}	1.13×10^{-5}	1.25×10^{-5}	kg·N-eqv.
	Production of potassium fertilizer	5.80×10^{-7}	8.92×10^{-7}	5.13×10^{-7}	8.14×10^{-7}	3.18×10^{-7}	3.51×10^{-7}	kg·N-eqv.
	Production of phosphate fertilizer	1.22×10^{-6}	1.87×10^{-6}	8.83×10^{-7}	1.40×10^{-6}	5.08×10^{-7}	5.60×10^{-7}	kg·N-eqv.
	Recycling of nutrients	-1.29×10^{-5}	-1.29×10^{-5}	-8.18×10^{-6}	-8.18×10^{-6}	-8.36×10^{-6}	-8.36×10^{-6}	kg·N-eqv.
	Herbicides	3.75×10^{-7}	5.76×10^{-7}	1.60×10^{-7}	2.54×10^{-7}	1.11×10^{-7}	1.22×10^{-7}	kg·N-eqv.
	Seeds/Rhizomes	1.89×10^{-6}	2.91×10^{-6}	1.06×10^{-6}	1.69×10^{-6}	1.64×10^{-6}	1.80×10^{-6}	kg·N-eqv.
Agricultural operations	Agricultural management	4.20×10^{-6}	6.46×10^{-6}	1.18×10^{-6}	1.87×10^{-6}	7.05×10^{-7}	7.77×10^{-7}	kg·N-eqv.
	Harvest	2.10×10^{-6}	3.23×10^{-6}	3.85×10^{-6}	6.11×10^{-6}	2.47×10^{-6}	2.72×10^{-6}	kg·N-eqv.
	Transport input substrates	2.97×10^{-7}	4.56×10^{-7}	2.16×10^{-7}	3.43×10^{-7}	1.50×10^{-7}	1.66×10^{-7}	kg·N-eqv.
	Transport biomass	2.97×10^{-6}	3.67×10^{-6}	2.86×10^{-6}	2.87×10^{-6}	2.72×10^{-6}	2.64×10^{-6}	kg·N-eqv.
	Ensilage	1.94×10^{-7}	2.98×10^{-7}	3.80×10^{-7}	6.03×10^{-7}	2.52×10^{-7}	2.78×10^{-7}	kg·N-eqv.
	Fertilizer-induced emissions	4.09×10^{-5}	6.29×10^{-5}	2.67×10^{-5}	4.25×10^{-5}	1.78×10^{-5}	1.96×10^{-5}	kg·N-eqv.
CHP	Biomass production system	6.78×10^{-5}	1.10×10^{-4}	4.67×10^{-5}	7.73×10^{-5}	2.96×10^{-5}	3.31×10^{-5}	kg·N-eqv.
	CHP-Direct emissions	4.58×10^{-6}	4.58×10^{-6}	4.58×10^{-6}	4.58×10^{-6}	4.58×10^{-6}	4.58×10^{-6}	kg·N-eqv.
	Credits heat utilization	-3.04×10^{-6}	-3.04×10^{-6}	-3.04×10^{-6}	-3.04×10^{-6}	-3.04×10^{-6}	-3.04×10^{-6}	kg·N-eqv.
Total	**Total with credits**	6.94×10^{-5}	11.2×10^{-5}	4.82×10^{-5}	7.88×10^{-5}	3.11×10^{-5}	3.47×10^{-5}	**kg·N-eqv.**
	Total without credits	7.24×10^{-5}	11.5×10^{-5}	5.13×10^{-5}	8.19×10^{-5}	3.42×10^{-5}	3.77×10^{-5}	**kg·N-eqv.**

3.3. Terrestrial Acidification

All scenarios led to higher terrestrial acidification than the fossil references. Maize without heat utilization performed worst and led to emissions of 3.5×10^{-3} kg·SO$_2$-eqv. (kWh$_{el}$.)$^{-1}$ (Figure 7). Miscanthus performed best with the lowest terrestrial acidification potential (Figure 7).

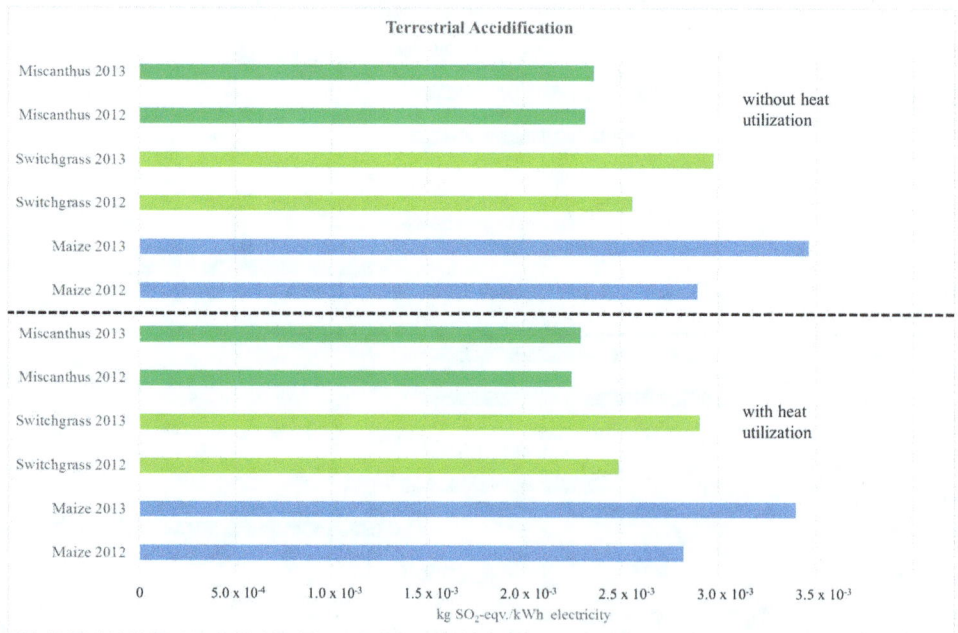

Figure 7. Assessment of the net benefits and impacts in kg·SO$_2$-eqv. of substituting 1 kWh$_{el}$. of the German electricity mix by power generated via combustion of the biogas in a CHP.

Fertilizer-induced emissions—especially ammonia—had the highest impact on terrestrial acidification for all crops and accounted on an average for around 20% of total emissions (Table 8). The second largest source of emissions responsible for terrestrial acidification was production of nitrogen fertilizer, followed by transport of the biomass (Table 8).

Table 8. Assessment of the terrestrial acidification in kg·SO_2-eqv. of 1 kWh$_{el}$. generated via the production and fermentation of dedicated energy crops and combustion of the biogas in CHP.

	Processes/Flows	Maize per FU 2012	Maize per FU 2013	Switchgrass per FU 2012	Switchgrass per FU 2013	Miscanthus per FU 2012	Miscanthus per FU 2013	Unit
Input substrates	Production of nitrogen fertilizer	7.34×10^{-5}	1.13×10^{-4}	4.80×10^{-5}	7.63×10^{-5}	3.19×10^{-5}	3.52×10^{-5}	kg·SO_2-eqv.
	Production of potassium fertilizer	8.25×10^{-6}	1.27×10^{-5}	7.30×10^{-6}	1.16×10^{-5}	4.53×10^{-6}	5.00×10^{-6}	kg·SO_2-eqv.
	Production of phosphate fertilizer	4.73×10^{-5}	7.28×10^{-5}	3.43×10^{-5}	5.45×10^{-5}	1.97×10^{-5}	2.18×10^{-5}	kg·SO_2-eqv.
	Recycling of nutrients	-6.22×10^{-5}	-6.22×10^{-5}	-4.71×10^{-5}	-4.71×10^{-5}	-4.88×10^{-5}	-4.88×10^{-5}	kg·SO_2-eqv.
	Herbicides	6.35×10^{-6}	9.77×10^{-6}	2.71×10^{-6}	4.31×10^{-6}	1.88×10^{-6}	2.08×10^{-6}	kg·SO_2-eqv.
	Seeds/Rhizomes	2.19×10^{-6}	3.36×10^{-6}	8.04×10^{-7}	1.28×10^{-6}	1.98×10^{-6}	2.18×10^{-6}	kg·SO_2-eqv.
Agricultural operations	Agricultural management	5.16×10^{-5}	7.95×10^{-5}	1.46×10^{-5}	2.32×10^{-5}	8.73×10^{-6}	9.62×10^{-6}	kg·SO_2-eqv.
	Harvest	2.58×10^{-5}	3.97×10^{-5}	4.72×10^{-5}	7.50×10^{-5}	3.03×10^{-5}	3.34×10^{-5}	kg·SO_2-eqv.
	Transport input substrates	8.90×10^{-7}	1.37×10^{-6}	6.47×10^{-7}	1.03×10^{-6}	4.51×10^{-7}	4.97×10^{-7}	kg·SO_2-eqv.
	Transport biomass	3.69×10^{-5}	4.57×10^{-5}	3.56×10^{-5}	3.57×10^{-5}	3.39×10^{-5}	3.28×10^{-5}	kg·SO_2-eqv.
	Ensilage	2.46×10^{-6}	3.78×10^{-6}	4.83×10^{-6}	7.67×10^{-6}	3.21×10^{-6}	3.54×10^{-6}	kg·SO2-eqv.
	Fertilizer-induced emissions	8.29×10^{-4}	1.28×10^{-3}	5.42×10^{-4}	8.61×10^{-4}	3.61×10^{-4}	3.97×10^{-4}	kg·SO_2-eqv.
CHP	Biomass production system	1.02×10^{-3}	1.60×10^{-3}	6.91×10^{-4}	1.10×10^{-3}	4.48×10^{-4}	4.95×10^{-4}	kg·SO_2-eqv.
	CHP - Direct emissions	2.61×10^{-3}	2.61×10^{-3}	2.61×10^{-3}	2.61×10^{-3}	2.61×10^{-3}	2.61×10^{-3}	kg SO_2-eqv.
	Credits heat utilization	-6.82×10^{-5}	-6.82×10^{-5}	-6.82×10^{-5}	-6.82×10^{-5}	-6.82×10^{-5}	-6.82×10^{-5}	kg·SO_2-eqv.
Total	**Total with credits**	3.57×10^{-3}	4.14×10^{-3}	3.24×10^{-3}	3.65×10^{-3}	2.99×10^{-3}	3.04×10^{-3}	**kg·SO_2-eqv.**
	Total without credits	3.64×10^{-3}	4.21×10^{-3}	3.31×10^{-3}	3.72×10^{-3}	3.06×10^{-3}	3.11×10^{-3}	**kg·SO_2-eqv.**

4. Discussion

Here the results of this study are considered in a broader context, also including other environmental aspects not modeled in the LCA. The discussion concludes with opportunities and challenges of the introduction of novel perennial C4 crops in the biogas sector.

4.1. Environmental Performance in Impact Categories Modelled in the LCA

The results of this study show that, as soon as more impact categories are assessed than climate change and fossil fuel depletion, the environmental performance of the bioenergy conversion route "biogas" is not so clear-cut. All three energy crops have a significantly better environmental profile than the fossil reference (German electricity mix) in the impact categories climate change (CC) and fossil fuel depletion (FFD). Similar findings have been reported in the literature [45,46]. However, all three energy crops showed significantly higher impacts than the fossil reference in the impact categories freshwater eutrophication (FE) and terrestrial acidification (TA). The results for marine eutrophication (ME) were more variable. Here, miscanthus (both years) and switchgrass (2012 only) had a significantly lower impact than the fossil reference, whereas maize had a significantly higher impact in 2013 due to the low yield. High biomass yields have been shown to be a crucial factor for favorable environmental performance [47]. Again, these results correspond to findings of other studies, which mainly also found a higher impact of energy-crop-derived biogas than the fossil reference in acidification and eutrophication potential [48–50].

4.1.1. Overall Impact of Process Steps in Impact Categories

The production of nitrogen fertilizer was identified as the most relevant process step in the impact categories FFD and CC and the second most relevant in ME. Fertilizer-induced emissions were identified as the most important flow in the categories FE and ME and second most important in CC and TA. Similar results have been reported in the literature and numerous studies have already described the strong impact of nitrogen fertilizer production and related direct and indirect emissions on FFD and CC (e.g., [39,46,50,51]). The present study also showed a strong impact of mineral nitrogen fertilizer application on eutrophication (FE and ME) and acidification potential of crop production. This seems logical, since nitrate is one of the major contributors to eutrophication and the nitrification process a major contributor to soil acidification [27].

In TA, direct CHP emissions were the most important flow. Rehl et al. [49] identified sulfur dioxide from the CHP as one of the most important contributors to the acidification potential. One possibility to reduce these emissions could be the upgrading of biogas to biomethane, because sulfur dioxide is almost completely removed during this process. In addition, new techniques for biomethane production (e.g., pressurized anaerobic digestion) could help reduce the carbon footprint of biomethane production in the near future, because the demand for energy-intensive compression is reduced in such approaches [52]. Lijó et al. [53] reported production of nitrogen fertilizer, fertilizer-induced emissions and emissions of agricultural management as important factors for the environmental performance of energy crops. In this study, emissions from agricultural management were found to be the third most relevant process in CC, FFD and ME for maize cultivation, but considerably less important for miscanthus and switchgrass.

4.1.2. Impact of the Process Steps for Each Crop

Emissions and fossil fuel depletion from production of nitrogen fertilizer and agricultural management and fertilizer-induced emissions were highest for maize in each of the considered impact categories. This is because maize production consumes more energy for soil cultivation and requires higher nitrogen fertilizer levels for high yields than the C4 perennial grasses. For maize, data from the treatment with the highest nitrogen fertilization (240 kg·N·ha^{-1}) were used, which on long-term average yielded significantly higher than the medium fertilization rate (120 kg·N·ha^{-1}). However, the high nitrogen fertilization is probably above the marginal revenue and a lower fertilization rate could reduce the environmental impact of maize. Nevertheless, the nitrogen demand of miscanthus and switchgrass are still lower than that of maize. In addition, for miscanthus and switchgrass, data from the treatment with the highest nitrogen fertilization rate (80 kg·N·ha^{-1}) were used, in order to consider the higher nutrient removal by the green harvested biomass. Although green harvest increases the withdrawal of nitrogen compared to a spring harvest, the biomass of miscanthus and switchgrass contained approximately 60% less nitrogen than maize biomass (Table 3).

The annual cultivation of maize led also to significantly higher emissions and fossil fuel depletion for agricultural management in CC, ME and FFD. For this reasons, changing the crop production system from annual crops with a high nitrogen demand to perennial C4 crops with improved nutrient efficiency seems to be a very promising option for increasing the environmental sustainability of the biogas sector and the bioeconomy, as already described by Lewandowski [54]. Compared to maize, miscanthus and switchgrass showed in the scenarios without heat utilization 59%–73% and 25%–28% lower CC potential, 68%–79% and 28%–30% lower FFD potential, 57%–69% and 25%–28% lower FE potential, 53%–67% and 29% lower ME potential and 16%–26% and 9%–12% lower TA potential, respectively.

Considering all impact categories, miscanthus performed best amongst the three assessed crops. Especially in 2013, the yield and thereby the environmental performance of miscanthus was much more stable compared to maize and switchgrass. Both crops reacted more sensitively to the unfavorable weather conditions in 2013. This resulted in lower yields and is also reflected by the performance in the environmental impact categories. The higher stress tolerance and yield stability of miscanthus is therefore not only favorable for the farmer, but also from an environmental point of view.

The nutrient recycling via fermentation residues led to a significant credit for all crops, especially in the impact categories CC, FFD and ME. However, fermentation residue application on the perennial grasses miscanthus and switchgrass and resulting emissions need to be further investigated. Since the fermentation residues cannot be incorporated into the soil in such perennials, higher ammonia emissions could occur, which could lead to higher eutrophication and acidification potentials [48]. This needs to be further investigated to allow consideration of such an effect in future assessments of the environmental performance.

4.2. Other Environmental Aspects

In the section above, the environmental performance was analyzed in five impact categories and it was shown that the perennial grasses, especially miscanthus, performed better than the annual crop maize. However, the five considered impact categories are not sufficient for a holistic assessment of the environmental performance. Therefore, other aspects relevant to environmental performance are discussed in the following section.

Intensive soil cultivation in annual maize is accompanied by an increased risk of soil erosion, due to the slow youth development of the crop [6]. For annual maize, there is also a low to medium risk for soil compaction [55]. However, for green-harvested miscanthus and switchgrass the risk of soil compaction may be lower due to its perennial nature, but needs to be assessed to allow comparison. The combination of intensive soil cultivation and low amount of crop residues in silage maize has a negative impact on content of soil organic carbon. Both environmental aspects could be improved by changing substrate supply of biogas plants from maize to perennial C4 grasses, since miscanthus and switchgrass generally lead to an increase in soil organic carbon compared to annual cropping systems [11,56,57]. Under miscanthus, the largest proportion of the soil organic carbon is found in the topsoil, which can be explained by the high proportion of roots in the top 0.35 m [58]. The sequestration of carbon in the soil can increase the GHG mitigation potential significantly, especially if the cropping system is changed from annual to perennial [56,59]. In this study, the sequestration effect was not considered, because the effect of the green harvest on the root and rhizome development and on the soil carbon sequestration potential is not yet known. Therefore, the development of the soil organic carbon under green harvested miscanthus and switchgrass needs to be further investigated to determine the sequestration potential of this harvest regime.

Agricultural land occupation is another important environmental aspect, due to limited expansion potential for agricultural land and negative impacts from the transformation of natural land. In this paper, agricultural land occupation was not directly assessed, but the data in Table 1 show that maize required the smallest area (173 ha) of agricultural land in 2012 to supply the biogas plant with the required biomass. Changing the input substrate from maize to miscanthus or switchgrass increased the agricultural land demand in 2012 by 12% or 68%, respectively. Under unfavorable weather conditions in 2013, the agricultural land demand for miscanthus cropping was 20% lower and for switchgrass 73% higher than for maize cultivation. Agricultural land occupation for biogas production can lead to indirect land-use change (iLUC), which can significantly reduce the GHG mitigation potential and even lead to higher GWP than the fossil reference [14]. For this reason, the comparatively high agricultural land demand of switchgrass to deliver the required biomass substrate is a clear disadvantage compared to the other crops. In contrast, the area demand of miscanthus was only slightly higher and even lower when unfavorable weather conditions occurred for maize production. Again, the higher abiotic stress tolerance and yield stability of miscanthus can be seen as environmental advantage. However, both perennial C4 crops could be grown in future mainly on marginal or contaminated land [9,23]. This could reduce the pressure on agricultural land and expand the area available for biomass production.

Biodiversity is difficult to assess just by the crop itself, because it strongly depends on other factors, e.g., the distribution of fields in a landscape and structural elements such as hedges. However, modern agriculture is assumed to have a negative impact on the biodiversity by simplification of agricultural landscapes, e.g., large field sizes, and small amount of crop varieties which are grown in monoculture [4]. An increased number of crop species and a higher proportion of perennial cropping systems in modern agriculture is seen as one option to promote biodiversity [4]. For this reason, replacing biogas maize with miscanthus or switchgrass could positively affect the biodiversity by adding novel, perennial crops to the agricultural landscapes. However, it should be noted that the impact on soil biodiversity may be influenced by the choice of the perennial biomass crop [60]. Furthermore, both perennials can be characterized by their comparatively low-input crop management, after their successful establishment in year one. For miscanthus, a higher abundance of insects, spiders and earthworms than in arable land is reported, as well as additional niches for birds and, provided a spring harvest is performed, over winter cover for small mammals in intensive arable regions [11,61]. For switchgrass, similar positive effects can be expected, which leads to the assumption, that both could increase the biodiversity and structure-richness of agricultural landscapes. Again, the effect of the pre-winter harvest, which clearly removes the winter cover for small mammals and reduces the mulch layer, is not yet known and needs to be investigated. However, both crops also induce risks for biodiversity because they are not native to Europe and could potentially appear as invasive species. *Miscanthus x giganteus* has a very low invasiveness risk, because it does not produce fertile seeds and no escapes were observed over more than two decades of *M. x giganteus* production in Europe. Current miscanthus breeding efforts aim to produce fertile genotypes that can be propagated by seeds [10], but several mechanisms to avoid seed escape are incorporated, including preferring candidates which require a very long vegetation period for seed production to avoid viable seeds being produced in regions of biomass cultivation [9]. It is also necessary to mention that miscanthus as well as switchgrass seedlings have a very low competitiveness compared to weeds and a slow youth development. For this reason it is quite unlikely that they become invasive species in Europe. Nonetheless, the invasiveness potential of novel miscanthus genotypes and switchgrass needs to be investigated and monitored.

Finally, the socioeconomic aspects of landscape appearance need to be considered. Crops such as maize are often criticized in the public, due to their height and monotony. The same could appear for miscanthus, due to its height and density in well-established commercial fields. Smaller and nicely flowering miscanthus genotypes or switchgrass could be experienced more favorably and might influence the appearance of landscapes more positively. However, this could compromise the yield and lead to a trade-off between yield and public acceptance. Public acceptance could also be positively influenced by using smaller fields or strip cropping instead large monoculture fields.

4.3. Implementation—Chances and Challenges

In this study, it is shown that implementation of perennial C4 grasses for biogas production can have significant environmental benefits. From an environmental point of view, miscanthus in particular would be a desirable crop for biogas production. The main weak point of switchgrass is clearly its lower yield potential than miscanthus and related to that its higher area demand, fossil fuel consumption and emissions. For the farmer, the implementation of miscanthus and switchgrass as biogas crops is accompanied by opportunities and challenges, which are discussed in the following section but require further research.

This study is based on methane yields measured in a batch test using milled biomass. In order to transfer these values to a full-scale biogas plant, a pre-treatment of the biomass was considered for miscanthus and switchgrass, which leads to a higher electricity demand for plant operation. For this reason, the electricity demand for miscanthus and switchgrass was assumed to be almost twice as high as that for maize. Before implementation, the methane yield, the necessity of a pre-treatment and the energy consumption of such a pre-treatment should be verified under more realistic conditions. Ensiling of miscanthus biomass, and presumably also switchgrass, appears possible [19], but also needs to be demonstrated in practice.

The long-term performance of green-harvested miscanthus is one of the major uncertainties for its biogas utilization, because miscanthus reacts sensitively to very early mid-season harvest, but tolerates green harvest in late autumn [18]. However, it is not yet known if green-harvested miscanthus is productive for as long as a spring-harvested crop (more than 20 years) and if recycling of fermentation residues is sufficient to maintain its productivity. In addition, the farmer has to dedicate arable land to miscanthus for several years to achieve return on investment, due to the high establishment costs. However, current research focuses on reducing establishment costs by developing seed-based genotypes, which may allow direct sowing in future [10]. Further, most biogas plants are designed for a minimum of 20 years' operation, which would fit in very well with the expected productive lifetime of miscanthus. Cost-effective miscanthus establishment offers the chance of significantly reducing biomass costs. As shown in this paper, the yield of miscanthus is not as sensitive as annual maize to unfavorable weather conditions, which may become more common in future due to climate change. One of the main reasons for the low maize yield was the very late sowing date and the early summer drought stress. In miscanthus, planting is only required once in 20–30 years and the established crop benefits from winter soil moisture. Therefore, miscanthus seems very suitable for risk mitigation of such weather conditions.

In contrast to miscanthus, switchgrass can be established cheaply via direct sowing of seeds. However, the establishment of switchgrass is difficult due to an often low germination rate, low competitiveness of seedlings and limited availability of herbicides. Current research focuses on the optimization of the establishment method and herbicide testing [62]. Nevertheless, early green harvest of switchgrass seems less problematic than in miscanthus and even a double cut is possible [21]. The shorter productive life of approximately 15 years, lower investment costs and the ability of direct sowing may increase farmers' willingness to adopt this crop. However, the lower yield potential limits its implementation to very poor and shallow soils, where it is likely to perform better than miscanthus [23].

From an environmental point of view, miscanthus cultivation for biogas production is generally recommended if the biogas plant technology is suitable for the digestion of fibrous substrates or adequate pre-treatment options are available.

Acknowledgments: This work was supported partly by a grant from the Ministry of Science, Research and the Arts of Baden-Württemberg (funding code: 7533-10-5-70) as part of the BBW ForWerts Graduate Program. The authors are grateful to Dagmar Mezger and Martin Zahner for their support in performing the laboratory analyses and to the staff of the research station Ihinger Hof, especially Thomas Truckses, for maintaining and managing the field trial. The authors would also like to thank to Nicole Gaudet for editing the manuscript.

Author Contributions: Andreas Kiesel collected the samples, performed the biogas batch test and analyzed the data as well as led the writing process and contributed mainly to the Introduction and the Discussion Sections. Moritz Wagner performed the LCA modeling and contributed mainly to writing Material and Methods and Results Sections. He also prepared the figures and tables. Iris Lewandowski added valuable contributions to each section and in manifold discussions.

Conflicts of Interest: The authors declare no conflict of interest. The funding sponsors had no role in the design of the study; in the collection, analyses, or interpretation of data; in the writing of the manuscript, and in the decision to publish the results.

References

1. Weiland, P. Production and Energetic Use of Biogas from Energy Crops and Wastes in Germany. *ABAB* **2003**, *109*, 263–274. [CrossRef]
2. FNR. *Basisdaten Bioenergie Deutschland 2015: Festbrennstoffe, Biokraftstoffe, Biogas*; FNR: Gülzow, Germany, 2015.
3. Zürcher, A. Dauerkulturen als Alternativen zu Mais: Wildartenmischungen, Topinambur, Durchwachsene Silphie, Virginiamalve und Riesenweizengras. Workshop "Pflanzliche Rohstoffe zur Biogasgewinnung" at LTZ Augustenberg on 16 October 2014. Available online: http://www.ltz-bw.de/pb/Lde/Startseite/Service/Nachlese#anker2300415 (accessed on 20 December 2016).
4. Altieri, M.A. The ecological role of biodiversity in agroecosystems. *Agric. Ecosyst. Environ.* **1999**, *74*, 19–31. [CrossRef]
5. Svoboda, N.; Taube, F.; Wienforth, B.; Kluß, C.; Kage, H.; Herrmann, A. Nitrogen leaching losses after biogas residue application to maize. *Soil Tillage Res.* **2013**, *130*, 69–80. [CrossRef]
6. Vogel, E.; Deumlich, D.; Kaupenjohann, M. Bioenergy maize and soil erosion—Risk assessment and erosion control concepts. *Geoderma* **2016**, *261*, 80–92. [CrossRef]

7. Herrmann, A. Biogas Production from Maize: Current State, Challenges and Prospects. 2. Agronomic and Environmental Aspects. *Bioenergy Res.* **2013**, *6*, 372–387. [CrossRef]

8. Clifton-Brown, J.; Schwarz, K.-U.; Hastings, A. History of the development of Miscanthus as a bioenergy crop: From small beginnings to potential realisation. *Biol. Environ. Proc. R. Irish Acad.* **2015**, *115B*, 1–13. [CrossRef]

9. Lewandowski, I.; Clifton-Brown, J.; Trindade, L.; van der Linden, G.; Schwarz, K.; Müller-Sämann, K.; Anisimov, A.; Chen, C.-L.; Dolstra, O.; Donnison, I.S.; et al. Progress on optimizing miscanthus biomass production for the European bioeconomy: Results of the EU FP7 project OPTIMISC. *Front. Plant Sci.* **2016**. [CrossRef] [PubMed]

10. Clifton-Brown, J.; Hastings, A.; Mos, M.; McCalmont, J.P.; Ashman, C.; Awty-Carroll, D.; Cerazy, J.; Chiang, Y.-C.; Cosentino, S.; Cracroft-Eley, W.; et al. Progress in upscaling Miscanthus biomass production for the European bio- economy with seed based hybrids. *GCB Bioenergy* **2016**. [CrossRef]

11. McCalmont, J.P.; Hastings, A.; McNamara, N.P.; Richter, G.M.; Robson, P.; Donnison, I.S.; Clifton-Brown, J. Environmental costs and benefits of growing Miscanthus for bioenergy in the UK. *GCB Bioenergy* **2015**. [CrossRef]

12. Lewandowski, I.; Schmidt, U. Nitrogen, energy and land use efficiencies of miscanthus, reed canary grass and triticale as determined by the boundary line approach. *Agric. Ecosyst. Environ.* **2006**, *112*, 335–346. [CrossRef]

13. Cadoux, S.; Riche, A.B.; Yates, N.E.; Machet, J.-M. Nutrient requirements of *Miscanthus x giganteus*: Conclusions from a review of published studies. *Biomass Bioenergy* **2012**, *38*, 14–22. [CrossRef]

14. Styles, D.; Gibbons, J.; Williams, A.P.; Dauber, J.; Stichnothe, H.; Urban, B.; Chadwick, D.R.; Jones, D.L. Consequential life cycle assessment of biogas, biofuel and biomass energy options within an arable crop rotation. *GCB Bioenergy* **2015**, *7*, 1305–1320. [CrossRef]

15. Felten, D.; Fröba, N.; Fries, J.; Emmerling, C. Energy balances and greenhouse gas-mitigation potentials of bioenergy cropping systems (Miscanthus, rapeseed, and maize) based on farming conditions in Western Germany. *Renew. Energy* **2013**, *55*, 160–174. [CrossRef]

16. Mayer, F.; Gerin, P.A.; Noo, A.; Lemaigre, S.; Stilmant, D.; Schmit, T.; Leclech, N.; Ruelle, L.; Gennen, J.; von Francken-Welz, H.; et al. Assessment of energy crops alternative to maize for biogas production in the Greater Region. *Bioresour. Technol.* **2014**, *166*, 358–367. [CrossRef] [PubMed]

17. Wahid, R.; Nielsen, S.F.; Hernandez, V.M.; Ward, A.J.; Gislum, R.; Jørgensen, U.; Møller, H.B. Methane production potential from Miscanthus sp: Effect of harvesting time, genotypes and plant fractions. *Biosyst. Eng.* **2015**, *133*, 71–80. [CrossRef]

18. Kiesel, A.; Lewandowski, I. Miscanthus as biogas substrate—Cutting tolerance and potential for anaerobic digestion. *GCB Bioenergy* **2015**. [CrossRef]

19. Whittaker, C.; Hunt, J.; Misselbrook, T.; Shield, I. How well does Miscanthus ensile for use in an anaerobic digestion plant? *Biomass Bioenergy* **2016**, *88*, 24–34. [CrossRef]

20. McLaughlin, S.B.; Adams Kszos, L. Development of switchgrass (*Panicum virgatum*) as a bioenergy feedstock in the United States. *Biomass Bioenergy* **2005**, *28*, 515–535. [CrossRef]

21. Masse, D.; Gilbert, Y.; Savoie, P.; Belanger, G.; Parent, G.; Babineau, D. Methane yield from switchgrass harvested at different stages of development in Eastern Canada. *Bioresour. Technol.* **2010**, *101*, 9536–9541. [CrossRef] [PubMed]

22. Heaton, E. A quantitative review comparing the yields of two candidate C4 perennial biomass crops in relation to nitrogen, temperature and water. *Biomass Bioenergy* **2004**, *27*, 21–30. [CrossRef]

23. Lewandowski, I.; Scurlock, J.M.; Lindvall, E.; Christou, M. The development and current status of perennial rhizomatous grasses as energy crops in the US and Europe. *Biomass Bioenergy* **2003**, *25*, 335–361. [CrossRef]

24. International Organization for Standardization (ISO). *ISO 14040: Environmental Management—Life Cycle Assessment—Principles and Framework*, 2nd ed.; ISO: Geneva, Switzerland, 2006.

25. International Organization for Standardization (ISO). *ISO 14044: Environmental Management—Life Cycle Assessment—Requirements and Guidelines*; ISO: Geneva, Switzerland, 2006.

26. Wagner, M.; Lewandowski, I. Relevance of environmental impact categories for perennial biomass production. *GCB Bioenergy* **2016**. [CrossRef]

27. Rice, K.C.; Herman, J.S. Acidification of Earth: An assessment across mechanisms and scales. *Appl. Geochem.* **2012**, *27*, 1–14. [CrossRef]

28. EEA. *Source Apportionment of Nitrogen and Phosphorus Inputs into the Aquatic Environment*; EEA Report No. 7; European Environment Agency: Copenhagen, Denmark, 2005; p. 48.

29. Boehmel, C.; Lewandowski, I.; Claupein, W. Comparing annual and perennial energy cropping systems with different management intensities. *Agric. Syst.* **2008**, *96*, 224–236. [CrossRef]

30. Iqbal, Y.; Gauder, M.; Claupein, W.; Graeff-Hönninger, S.; Lewandowski, I. Yield and quality development comparison between miscanthus and switchgrass over a period of 10 years. *Energy* **2015**, *89*, 268–276. [CrossRef]

31. GaBi Database. *Service Pack*; GaBi Software System; Thinkstep AG: Leinfelden-Echterdingen, Germany, 2016.

32. Bouwman, A.F.; Boumans, L.J.M.; Batjes, N.H. Modeling global annual N_2O and NO emissions from fertilized fields. *Glob. Biogeochem. Cycles* **2002**, *16*, 28-1–28-9. [CrossRef]

33. Intergovernmental Panel on Climate Change (IPCC). *Guidelines for National Greenhouse Gas Inventories*; Prepared by the National Greenhouse Gas Inventories Programme; Eggleston, H.S., Buendia, L., Miwa, K., Ngara, T., Tanabe, K., Eds.; IGES: Hayama, Japan, 2006.

34. Faist Emmenegger, M.; Reinhard, J.; Zah, R. *Sustainability Quick Check for Biofuels—Intermediate Background Report*; Agroscope Reckenholz-Tänikon Research Station ART: Dübendorf, Switzerland, 2009.

35. EMEP/CORINAIR. *Joint EMEP/CORINAIR Atmospheric Emission Inventory Guidebook*, 3rd ed.; European Environment Agency: Copenhagen, Denmark, 2001.

36. Van der Werf, H.M.; Petit, J.; Sanders, J. The environmental impacts of the production of concentrated feed: The case of pig feed in Bretagne. *Agric. Syst.* **2005**, *83*, 153–177. [CrossRef]

37. Walla, C.; Schneeberger, W. The optimal size for biogas plants. *Biomass Bioenergy* **2008**, *32*, 551–557. [CrossRef]

38. Bacenetti, J.; Fusi, A.; Negri, M.; Guidetti, R.; Fiala, M. Environmental assessment of two different crop systems in terms of biomethane potential production. *Sci. Total Environ.* **2014**, *466–467*, 1066–1077. [CrossRef] [PubMed]

39. Gützloe, A.; Thumm, U.; Lewandowski, I. Influence of climate parameters and management of permanent grassland on biogas yield and GHG emission substitution potential. *Biomass Bioenergy* **2014**, *64*, 175–189. [CrossRef]

40. *Faustzahlen Biogas: 3. Ausgabe*; Kuratorium für Technik und Bauwesen in der Landwirtschaft: Darmstadt, Germany, 2013.

41. Bachmaier, J.; Effenberger, M.; Gronauer, A. Greenhouse gas balance and resource demand of biogas plants in agriculture. *Eng. Life Sci.* **2010**, *10*, 560–569. [CrossRef]

42. Weidema, B.P.; Bauer, C.; Hischier, R. The Ecoinvent Database: Overview and Methodology. Data Quality Guideline for the Ecoinvent Database Version 3. 2013. Available online: http://www.ecoinvent.org/files/dataqualityguideline_ecoinvent_3_20130506.pdf (accessed on 20 December 2016).

43. Naumann, C.; Bassler, R. *Die Chemische Untersuchung von Futtermitteln*; VDLUFA-Verl.: Darmstadt, Germany, 1976/2012.

44. Goedkoop, M.; Heijungs, R.; Huijbregts, M.; De, S.A.; Struijs, J.; Van, Z.R. *ReCiPe 2008. A life Cycle Impact Assessment Method Which Comprises Harmonised Category Indicators at the Midpoint and the Endpoint Level, First Edition Report I. Characterisation*; VROM: Den Haag, The Netherlands, 2008.

45. Gerin, P.A.; Vliegen, F.; Jossart, J.-M. Energy and CO_2 balance of maize and grass as energy crops for anaerobic digestion. *Bioresour. Technol.* **2008**, *99*, 2620–2627. [CrossRef] [PubMed]

46. Hijazi, O.; Munro, S.; Zerhusen, B.; Effenberger, M. Review of life cycle assessment for biogas production in Europe. *Renew. Sustain. Energy Rev.* **2016**, *54*, 1291–1300. [CrossRef]

47. Meyer, F.; Wagner, M.; Lewandowski, I. Optimizing GHG emission and energy-saving performance of miscanthus-based value chains. *Biomass Conv. Bioref.* **2016**. [CrossRef]

48. Hartmann, J.K. *Life-Cycle-Assessment of Industrial Scale Biogas Plants*; eDiss: Göttingen, Germany, 2006.

49. Rehl, T.; Lansche, J.; Müller, J. Life cycle assessment of energy generation from biogas—Attributional vs. consequential approach. *Renew. Sustain. Energy Rev.* **2012**, *16*, 3766–3775. [CrossRef]

50. González-García, S.; Bacenetti, J.; Negri, M.; Fiala, M.; Arroja, L. Comparative environmental performance of three different annual energy crops for biogas production in Northern Italy. *J. Clean. Prod.* **2013**, *43*, 71–83. [CrossRef]

51. Cherubini, F. GHG balances of bioenergy systems—Overview of key steps in the production chain and methodological concerns. *Renew. Energy* **2010**, *35*, 1565–1573. [CrossRef]

52. Budzianowski, W.M.; Postawa, K. Renewable energy from biogas with reduced carbon dioxide footprint: Implications of applying different plant configurations and operating pressures. *Renew. Sustain. Energy Rev.* **2016**. [CrossRef]

53. Lijó, L.; González-García, S.; Bacenetti, J.; Fiala, M.; Feijoo, G.; Lema, J.M.; Moreira, M.T. Life Cycle Assessment of electricity production in Italy from anaerobic co-digestion of pig slurry and energy crops. *Renew. Energy* **2014**, *68*, 625–635. [CrossRef]

54. Lewandowski, I. Securing a sustainable biomass supply in a growing bioeconomy. *Glob. Food Secur.* **2015**, *6*, 34–42. [CrossRef]

55. Goetze, P.; Ruecknagel, J.; Jacobs, A.; Marlander, B.; Koch, H.-J.; Christen, O. Environmental impacts of different crop rotations in terms of soil compaction. *J. Environ. Manag.* **2016**, *181*, 54–63. [CrossRef] [PubMed]

56. Zeri, M.; Anderson-Teixeira, K.; Hickman, G.; Masters, M.; De Lucia, E.; Bernacchi, C.J. Carbon exchange by establishing biofuel crops in Central Illinois. *Agric. Ecosyst. Environ.* **2011**, *144*, 319–329. [CrossRef]

57. Gauder, M.; Billen, N.; Zikeli, S.; Laub, M.; Graeff-Hönninger, S.; Claupein, W. Soil carbon stocks in different bioenergy cropping systems including subsoil. *Soil Tillage Res.* **2016**, *155*, 308–317. [CrossRef]

58. Gioacchini, P.; Cattaneo, F.; Barbanti, L.; Montecchio, D.; Ciavatta, C.; Marzadori, C. Carbon sequestration and distribution in soil aggregate fractions under Miscanthus and giant reed in the Mediterranean area. *Soil Tillage Res.* **2016**, *163*, 235–242. [CrossRef]

59. Meyer-Aurich, A.; Lochmann, Y.; Klauss, H.; Prochnow, A. Comparative Advantage of Maize- and Grass-Silage Based Feedstock for Biogas Production with Respect to Greenhouse Gas Mitigation. *Sustainability* **2016**, *8*, 617. [CrossRef]

60. Schrama, M.; Vandecasteele, B.; Carvalho, S.; Muylle, H.; van der Putten, W.H. Effects of first- and second-generation bioenergy crops on soil processes and legacy effects on a subsequent crop. *GCB Bioenergy* **2016**, *8*, 136–147. [CrossRef]

61. Clapham, S.J.; Slater, F.M. The biodiversity of established biomass grass crops. *Aspects of Appl. Biol.* **2008**, 325–330.

62. Sadeghpour, A.; Hashemi, M.; DaCosta, M.; Gorlitsky, L.E.; Jahanzad, E.; Herbert, S.J. Switchgrass Establishment and Biomass Yield Response to Seeding Date and Herbicide Application. *Agronomy J.* **2015**, *107*, 142. [CrossRef]

12

Be Sustainable to Be Innovative: An Analysis of Their Mutual Reinforcement

Sarah Behnam [1,2],* and Raffaella Cagliano [1]

[1]   Department of Management, Economics and Industrial Engineering, Polytechnic University of Milan, 20133 Milano, Italy; raffaella.cagliano@polimi.it
[2]   Department of Industrial Engineering, Business Administration and Statistics, Universidad Politécnica de Madrid, 28006 Madrid, Spain
*   Correspondence: sarah.behnam@gmail.com

Academic Editor: Marc Rosen

Abstract: Sustainable development has attracted the increasing attention of both researchers and practitioners. While academicians and practitioners' focus towards sustainability has shifted to innovation, there is a need to understand how sustainability and innovation are interlinked. Thus, this paper attempts to analyze, first, the bidirectional impact of the firms' pursuit of sustainability and innovation as the priority, second, the bidirectional impact of the adoption of sustainability innovation action programs and, third, to discern the bidirectional influence of sustainability and innovation performances. The evidence is drawn from a sample of 860 manufacturing plants in 22 countries from the sixth edition of the International Manufacturing Strategy Survey 2013. The survey was conducted using a self-administered questionnaire. Structural equation modelling has been employed to test the model. The results show that sustainability and innovation positively and significantly impact each other in terms of the adoption of their relevant action programs and performance. However, the pursuit of sustainability priority acts as an antecedent of innovation priority.

Keywords: sustainability; innovation; mutual influence; priority; action programs; performance

---

1. Introduction

Since the release of World Commission on Environment and Development Report in 1987 (commonly known as the 'Brundtland Commission Report'), research on sustainability has attracted increasing attention. The main concern of sustainability is meeting the needs of the present without compromising the ability of future generations to meet their own needs [1]. Sustainable development encompasses a triple bottom line, a concept developed by [2], which integrates economic, environmental and social issues in operations. While environmental sustainability refers to consuming natural resources in a more advantageous manner and producing less emission to preserve the ecosystem [3], social sustainability refers to the skills' preservation and enhancing health and quality of life [4]. This study, by adopting the sustainability concept considering both environmental and social perspectives simultaneously, primarily attempts to provide important value to the literature, especially in the operations management literature [5].

Studying sustainability from the Operations Management (OM) field is essential since companies have to consider the footprint left behind in terms of the resources used (e.g., energy) [3]. Moreover, companies are required to operate responsibly for employees' health and safety and society's welfare. To conclude, "given the impact of the manufacturing industry on the environment, people and economy, OM gives new opportunities to significantly contribute to sustainability" [6] (p. 1).

Despite the fact that sustainability has been studied extensively for decades, practitioners, as well as academics still interrogate whether or not existing paths of businesses are sustainable [7]. As a result, recently, scholars suggest innovation as one of the possible ways to achieve sustainability, because in order to develop more sustainable products/services, companies need to change their way of doing things: innovate [7]. Accordingly, scholars claim that the incorporation of sustainability issues in innovation is needed [8,9]. Innovation is defined in any form of new products, processes, methods, markets and supply sources [10]. The core perspective of innovation is the novelty and change that must be achieved through a noticeable change from the previous product, process, service or business model [11]. Similar to several previous studies, (e.g., [12]), our notion of innovation is related to the change and novelty of the firm as an institutional context.

Companies require innovative responses, either incremental (doing better) or radical (doing different) [13]. In a similar vein, a broad understanding exits that the sustainability challenges propose meaningful opportunities for innovation [14]. Thus, sustainability challenges act as the source of change and the driver for innovation [15]. Thus, several pieces of evidence are documented on cases where sustainability orientation influences firms to see innovation as a priority or vice versa; the innovation orientation leads firms to consider sustainability as a priority, (e.g., [13,14]). This paper aims at expanding this debate by testing the mutual influence of sustainability and innovation as a business priority in a large sample of firms. To the best of our knowledge, such a large-scale empirical test has not been investigated in the previous studies.

When firms consider sustainability and/or innovation as a priority, they adopt the relevant action programs, (e.g., [14,16]). Moreover, scholars show the positive relationship between the internal sustainability action programs and managing new external insights for sustainability (referred to as external sustainability programs) [3,17]. In this regard, it is argued that environmental and social sustainability development relates to the exploration of product alternatives and, thus, to actions coordinating new product development, (e.g., [18]). The results of the studies propose the impact of the adoption of sustainability programs on innovation programs or vice versa. However, there is a lack of investigation on their simultaneous interaction to shed light on whether firms should focus on one before the other or if they should adopt them at the same time. Additionally, the majority of studies lie in showing successful cases or projects, and there is a lack of generalizable large-scale empirical investigations. That is why we further aim at testing the bidirectional influence of the adoption of sustainability and innovation action programs.

Finally, the adoption of sustainability or innovation programs enhances the business performance, whereas prior investigations on the inter-relationship between sustainability and innovation either lack taking into account the performance perspective or investigate the unidirectional relationship (through showing cases where either one impacts the other) [19]. However, investigating the bidirectional influence of sustainability and innovation performance is important because it can suggest the synergetic effect between them. To satisfy this need, we extend the course of this paper to test empirically the bidirectional influence of innovation and sustainability performance.

To do so, we target firms globally because of the need for the generalizability of the results. Moreover, aligned with Pagell and Gobeli's recommendation [20], we analyze the data gathered from a sample of individual plants unlike the majority of the previous research that considers a sample of companies [6]. The advantage of scrutinizing individual plants is the ability to study day-to-day decisions and the exact implementation of sustainable and innovative action programs.

To conclude, since the attention of recent sustainability studies has shifted to innovation management as a priority [21,22], the purpose of this study is to investigate on a large and generalizable empirical basis the relationship between sustainability and innovation on three different levels: the firms' pursuit of sustainability and innovation as the priority, the implementation of their relevant action programs and the performance achieved.

Through this analysis, this paper contributes not only to the sustainability research field, but also to the operation management and innovation fields. Moreover, the study aims at proposing new foundations in further understanding the determinants for sustainable development and innovation management. In particular, while there are fruitful investigations addressing the relationships between sustainability and innovation priorities, programs and performance, they are either focused on a few constructs or demonstrating cases/projects and do not investigate the bidirectional relationships in a generalizable sample of firms. We address this gap and present a more comprehensive study of the relationships between and among sustainability and innovation priorities, action programs and performance.

Moreover, we provide some guidance to managers when deploying sustainability and/or innovation priorities. In particular, the paper gives insights to practitioners on whether they can generally leverage superior sustainability and innovation action programs/performance by taking advantage of their mutual influence.

The structure of this paper is as follows: First, we review the literature on the link between sustainability and innovation. We then present the methodology used. Next, we present and discuss our results. Lastly, we present our conclusions, the limitations of this study, its managerial implications and lines for further research.

2. Literature Review

Scholars propose that sustainability has been emerging in addition to the traditional competitive priorities, (e.g., [23,24]). It is suggested that firms increasingly are integrating sustainability priorities in their business [25]. Furthermore, according to several surveys, the majority of firms take sustainability into account in developing and marketing new products, (e.g., [26]). The fact that sustainable firm pursue innovation as a priority suggests that a sustainability orientation triggers firms to acknowledge innovation as a business priority [14,15]. This argument emerges from two lines of reasoning. First, sustainability introduces additional contemporary visions for new business opportunities [27]; second, sustainability regulations push businesses to be innovative in order to find ways to comply with them without losing money (or even by increasing economic performance) [22,28]. In this regard, sustainability has been seen as a change force that generates new products and processes challenging existing practices in firms [29]. Porter contends that vigorous sustainability strategies would trigger innovation, stating: "properly constructed regulatory standards, which aim at outcomes and not methods, will encourage companies to re-engineer their technology" [30] (p. 96). Porter and van der Linde demonstrate multiple cases where a sustainability orientation leads to enhanced innovativeness in products and processes [31]. Indeed, the quest for sustainability has been shifting the competitive landscape, leading firms to re-think products, processes, if not business models. Particularly in times of economic crisis, the key to re-think products and processes is innovation. Nidumolu, Prahalad and Rangaswami, by scrutinizing longitudinally the sustainability initiatives of 30 large corporations, demonstrate that sustainability is a mother lode of innovations that generates both bottom-line and top-line returns [7], as environmental initiatives are able to decrease costs while generating additional revenues [32]. By pursuing sustainability as a priority, early movers are proven to view compliance as opportunity, make their value chain sustainable, design sustainable new products and processes, develop new business models and create next-practice platforms [7]. This would support also innovation strategies to be sustained in the long-term [33].

On the other hand, extensive practical illustrations are documented where firms' innovativeness motivates businesses to perceive sustainability as a priority, (e.g., [14,34,35]). In this regard, it is argued that, as innovations are able to target beyond economic goals, through aiming at social and environmental purposes, they trigger business towards sustainability priorities [36]; second, because innovation strategies cannot be sustained long term without being merged with sustainability strategic orientation [37]. We can conclude that, whatever the perspective adopted, it is clear that innovation and sustainability orientation influence each other [13,38–40]. What still remains as a gap is whether, on a generalizable sample of firms, the pursuits of sustainability and innovation priority influence each other. Thus, the following hypothesis is formulated:

H1a: *The pursuit of sustainability priority leads to the pursuit of innovation priority.*
H1b: *The pursuit of innovation priority leads to the pursuit of sustainability priority.*

Business priorities and programs have been proposed as a coupled bundle for strategy deployment [41]. In this regard, there is clear evidence of the pursuit of sustainability priority, which urges firms to adopt relevant action programs, (e.g., [14,16]). However, scholars differentiated between external vs. internal and environmental vs. social action programs, (e.g., [6,42]). In particular, internal environmental action programs are classified into: (1) environmental certifications; (2) energy and water consumption reduction programs; (3) pollution emission reduction and waste recycling programs; and internal social action programs include: (1) social certification; (2) formal sustainability oriented communication, training programs and involvement; (3) a formal occupational health and safety management system; (4) work/life balance policies, (e.g., [6,43]).

Recently, scholars show the positive relationship between the general internal sustainability action programs with external sustainability programs [3]. The majority of these investigations focus on sustainability-oriented supply chain programs [44]. However, fewer studies relate the external sustainability programs to managing new external insights for sustainability, (e.g., [17,45]). It is argued that environmental and social sustainability development relates to the exploration of alternatives and to actions related to the new product development [7,17]. To do so, one common action program is argued to be new product development coordination with the manufacturing processes [18,46–49]. The results of the studies propose that the adoption of sustainability programs impacts the adoption of innovation programs and vice versa; because coordination programs for product innovation include cross-functional integration, employee involvement and leveraging firms' ability to generate information, which in return would enhance the adoption of sustainability programs, (e.g., [50–52]). In this regard, case-based investigations illustrate that integrating innovation supports firms coping with the challenges of operational processes, particularly of implementing sustainability programs [53,54]. While, the majority of the studies lie in showing successful cases and/or projects, we are not aware of a simultaneous test on a large scale of firms on the bidirectional impact of the adoption of sustainability and innovation action programs. Thus, we hypothesize the following:

H2a: *The adoption of sustainability action programs leads to the adoption of innovation action programs.*
H2b: *The adoption of innovation action programs leads to the adoption of sustainability action programs.*

The adoption of sustainability and/or innovation action programs enhances the business performance in different ways [55] and plays a role in achieving business success, (e.g., [49,56–58]). Whereas prior investigations mostly lie on how sustainability performance impacts economic performance or organizational performance, (e.g., [17]), there are few investigations on the link between sustainability and innovation performance. It is argued that sustainability performance impacts product innovations [59] because ultimately, the sustainability contribution will be achieved when a viable new product is provided [40,60,61]. In this regard, for improving sustainability performance, a regular plan for all of the products is needed, leading to enhanced new product development performance [62].

On the other hand, it is argued that firms with higher innovation performance tend to achieve higher sustainability performance [26,48,63]. To conclude, while the literature highlights the critical role of examining performance in operational studies, prior research on the inter-relationship between sustainability and innovation lacks taking into account the performance dimension or investigating the unidirectional relationship [19]. In other words, few prior research that considers the performance dimension of sustainability and innovation focused on examples where either one impacts the other. However, we argue that taking into account the bidirectional influence of sustainability and innovation performance brings an understanding of the possible existence of the synergetic effect between them. Since there is a need to discern how innovation and sustainability performance would mutually contribute to each other on a generalizable scale [39], we hypothesize:

H3a: *Sustainability performance positively impacts innovation performance.*
H3b: *Innovation performance positively impacts sustainability performance.*

3. Methodology

A survey research methodology is congruent with the aim of this study because although the operation management literature is quite rich on each of the constructs of sustainability and innovation, previous studies lack generalizable investigations on their mutual inter-relationship. Moreover, a survey research methodology is aligned with other studies published in the managerial literature on sustainability action programs and performance [6,64,65].

To test the above research hypothesis, we used data collected in the sixth edition of the International Manufacturing Strategy Survey (IMSS VI), a research project carried out in 2013–2014 by a global network. The IMSS project, originally launched in 1992 by the London Business School and Chalmers University of Technology, studies manufacturing and supply chain strategies within the assembly industry (25–30 classifications of International Standard Industrial Classification (ISIC)—25, manufacture of rubber and plastics products; 26, manufacture of other non-metallic mineral products; 27, manufacture of basic metals; 28, manufacture of fabricated metal products, except machinery and equipment; 29, manufacture of machinery and equipment not elsewhere classified; 30, manufacture of office, accounting and computing machinery). It is carried out through a detailed questionnaire administered simultaneously in many countries by local research groups. The first section of the questionnaire is related to the general information about the business unit, while the other sections advise on the dominant activities of the plant (particularly focusing on business, strategies, action programs and performance). The unit of analysis is the plant to avoid issues related to diverse working manners of multiple plants of a business unit. To ensure alignment with the most recent trends in operations strategies, part of the questionnaire is redesigned in each edition. The update is performed by a design team composed of a pool of international researchers bypassing country biases of the team [66]. Aligned with the majority of the studies on priorities, programs and performances, we used responses from a single manager within each plant (e.g., operations, manufacturing, general or technical) [60]. This implies the assumption that such a manager has adequate and precise related information [67]. Responses have been gathered in a unique global database [68]. The sample consists of 931 manufacturing plants from 22 different countries, with an average response rate of 36 percent (Table 1). Data have been collected from May 2013–March 2014. Firms have been selected through three different ways: convenience sampling, random sampling and firms that participated in previous versions of the surveys.

Table 1. Firms' descriptive data.

	Number of Firms Contacted	Number of Firms Agreeing to Participate	Number of Responses	Valid Responses in the Final Release	Agreement Rate	Valid Response Rate (on Contacted Firms)	Valid Response Rate
TOTAL	7167	2586	1003	931	36.1%	13.0%	36.0%

The quality of the global database has been checked for all respondents. Cases with more than 60% of answers missing were deleted. Non-response and late response bias have been checked for all of the countries' database, except for a few cases: Hungary, around half of the Norway cases and Germany. Two procedures are carried out on the country level for checking the non-response bias (objective measures (obtain sales, number of employees and Standard Industrial Classification code figures for respondents and non-respondents; these figures are normally available in most databases; t-test comparing sales_respondents and sales_nonrespondents; t-test comparing employees_respondents and employees_nonrespondents; chi^2 test comparing SICCode_respondents and SICCode_nonrespondents) and contact non-respondents (contact non-respondents and ask them a couple of questions that can be important for checking non-response bias, e.g., existence of a formalized manufacturing strategy, manufacturing performance, strategic relevance of the manufacturing function, ongoing restructuring process, etc.; if possible, try to ask for the reason why they do not want to answer the questionnaire)) and one procedure for late-response bias (obtain sales, number of employees and SIC code figures for early respondents and late respondents). In addition, for questionnaires gathered on paper, fault-proof methods have been used for data consistency. Furthermore, in some countries, tests were performed for distinguishing between those answers collected on "paper" and those answers collected via "electronic survey". No noticeable pattern suggesting a bias was found. Moreover, "since data were collected from a single person at a single point in time, common method variance (CMV) might be a threat to the validity of our results" [6] (p. 153). To control the CMV, in the questionnaire and research design, the following measures were considered [69]: (1) respondent anonymity/confidentiality was protected; (2) the questions are designed to be as clear and concise as possible; (3) constructs' questions were distributed in different sections of the questionnaire [70–72]. In particular, questions on action programs are asked in different sections of the questionnaire, and these programs are separated from priorities and performances.

To ensure the validity of the survey variables, the project team assessed content validity. The content validity analysis is carried out through: (1) a review of questions for face validity; (2) the process of variable construction [73,74]. The team thoroughly reviewed the existing literature to establish appropriate domains and to extend them (the data were collected in 1992, 1996, 2001, 2005, 2009 and 2013–2014; the sixth version). Thus, the questions are formulated by a discussion of several academicians.

In addition, the confirmatory factor analysis of the SEM method is applied to confirm the convergent and discriminant validity (through testing the relations among the observables and the uni-dimensionality of the constructs) [73].

In the next step, the quality of the part of the database used particularly for this study has been checked. Missing data replacement was carried out in two steps: first, responses lacking one or more complete construct are eliminated (860 valid respondents remained); second, within these responses, the missing values were replaced with the mean of the series [75]; finally, the coherency of the replaced values with the other responses was checked. The distribution of the sample, in terms of country and valid response, rates is shown in Table 2.

Table 2. Sample descriptive data.

Country	N	%	Valid Response Rate (on Contacted Firms)	Valid Response Rate
Belgium	27	3%	25.9%	65.9%
Canada	23	3%	20.4%	33.7%
China	113	13%	26.1%	79.2%
Denmark	36	4%	12.6%	30.2%
Finland	31	4%	6.2%	40.5%
Germany	14	2%	9.7%	23.4%
Hungary	55	6%	17.6%	26.8%
India	87	10%	18.2%	19.9%
Italy	44	5%	17.0%	35.0%
Malaysia	13	2%	5.6%	46.7%
Netherlands	48	6%	14.8%	51.0%
Norway	26	3%	23.6%	53.5%
Portugal	30	3%	26.0%	65.4%
Romania	39	5%	8.0%	21.5%
Slovenia	17	2%	6.7%	34.0%
Spain	26	3%	11.3%	12.9%
Switzerland	24	3%	15.8%	37.0%
Taiwan	26	3%	5.6%	48.3%
USA	37	4%	3.2%	75.0%
Japan	82	10%	48.0%	82.0%
Sweden	32	4%	17.3%	19.9%
Brazil	29	3%	15.9%	73.8%
TOTAL	**860**	**100.0%**	**13.0%**	**36.0%**

Data have been tested primarily for the normality check by the skewness and kurtosis tests. The results reveal that the normality assumption is rejected for almost all of the variables. Even though the samples are ordinal and shown to be non-normal, maximum likelihood (ML) is considered despite its assumption of a continuous and normal sample, since previous scholars contend that treating ordinal data as continuous will result in negligible the underestimation of path coefficients, factor loadings and correlations [76]. We did not control for industry, because the sample of the IMSS survey is being restricted to what is called as assembly industries (ISIC 25–30), meaning that the examined industries are already homogeneous in nature. However, we controlled for size, perceived trend of environmental/social pressure and technological change. Following, the summary statistics of the dataset are reported (Table 3).

Table 3. Summary statistics of the dataset.

Variable	Obs	Mean	SD	Min	Max
Control Variables					
Industry	860	26.83605	1.559037	25	30
Country	860	10.77907	6.511209	1	22
Age	807	42.75341	31.5413	3	243
Size	860	2.532558	0.5523665	1	3
Environmental pressure	860	3.310744	1.046832	1	5
Social pressure	860	3.236558	1.06932	1	5
Technological change	860	3.306174	0.9773198	1	5
Latent variables of sustainability as priority					
Environmental sustainability as priority					
More environmentally-sound products and processes	860	3.257326	1.044634	1	5
Social sustainability as priority					
1. Higher contribution to the development and welfare of society	860	3.0315	1.108833	1	5
2. More safe and health respectful processes	860	3.407244	1.117578	1	5
Latent variables of innovation as priority					
Incremental: offering new products more frequently	860	3.253198	1.078175	1	5
Radical: offering products that are more innovative	860	3.614267	1.040566	1	5
Latent variables of sustainability action programs					
Environmental action programs					
1. Environmental certifications	860	3.264	1.412703	1	5
2. Energy and water consumption reduction programs	860	3.078279	1.157934	1	5
3. Pollution emission reduction and waste recycling programs.	860	3.11136	1.202645	1	5
Social action programs					
1. Social certifications	860	2.623488	1.470521	1	5
2. Formal sustainability oriented communication, training programs and involvement	860	2.871291	1.185365	1	5
3. Formal occupational health and safety management system	860	3.375023	1.125752	1	5
4. Work/life balance policies	860	2.751453	1.167722	1	5
Latent variables of innovation action programs					
1. Informal mechanisms	860	3.3275	1.013474	1	5
2. Design integration between product development and manufacturing	860	3.189477	1.082889	1	5
3. Organizational integration between product development and manufacturing	860	3.064302	1.064542	1	5
4. Technological integration between product development and manufacturing	860	3.020047	1.167397	1	5
5. Integrating tools and techniques	860	2.977721	1.236134	1	5
6. Communication technologies	860	3.245058	1.148234	1	5
7. Forms of process standardization	860	3.137674	1.140275	1	5
Latent variables of innovation performance					
Incremental: product customization ability	860	3.076233	1.00641	1	5
Radical: new product introduction ability	860	3.207767	1.009979	1	5
Latent variables of sustainability performance					
Social sustainability performance					
1. Workers' motivation and satisfaction	860	2.889023	0.9458301	1	5
2. Health and safety conditions	860	3.258186	0.9442817	1	5
Environmental sustainability performance					
1. Materials, water and/or energy consumption levels	860	2.575907	0.9307496	1	5
2. Pollution emission and waste production	860	2.806709	0.9315846	1	5

4. Operationalization of the Variables

The variables (observed variables) of all constructs (latent variables) were grounded based on previous research studies. All attributes are measured by a Likert scale of 1–5.

Business priorities: Respondents were asked to rate the importance of the competitive priorities for winning orders from the major customers, on a five-point Likert scale (1 = not important to 5 = very important). According to the literature, sustainability encompasses both environmental [6] and social [77] perspectives [3]. In IMSS, both variables are measured. We categorized these priorities into two groups: first, sustainability, which is composed by more environmentally-sound products and processes, a higher contribution to the development and welfare of society, safer and health respectful processes.

According to the literature, innovation has been differentiated mainly between product/process/ organizational [78] or radical/incremental [79,80]. However, in IMSS, the attributes relate to the radical/incremental innovations, but are limited to product innovation. The reason lies within the inability to assess process/organizational innovation with measuring the importance of winning the order from the customers' point of view, (e.g., [81]). Thus, innovation priority has been explored through the priority of offering new products more frequently and offering products that are more innovative, aligned with the way the innovator strategy is generally measured [82].

Action programs: In a similar vein to the course of this paper, we categorized the sets of action programs into two groups:

(1) Sustainability:

Respondents were asked to rate the level of effort put into the implementation of the action programs in the last three years (1 = none to 5 = high), (e.g., [6,83,84]). Following the framework developed in [5], both environmental and social internal programs are assessed for measuring sustainability action programs.

Environmental programs include, (e.g., [6,20,85–89]):

- Environmental certifications (e.g., Eco-Management and Audit Scheme (EMAS)or International Organization for Standardization-ISO 14001)
- Energy and water consumption reduction programs
- Pollution emission reduction and waste recycling programs.

Social programs include [6,90–92] (an adaptation of the scale used by [93]):

- Social certifications (e.g., SA8000 or OHSAS 18000)
- Formal sustainability-oriented communication, training programs and involvement
- Formal occupational health and safety management system
- Work/life balance policies

(2) Innovation:

Respondents were asked to indicate the effort in the last three years put into implementing action programs to coordinate the new product development and manufacturing processes [94]. Following prior investigations [95,96], the innovation programs investigated in this survey are:

- Informal mechanisms, such as direct, face-to-face communication, informal discussions and ad hoc meetings
- Design integration between product development and manufacturing through, e.g., platform design, standardization and modularization, design for manufacturing, design for assembly
- Organizational integration between product development and manufacturing through, e.g., cross-functional teams, job rotation, co-location, role combination, secondment and coordinating managers

- Technological integration between product development and manufacturing through, e.g., CAD-CAM, CAPP, CAE, product lifecycle management
- Integrating tools and techniques, such as failure mode and effect analysis, quality function deployment and rapid prototyping
- Communication technologies, such as teleconferencing, web meetings, intranet and social media

Forms of process standardization, such as a stage-gate process, design reviews and performance management.Performance: Respondents were asked to rate the level of change in manufacturing performance over the last three years (1 = decrease −5% or worse, 2 = stayed about the same −5%–+5%, 3 = slightly increased +5%–+15%, 4 = increased = 15%–25%, 5 = strongly increased +25% or better). Aligned with the previous measurements of sustainability and innovation, their performances are measured. In particular, sustainability performance is measured in terms of environmental performance (materials, water and/or energy consumption, pollution emission and waste production levels) [83,97,98] and social performance (workers' motivation and satisfaction and health and safety conditions) [4,6,20,77]. Innovation performance is measured through a bundle of product customization ability and new product introduction ability [95,99].

5. Analysis and Findings

The structural equation modelling (SEM) approach has been employed to test the hypotheses [100]. It is a statistical method with a confirmatory approach [101] in which constructs' validity (measurement model) and relationships (structural model) are tested in a hypothesized model concurrently. Moreover, the advantage of the SEM method compared to traditional statistical techniques is the fact that it allows the measurement of several variables and their inter-relationship simultaneously, thanks to an indication of model fitness [102,103]. SEM has attracted increasing attention because of the need for testing a complex phenomenon, which consists of several items, respondents, relationships and combined effects (synergies) [104]. Accordingly, in this paper, this method is adopted for its capability in testing the combination of the hypotheses simultaneously with a large number of respondents.

First, the necessary conditions for model identification were assessed. There is p = 25 observable variables; thus, the non-redundant data points in the sample variance covariance matrix is given by: p = (p(p + 1))/2 = 325. The number of parameters to be estimated is 62, so the model has the necessary condition and can be over identified. The second tested condition is related to the measurement portions of the model. The first regression coefficient of each construct is constrained to one for establishing the scale of factors.

Finally, we specified a model based on our hypothesized framework providing also the parameter values for significant paths (Figure 1 and Table 4). The overall fit for the path model was acceptable (Chi^2 = 263). Even though the overall fitness of the model was acceptable, we used modification indices to enhance the model with respect to the theoretical considerations. Significantly, covariance between two measurements for each sustainability action program and performance were suggested: energy/material and water consumption (reduction programs) and pollution emission reduction and waste recycling programs/production level). Subsequently, further theoretical investigation of the variables shows that the correlations are justifiable based on the theoretical background due to the fact that both measurements are targeting environmental sustainability programs. Based on previous studies, the environmental programs may be correlated in a sense that firms with proactive environmental strategies engage typically to enhance diverse type of environmental programs [105,106]. As a consequence, the covariance of the environmental programs is considered in the final model with improved overall fit (Chi^2 = 262; Table 4).

Table 4. Overall fit of the model and values of the hypothesized relationships.

Fit Statistic	Value	Description
Likelihood ration		
Chi^2_ms (262)	1051.943	Model vs. saturated
$p > Chi^2$	0.000	
Chi^2_bs (300)	10,451.881	Baseline vs. saturated
$p > Chi^2$	0.000	
Population error		
RMSEA	0.059	Root mean squared error of approximation
90% CI, lower bound	0.055	
Upper bound	0.063	
p close	0.000	Probability RMSEA ≤ 0.05
Baseline comparison		
CFI	0.922	Comparative fit index
TLI	0.911	Tucker–Lewis index
Size of residuals		
SRMR	0.044	Standardized root mean squared residual
CD	0.929	Coefficient of determination

Values of the Hypothesized Relationships (Structural Model)	Coef.	OIM Std. Error	$p > IZI$
Innovation as priority <-			
Sustainability as priority	0.567	0.046	0.000
Sustainability action program <-			
Innovation action program	0.497	0.131	0.000
Sustainability as priority	0.497	0.053	0.000
Sustainability Performance <-			
Sustainability action programs	0.138	0.028	0.000
Sustainability as priority	0.072	0.029	0.012
Innovation performance	0.348	0.076	0.000
Innovation action program <-			
Innovation as priority	0.133	0.027	0.000
Sustainability action program	0.332	0.042	0.000
Innovation performance <-			
Innovation as priority	0.215	0.039	0.000
Innovation action program	0.237	0.061	0.000
Sustainability performance	0.335	0.116	0.004

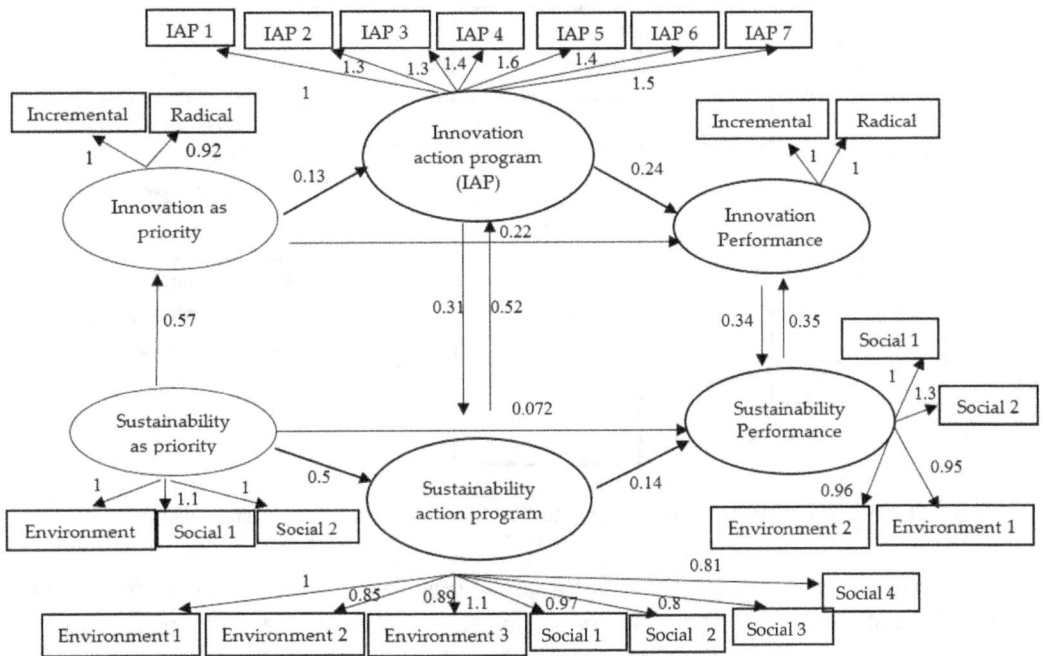

Figure 1. Structural equation model based on our second hypothesized model.

As already widely proven by the literature, the pursuit of priorities positively impacts firms' performance with the mediating role of the adoption of action programs. This is supported, in the model, with the proof of a partial mediation role for innovation and a fully mediating role for sustainability (considering a 99% confidence interval) (partial mediating role for both sustainability and innovation considering a 95% confidence interval).

The results of the analyses provide mixed support to the hypotheses.

Hypothesis 1a, stating that the pursuit of sustainability priority positively impacts innovation priority, is fully supported. However, Hypothesis 1b, stating that the pursuit of innovation priority positively impacts sustainability priority, is not supported. In other words, if the bidirectional relationship is considered, both fall into insignificant. However, when unidirectional relationships are considered, only the impact of sustainability as priority on innovation priority falls into significant and positive.

Hypothesis 2a,b stating that the adoption of sustainability action programs impacts the adoption of innovation action programs to coordinate new product development and manufacturing processes is fully supported. This means that on the level of effort firms provide for the adoption of innovation and sustainability action programs, not only innovation programs are functional to adopt more sustainability programs, but also sustainability programs enhance the adoption of innovation programs.

In a similar vein, Hypothesis 3a,b, stating that sustainability performance positively impacts innovation performance and vice versa, is also fully supported: innovation and sustainability performance improvement are highly correlated and show combined (synergetic) effects.

6. Discussion and Future Research

The results obtained by testing of the research hypotheses can be summarized in the framework depicted in Figure 2.

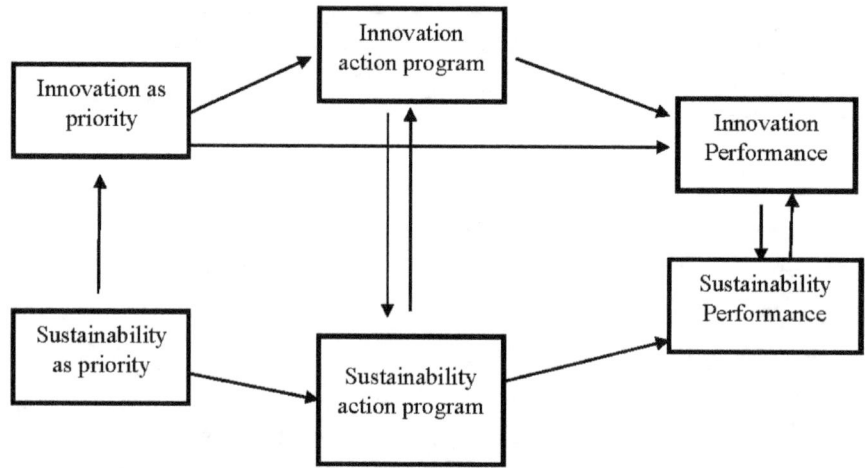

Figure 2. Framework for the relationship of innovation and sustainability.

The results reveal appealing insight about the relationship between sustainability and innovation as priorities. This study shows that the pursuit of sustainability as a priority predicts pursuing innovation as a priority according to a generalizable sample of firms; meaning that innovation becomes an order winner when it is driven by sustainable orientation. Our results added value, thanks to its generalizability, to previous proposals of scholars through some cases that sustainability acts a driver of innovation (e.g., based on 30 big corporations in [7]). Our results suggest that innovating without sustainability consideration is no longer a strong competitive priority for firms. Thus, we propose that the sustainable product innovation (as a result of interaction between sustainability and innovation) is emerging as a competitive priority in practice. In this regard, the environmental (green/eco) product innovation is well-studied in the literature [107–109]. However, we argue that the social product innovations also turn out to be a business priority [110]. Accordingly, we suggest further research extending the debate of social product innovations as a business priority. The results, suggesting that the innovation priorities do not directly impact sustainability priorities, imply the fact that firms should anticipate sustainability priorities over innovation priorities.

Second, looking at the adoption of action programs, the results of our study confirm on a wider scale the positive correlation between sustainability and innovation action programs [9]. Therefore, we argue that the adoption of sustainability and innovation action programs positively and significantly impacts one another [53]; because coordination programs for product innovation include organizational integration (and cross-functional teams), which in return would enhance the adoption of sustainability programs, (e.g., [50,111,112]). Moreover, the coordination programs for product innovation enhance the employee involvement which in return is a critical element for implementing sustainability programs [51,87,92,113–115]. Pagell and Wu contributed to this debate by arguing that innovative firms leverage their ability to generate beneficial information, which supports establishing new sustainability programs [52]. To conclude, our results confirm many case-based studies arguing that innovative firms integrating innovation with manufacturing processes allow businesses to face fewer obstacles to the operational processes of implementing sustainability programs and are better equipped to cope with the challenges [53,54].

To conclude, the adoption of innovation coordination programs enhances the adoption and implementation of sustainability programs [7,30,52,54,116]. However, the majority of the above studies focuses on environmental programs. This study shows that innovation programs act as a predictor also for social sustainability management ([116], based on case data). This preliminary evidence call for further empirical investigation on the relationship between innovation and social sustainability programs may be from the process innovation perspective [54].

This study also shows the significant impact of the adoption of sustainability programs on innovation programs, which has been rarely investigated [117]. This finding can be justified as sustainability is proposed to be a facilitator of the product design integration [117]. Alternatively, the implementation of sustainability programs is argued to be more than just a technical process and also requires organizational redesign [118,119]. Moreover, it is argued that through adopting sustainability programs (e.g., certifications), firms can establish innovative technologies [119].

Finally, this study contributes to the literature on the interrelationship between sustainability and innovation performance, (e.g., [6,83,97,98]). The results demonstrate the positive and significant correlation of these two performance dimensions. Thus, we argue that organizations with higher sustainability performance are more innovative and vice versa [120]. Accordingly, plants that simultaneously pursue exploratory innovative programs and exploitative sustainability programs are able to enhance sustainability in existing systems and are able to develop new products [17]. This is aligned with previous scholars suggesting that innovative companies perform best at sustainability [7,52]. Gualandris and Kalchschmidt contributed to this debate by suggesting that high innovation performance may not be sufficient to guarantee high sustainability performance, but its absence may hinder it [53]. Our results, based on large-scale and multi-national empirics, confirm that innovation is not only critical for operational performances (e.g., cost and/or quality), but also it is valuable for environmental and social benefits.

We believe this study drives several valuable contributions: first, we analyze the bidirectional impact of sustainability and innovation; second, by considering both environmental and social perspectives of sustainability, as well as the performances of sustainability and innovation (we are not aware of any clear test that simultaneously does so), so filling a gap in the existing literature. In addition, by scrutinizing the impact of sustainability performance on innovation performance, we propose a new area where enhancing sustainability performance is paid back to the firms.

The paper provides managers with clear evidence of the necessity for sustainability as a priority and endeavors for enhanced innovation at the operational level. Moreover, the study proposes new foundations in further understanding the determinants for sustainable development and innovation. It can support firms to implement sustainability action programs within their operational processes by empirically analyzing how innovation, and in particular, innovation coordination programs, can increase environmental and social sustainability performance.

While this study provides a compelling contribution to the sustainability and innovation literature and possesses valuable implications for practice, there are some limitations and opportunities for future studies. First, the applied measurement items to measure environmental, social and innovation as the priority, program and performance may be considered as a limitation of this study, since the team was required to keep the IMSS questionnaire to a reasonable length, which made the researchers select only the most relevant items. Thus, we suggest using different measurements. Moreover, since sustainability and innovation interlinks evolve over time, examining the patterns in a longitudinal study will be advantageous. Future studies can also extend this paper's scope by considering other industries. Moreover, we suggest future research to investigate possible differences between plants placed in various continents when it comes to sustainable development and innovation management. Finally, innovation measurements in this study are limited to new product developments, while future research needs to take into account other taxonomies of innovation, including new processes, management systems or business models.

Acknowledgments: This paper is produced as part of the Erasmus Mundus Joint Doctorates (EMJDs) Programme, European Doctorate in Industrial Management (EDIM), funded by the European Commission, Erasmus Mundus Action 1. For their enormous effort in gathering the data, we acknowledge all members of the IMSS project, sixth version.

Author Contributions: All authors have designed the research and conclusions. Both authors contributed to performing the IMSS project as part of the Politecnico di Milano group, with Raffaella Cagliano as the main contributor. Sarah Behnam was the main contributor in reviewing the literature and analyzing the data. Raffaella Cagliano guided the paper editing. All authors read and approved the final manuscript.

Conflicts of Interest: The authors declare no conflict of interest.

References

1. WCED (World Commission on Environment and Development), United Nations. *Our Common Future*; World Commission on Environment and Development Oxford University Press: Oxford, UK, 1987.
2. Elkington, J. *Cannibals with Forks: The Triple Bottom Line of 21st Century*; Capstone Publishing Ltd.: Oxford, UK, 1997.
3. Kleindorfer, P.R.; Kalyan, S.; Luk, N.W. Sustainable operations management. *Prod. Oper. Manag.* **2005**, *14*, 482–492. [CrossRef]
4. McKenzie, S. *Social Sustainability: Towards Some Definitions*; Hawke Research Institute: Magill, Australia, 2004.
5. Gimenez, C.; Sierra, V.; Rodon, J. Sustainable operations: Their impact on the triple bottom line. *Int. J. Prod. Econ.* **2012**, *140*, 149–159. [CrossRef]
6. Cagliano, R.; Golini, R.; Longoni, A. *The Role of NFWO in Sustainability Strategies: An OM Perspective*; Sousa, R., Ed.; Catholic University of Portugal: Porto, Portugal, 2010; pp. 1–10.
7. Nidumolu, R.; Prahalad, C.K.; Rangaswami, M.R. Why sustainability is now the key driver of innovation. *Harv. Bus. Rev.* **2009**, *87*, 56–64.
8. De Medeiros, J.F.; Ribeiro, J.L.D.; Cortimiglia, M.N. Success factors for environmentally sustainable roduct innovation: A systematic literature review. *J. Clean. Prod.* **2014**, *65*, 76–86. [CrossRef]
9. Gmelin, H.; Seuring, S. Determinants of a sustainable new product development. *J. Clean. Prod.* **2014**, *69*, 1–9. [CrossRef]
10. Schumpeter, J.A. *Business Cycles*; McGraw-Hill: New York, NY, USA, 1939; Volume 1.
11. Harper, S.M.; Selwyn, W.B. On the leading edge of innovation: A comparative study of innovation practices. *South. Bus. Rev.* **2004**, *29*, 1.
12. Kemp, R.; Peter, P. *Measuring Eco-Innovation*; United Nations University: Maastricht, The Netherlands, 2008.
13. Seebode, D.; Sally, J.; John, B. Managing innovation for sustainability. *R&D Manag.* **2012**, *42*, 195–206.
14. Hansen, E.G.; Friedrich, G.-D.; Ralf, R. Sustainability innovation cube—A framework to evaluate sustainability-oriented innovations. *Int. J. Innov. Manag.* **2009**, *13*, 683–713. [CrossRef]
15. Noci, G.; Verganti, R. Managing 'green' product innovation in small firms. *R&D Manag.* **1999**, *29*, 3–15.
16. Adamczyk, S.; Hansen, E.G.; Reichwald, R. Measuring Sustainability by Environmental Innovativeness: Results from Action Research at a Multinational Corporation in Germany. In Proceedings of the International Conference and Doctoral Consortium on Evaluation Metrics of Corporate Social and Environmental Responsibility, Lyon, France, 8–10 June 2009.
17. Maletic, M.; Maletic, D.; Dahlgaard, J.J.; Dahlgaard-Park, S.M.; Gomiscek, B. Sustainability exploration and sustainability exploitation: From a literature review towards a conceptual framework. *J. Clean. Prod.* **2014**, *79*, 182–194. [CrossRef]
18. Schaltegger, S.; Beckmann, M.; Hansen, E.G. Transdisciplinarity in corporate sustainability: Mapping the field. *Bus. Strategy Environ.* **2013**, *22*, 219–229. [CrossRef]
19. Pujari, D. Eco-innovation and new product development: Understanding the influences on market performance. *Technovation* **2006**, *26*, 76–85. [CrossRef]
20. Pagell, M.; David, G. How plant managers' experiences and attitudes toward sustainability relate to operational performance. *Prod. Oper. Manag.* **2009**, *18*, 278–299. [CrossRef]
21. Hart, S.L.; Sharma, S. Engaging fringe stakeholders for competitive imagination. *Acad. Manag. Executive* **2004**, *18*, 7–18. [CrossRef]
22. Hockerts, K.; Morsing, M. *A Literature Review on Corporate Social Responsibility in the Innovation Process*; Center for Corporate Social Responsibility, Copenhagen Business School (CBS): Frederiksberg, Denmark, 2008; pp. 1–28.
23. De Burgos, J.J.; Jose, J.C.L. Environmental performance as an operations objective. *Int. J. Oper. Prod. Manag.* **2001**, *21*, 1553–1572. [CrossRef]
24. Porter, M.; Kramer, M.R. The link between competitive advantage and corporate social responsibility. *Harv. Bus. Rev.* **2006**, *84*, 1–24.
25. McKinsey Global Institute. *The Business of Sustainability*; McKinsey Global Institute: New York, NY, USA, 2013.

26. May, G.; Taisch, M.; Kerga, E. Assessment of sustainable practices in new product development. In *IFIP International Conference on Advances in Production Management Systems*; Springer: Berlin/Heidelberg, Germany, 2011; pp. 437–447.

27. Hart, S.L. Beyond greening: Strategies for a sustainable world. *Harv. Bus. Rev.* **1997**, *75*, 66–76.

28. Preuss, L. Contribution of purchasing and supply management to ecological innovation. *Int. J. Innov. Manag.* **2007**, *11*, 515–537. [CrossRef]

29. Martina, B.-K.; Hussain, S.S. Innovation and corporate sustainability: An investigation into the process of change in the pharmaceuticals industry. *Bus. Strategy Environ.* **2001**, *10*, 300.

30. Porter, M.E. Towards a dynamic theory of strategy. *Strateg. Manag. J.* **1991**, *12*, 95–117. [CrossRef]

31. Porter, M.E.; Van der Linde, C. Toward a new conception of the environment-competitiveness relationship. *J. Econ. Perspect.* **1995**, *9*, 97–118. [CrossRef]

32. Bos-Brouwers, H.E.J. Corporate sustainability and innovation in SMEs: Evidence of themes and activities in practice. *Bus. Strategy Environ.* **2010**, *19*, 417–435. [CrossRef]

33. Hall, J. Environmental innovation (Editorial). *J. Clean. Prod.* **2003**, *11*, 343–346. [CrossRef]

34. Hart, S.L.; Mark, B.M. Global sustainability and the creative destruction of industries. *MIT Sloan Manag. Rev.* **1999**, *41*, 23.

35. Westley, F.; Harrie, V. Interorganizational collaboration and the preservation of global biodiversity. *Organ. Sci.* **1997**, *8*, 381–403. [CrossRef]

36. Adams, R.; Jeanrenaud, S.; Besant, J.; Denyer, D.; Overy, P. Sustainability-oriented innovation: A systematic review. *Int. J. Manag. Rev.* **2015**, *18*, 180–205. [CrossRef]

37. Steiner, G. Supporting sustainable innovation through stakeholder management: A systems view. *Int. J. Innov. Learn.* **2008**, *5*, 595–616. [CrossRef]

38. Pujari, D.; Wright, G.; Peattie, K. Green and competitive: Influences on environmental new product development performance. *J. Bus. Res.* **2003**, *56*, 657–671. [CrossRef]

39. Longoni, A.; Cagliano, R. Environmental and social sustainability priorities: Their integration in operations strategies. *Int. J. Oper. Prod. Manag.* **2015**, *35*, 216–245. [CrossRef]

40. Gomez-Conde, J. Examining the link between outsourcing and performance: The leverage effect of the interactive use of management accounting and control systems. *Span. J. Financ. Account./Rev. Esp. Financ. Contab.* **2015**, *44*, 298–325. [CrossRef]

41. Longoni, A.; Golini, R.; Cagliano, R. The role of New Forms of Work Organization in developing sustainability strategies in operations. *Int. J. Prod. Econ.* **2014**, *147*, 147–160. [CrossRef]

42. Quesada, G.; Rachamadugu, R.; Gonzalez, M.; Luis Martinez, J. Linking order winning and external supply chain integration strategies. *Supply Chain Manag.* **2008**, *13*, 296–303. [CrossRef]

43. Van Kleef, J.A.G.; Roome, N.J. Developing capabilities and competence for sustainable business management as innovation: A research agenda. *J. Clean. Prod.* **2007**, *15*, 38–51. [CrossRef]

44. Biondi, V.; Iraldo, F.; Meredith, S. Achieving sustainability through environmental innovation: The role of SMEs. *Int. J. Technol. Manag.* **2002**, *24*, 612–626. [CrossRef]

45. Alblas, A.A.; Peters, K.; Wortmann, J.C. Fuzzy sustainability incentives in new product development: An empirical exploration of sustainability challenges in manufacturing companies. *Int. J. Oper. Prod. Manag.* **2014**, *34*, 513–545. [CrossRef]

46. Endris, K.; Marco, T.; Sergio, T.; Gokan, M. Integration of sustainability in NPD process: Italian Experiences. In Proceedings of the IFIP WG 5.1 8th International Conference on Product lifecycle Management (PLM2011), Eindhoven, The Netherlands, 11–13 July 2011.

47. Fish, L. *Recommendations for Implementing Sustainability in New Product Development for Supply Chain Management*; Business Research Consortium of Western New York: New York, NY, USA, 2015; p. 119.

48. Schaltegger, S.; Marcus, W. *Managing the Business Case for Sustainability*; Greenleaf Publishing: Sheffield, UK, 2006.

49. Wagner, M.; Schaltegger, S. How does sustainability performance relate to business competitiveness? *Greener Manag. Int.* **2003**, *44*, 5–16. [CrossRef]

50. Koufteros, X.A.; Nahm, A.Y.; Edwin Cheng, T.C.; Lai, K. An empirical assessment of a nomological network of organizational design constructs: From culture to structure to pull production to performance. *Int. J. Prod. Econ.* **2007**, *106*, 468–492. [CrossRef]

51. Hanna, M.D.; Newman, W.R.; Pamela, J. Linking operational and environmental improvement through employee involvement. *Int. J. Oper. Prod. Manag.* **2000**, *20*, 148–165. [CrossRef]

52. Pagell, M.; Wu, Z. Building a more complete theory of sustainable supply chain management using case studies of 10 exemplars. *J. Supply Chain Manag.* **2009**, *45*, 37–56. [CrossRef]

53. Gualandris, J.; Kalchschmidt, M. Customer pressure and innovativeness: Their role in sustainable supply chain management. *J. Purch. Supply Manag.* **2014**, *20*, 92–103. [CrossRef]

54. Christmann, P. Effects of "best practices" of environmental management on cost advantage: The role of complementary assets. *Acad. Manag. J.* **2000**, *43*, 663–680. [CrossRef]

55. Schaltegger, S.; Marcus, W. Managing the business case for sustainability. In Proceedings of the EMAN-EU 2008 Conference, Budapest, Hungary, 6–7 October 2008; Volume 7.

56. Wagner, M.; Schaltegger, S.; Wehrmeyer, W. The relationship between the environmental and economic performance of firms: What does theory propose and what does empirical evidence tell us? *Greener Manag. Int.* **2001**, *34*, 95–108. [CrossRef]

57. Totterdell, P.; Leach, D.; Birdi, K.; Clegg, C.; Wall, T. An investigation of the contents and consequences of major organizational innovations. *Int. J. Innov. Manag.* **2002**, *6*, 343–368. [CrossRef]

58. Zhang, Q.; Doll, W.J. The fuzzy front end and success of new product development: A causal model. *Eur. J. Innov. Manag.* **2001**, *4*, 95–112. [CrossRef]

59. Ottman, J.A. *Green Marketing: Challenges and Opportunities*; NTC Business Books: Linclonwood, IL, USA, 1994.

60. Pujari, D.; Wright, G. Integrating environmental issues into product development: Understanding the dimensions of perceived driving forces. *J. Eur. Mark.* **1999**, *7*, 43–63. [CrossRef]

61. Pujari, D.; Wright, G. Management of environmental new product development in charter. In *Greener Marketing*, 2nd ed.; Charter, M., Polonsky, M., Eds.; Greenleaf Publishing: London, UK, 1999.

62. Charter, M. Sustainable Consumption & Production, Business and Innovation. In *Perspectives on Radical Changes to Sustainable Consumption and Production (SCP)*; Sustainable Consumption Research Exchange: Copenhagen, Denmark, 2006; Volume 20, p. 243.

63. Boons, F.; Montalvo, C.; Quist, J.; Wagner, M. Sustainable innovation, business models and economic performance: An overview. *J. Clean. Prod.* **2013**, *45*, 1–8. [CrossRef]

64. Searcy, C. Corporate sustainability performance measurement systems: A review and research agenda. *J. Bus. Ethics* **2012**, *107*, 239–253. [CrossRef]

65. Crowe, D.; Brennan, L. Environmental considerations within manufacturing strategy: An international study. *Bus. Strategy Environ.* **2007**, *16*, 266–289. [CrossRef]

66. Van de Vijver, F.J.R.; Kwok, L. *Methods and Data Analysis for Cross-Cultural Research*; Sage: Riverside County, CA, USA, 1997; Volume 1.

67. Szwejczewski, M.; Mapes, J.; New, C. Delivery and trade-offs. *Int. J. Prod. Econ.* **1997**, *53*, 323–330. [CrossRef]

68. Lindberg, P.; Voss, C.A.; Blackmon, K. *International Manufacturing Strategies. Context, Content, and Change*; Kluwer Academic Publishers: Dordrecht, The Netherlands, 1998.

69. Conway, J.M.; Charles, E.L. What reviewers should expect from authors regarding common method bias in organizational research. *J. Bus. Psychol.* **2010**, *25*, 325–334. [CrossRef]

70. Chang, S.J.; Van Witteloostuijn, A.; Eden, L. From the editors: Common method variance in international business research. *J. Int. Bus. Stud.* **2010**, *41*, 178–184. [CrossRef]

71. Malhotra, N.K.; Kim, S.S.; Patil, A. Common method variance in IS research: A comparison of alternative approaches and a reanalysis of past research. *Manag. Sci.* **2006**, *52*, 1865–1883. [CrossRef]

72. Podsakoff, P.M.; MacKenzie, S.B.; Lee, J.Y.; Podsakoff, N.P. Common method biases in behaveral research: A critical review of the literature and recommended remedies. *J. Appl. Psychol.* **2003**, *88*, 879–903. [CrossRef] [PubMed]

73. Nunnally, J. *Psychometric Methods*; McGraw-Hill: New York, NY, USA, 1978.

74. Widener, S.K. An empirical analysis of the levers of control framework. *Account. Organ. Soc.* **2007**, *32*, 757–788. [CrossRef]

75. Downey, R.G.; King, C.V. Missing data in Likert ratings: A comparison of replacement methods. *J. Gen. Psychol.* **1998**, *125*, 175–191. [CrossRef] [PubMed]

76. West, S.G.; Finch, J.F.; Curran, P.J. *Structural Equation Models with Nonnormal Variables: Problems and Remedies*; Sage Publications: Riverside County, CA, USA, 1995.

77. Maxwell, D.; van der Vorst, R. Developing sustainable products and services. *J. Clean. Prod.* **2003**, *11*, 883–895. [CrossRef]

78. Gritti, P.; Leoni, R. High Performance Work Practices, Industrial Relations and Firm Propensity for Innovation. In *Advances in the Economic Analysis of Participatory and Labor-Managed Firms*; Bryson, A., Ed.; Emerald Group Publishing Limited: Bingley, UK, 2012; Volume 13, pp. 267–309.

79. Hult, G.; Tomas, M.; Robert, F.H.; Gary, A.K. Innovativeness: Its Antecedents and Impact on Business Performance. *Ind. Mark. Manag.* **2004**, *33*, 429–438. [CrossRef]

80. Jay, S.K.; Peter, A. Manufacturing Competence and Business Performance: A Framework and Empirical Analysis. *Int. J. Oper. Prod. Manag.* **1993**, *13*, 4–25.

81. Bisbe, J.; Otley, D. The effects of the interactive use of management control systems on product innovation. *Account. Organ. Soc.* **2004**, *29*, 709–737. [CrossRef]

82. Miller, J.; Roth, A. A taxonomy of manufacturing strategies. *Manag. Sci.* **1994**, *40*, 285–304. [CrossRef]

83. Zhu, Q.; Joseph, S. Relationships between operational practices and performance among early adopters of green supply chain management practices in Chinese manufacturing enterprises. *J. Oper. Manag.* **2004**, *22*, 265–289. [CrossRef]

84. Zhu, Q.; Joseph, S.; Kee-hung, L. Confirmation of a measurement model for green supply chain management practices implementation. *Int. J. Prod. Econ.* **2008**, *111*, 261–273. [CrossRef]

85. Berkhout, F.; Verbong, G.; Wieczorek, A.J.; Raven, R.; Lebel, L.; Bai, X. Sustainability experiments in Asia: Innovations shaping alternative development pathways? *Environ. Sci. Policy* **2010**, *13*, 261–271. [CrossRef]

86. Klassen, R.D.; Whybark, D.C. The Impact of Environmental Technologies on Manufacturing Performance. *Acad. Manag. J.* **1999**, *42*, 599–615. [CrossRef]

87. Kitazawa, S.; Sarkis, J. The relationship between ISO 14001 and continuous source reduction programs. *Int. J. Oper. Prod. Manag.* **2000**, *20*, 225–248. [CrossRef]

88. Russo, M.V. Explaining the impact of ISO 14001 on emission performance: A Dynamic Capabilities Perspective on process learning. *Bus. Strategy Environ.* **2009**, *18*, 307–319. [CrossRef]

89. Sarkis, J. Evaluating environmentally conscious business practices. *Eur. J. Oper. Res.* **1998**, *107*, 159–174. [CrossRef]

90. Florida, R. Lean and green: The move to environmentally conscious manufacturing. *Calif. Manag. Rev.* **1996**, *39*, 80–105. [CrossRef]

91. Longo, M.; Mura, M.; Bonoli, A. Corporate Social Responsibility and Corporate Performance: The Case of Italian SMEs. *Corp. Gov.* **2005**, *5*, 28–42. [CrossRef]

92. Daily, B.F.; Huang, S. Achieving sustainability through attention to human resource factors in environmental management. *Int. J. Oper. Prod. Manag.* **2001**, *21*, 1539–1552. [CrossRef]

93. Pullman, M.E.; Michael, J.M.; Craig, R.C. Food for thought: Social versus environmental sustainability practices and performance outcomes. *J. Supply Chain Manag.* **2009**, *45*, 38–54. [CrossRef]

94. Paashuis, V.; Boer, H. Organizing for concurrent engineering: An integration mechanism framework. *Integr. Manuf. Syst.* **1997**, *8*, 79–89. [CrossRef]

95. Boer, H.E.; Boer, H. Modularization, inter-functional integration and operational performance. In Proceedings of the 15th International CINet Conference on Operating Innovation—Innovating Operations. Continuous Innovation Network (CINet), Budapest, Hungary, 7–9 September 2014.

96. Boer, H.; Kuhn, J.; Gertsen, F. Continuous innovation: Managing dualities through co-ordination. In *Continuous Innovation Network*; CiteseerX: Princeton, NJ, USA, 2006.

97. Rao, P. Greening the supply chain: A new initiative in South East Asia. *Int. J. Oper. Prod. Manag.* **2002**, *22*, 632–655. [CrossRef]

98. Rao, P.; Holt, D. Do green supply chains lead to competitiveness and economic performance? *Int. J. Oper. Prod. Manag.* **2005**, *25*, 898–916. [CrossRef]

99. Efeoglu, A.; Moller, C.; Serie, M. Corporate Innovation Management Framework Based on Design Thinking. In Proceedings of the International CINet Conference on Operating Innovation—Innovating Operations, Budapest, Hungary, 7–9 September 2014.

100. Anderson, J.C.; Gerbing, D.W. Structural equation modeling in practice: A review and recommended two-step approach. *Psychol. Bull.* **1998**, *103*, 411. [CrossRef]

101. Byrne, B.M. Structural equation modeling with AMOS, EQS, and LISREL: Comparative approaches to testing for the factorial validity of a measuring instrument. *Int. J. Test.* **2001**, *1*, 55–86. [CrossRef]

102. Hoe, S.L. Issues and procedures in adopting structural equation modeling technique. *J. Appl. Quant. Methods* **2008**, *3*, 76–83.

103. Schreiber, J.B.; Nora, A.; Stage, F.K.; Barlow, E.A.; King, J. Reporting structural equation modeling and confirmatory factor analysis results: A review. *J. Educ. Res.* **2006**, *99*, 323–338. [CrossRef]

104. Dell'Anno, R.; Schneider, F. A complex approach to estimate shadow economy: The structural equation modelling. In *Coping with the Complexity of Economics*; Springer: Milan, Italy, 2009; pp. 111–130.

105. Arbuckle, J.G.; James, M.A.; Miller, M.L.; Sullivan, T.F. *Environmental Law Handbook*; Government Institutes, Inc.: Washington, DC, USA, 1976.

106. Brownell, F.W.; Case, D.R.; Cardwell, R.E. *Environmental Law Handbook*; Government Institutes: Washington, DC, USA, 2011.

107. Dangelico, R.M.; Pujari, D. Mainstreaming green product innovation: Why and how companies integrate environmental sustainability. *J. Bus. Ethics* **2010**, *95*, 471–486. [CrossRef]

108. Rennings, K. Redefining innovation—Eco-innovation research and the contribution from ecological economics. *Ecol. Econ.* **2000**, *32*, 319–332. [CrossRef]

109. Carrillo-Hermosilla, J.; del Gonzalez, P.R.; Konnola, T. What is eco-innovation? In *Eco-Innovation*; Palgrave Macmillan: London, UK, 2009; pp. 6–27.

110. Piller, F.T.; Vossen, A.; Ihl, C. From social media to social product development: The impact of social media on co-creation of innovation. *Die Unternehm.* **2012**, *65*, 1. [CrossRef]

111. Fawcett, S.E.; Myers, M.B. Product and employee development in advanced manufacturing: Implementation and impact. *Int. J. Prod. Res.* **2001**, *39*, 65–79. [CrossRef]

112. Russo, M.V.; Paul, A.F. A resource-based perspective on corporate environmental performance and profitability. *Acad. Manag. J.* **1997**, *40*, 534–559. [CrossRef]

113. Bunge, J.; Edward, C.-R.; Antonio, R.-Q. Employee participation in pollution reduction: Preliminary analysis of the Toxics Release Inventory. *J. Clean. Prod.* **1996**, *4*, 9–16. [CrossRef]

114. Hui, I.K.; Alan, H.S.C.; Pun, K.F. A study of the environmental management system implementation practices. *J. Clean. Prod.* **2001**, *9*, 269–276. [CrossRef]

115. Daily, B.F.; James, W.B.; Robert, S. The mediating role of EMS teamwork as it pertains to HR factors and perceived environmental performance. *J. Appl. Bus. Res.* **2011**, *23*. [CrossRef]

116. Klassen, R.D.; Ann, V. Social issues in supply chains: Capabilities link responsibility, risk (opportunity), and performance. *Int. J. Prod. Econ.* **2012**, *140*, 103–115. [CrossRef]

117. Hong, P.; Kwon, H.B.; Jungbae Roh, J. Implementation of strategic green orientation in supply chain: An empirical study of manufacturing firms. *Eur. J. Innov. Manag.* **2009**, *12*, 512–532. [CrossRef]

118. Mohrman, S.A.; Worley, C.G. The organizational sustainability journey: Introduction to the special issue. *Organ. Dyn.* **2010**, *4*, 289–294. [CrossRef]

119. Golini, R.; Longoni, A.; Cagliano, R. Developing sustainability in global manufacturing networks: The role of site competence on sustainability performance. *Int. J. Prod. Econ.* **2014**, *147*, 448–459. [CrossRef]

120. He, Z.-L.; Wong, P.-K. Exploration vs. exploitation: An empirical test of the ambidexterity hypothesis. *Organ. Sci.* **2004**, *15*, 481–494. [CrossRef]

13

Using Goal-Programming to Model the Effect of Stakeholder Determined Policy and Industry Changes on the Future Management of and Ecosystem Services Provision by Ireland's Western Peatland Forests

Edwin Corrigan *,† **and Maarten Nieuwenhuis** †

UCD Forestry, School of Agriculture & Food Science, UCD, Belfield, Dublin 4, Ireland;
maarten.nieuwenhuis@ucd.ie
* Correspondence: edwintcorrigan@gmail.com
† These authors contributed equally to this work.

Academic Editor: Jose G. Borges

Abstract: Recent studies have highlighted land-use conflicts between stakeholder groups in Ireland. Some of these conflicts can be attributed to European directives, designed with sustainable forest management principles in mind, but imposing incoherencies for land-owners and stakeholders at the local level. This study, using Ireland's Western Peatland forests as a case study area, focused on the development and implementation of a goal programming model capable of analysing the long term impact of policy and industry changes at the landscape level. The model captures the essential aspects of the changes identified by local level stakeholders as influencing forest management in Ireland and determines the future impact of these changes on ecosystem services provisions. Initially, a business as usual potential future is generated. This is used as a baseline against which to compare the impact of industry and policy changes. The model output indicated that the current forest composition is only really suited to satisfy a single, financial objective for forest management. The goal programming model analysed multiple objectives simultaneously and the results indicated that the stakeholders' desired ecosystem service provisions in the future will be more closely met by diversifying the forest estate and/or by changing to an alternative, non-forest land-use on less productive areas.

Keywords: Remsoft; optimisation; land-use change; forest management; scenario analysis; sustainable forest management

---

1. Introduction

Proper implementation of Sustainable Forest Management (SFM) will depend on an acceptable balance between the three pillars of SFM, i.e., the economic, ecological and the social, as quantified by their indicators [1–3]. Each pillar provides functional support to individuals at the local, national and global scale [4]. It has been suggested that Ecosystem Services (ESs), a utilitarian concept, may prove useful to identify this balance [5]. A full history of the evolution of the ES concept is described by Gómez-Baggethun et al. [6]. Briefly, the origin of the concept dates back to the late 1960s [7]. It was used in the 1970s to capture public interest in terms of biodiversity conservation [8]. Since then, the number of academic publications referring to the term ESs has gradually increased prompting the Millennium Ecosystem Assessment; a major piece of work which provided an indication of the level of degradation of the world's ESs [9]. The concept of ESs makes it possible to evaluate the trade-offs and the compatibilities between different management scenarios. However, debate continues about the most appropriate method to define and quantify ESs with consistency. De Groot et al. [10] have suggested indicators that link ESs to human well-being. In Ireland, political attention has focused on specific ESs, and land-use disputes have identified ESs that are strongly related to human welfare. A certain amount of research into these ESs has been carried out [11–14] and most of the ESs analysed can be categorised as outlined by De Groot, Alkemade, Braat, Hein and Willemen [10]: food, raw materials, climate regulation, gene pool protection, water regulation, and recreation.

Since the 1960s, many forests in Ireland have been established with a focus on generating revenue and this remains their main objective. This single financially oriented objective has guided early forest Decision Support System (DSS) research in Ireland in the past. Nieuwenhuis and Williamson [15] developed a system for timber harvesting and sawmill delivery while minimising costs. More recently, SFM has come to the fore in Irish forest management and research effort has shifted, becoming less focused on timber production only and more on the integration of non-timber benefits with a sustainable timber supply, using the ecosystem services concept. The potential to group sub-compartments (the smallest unit of management used by Coillte, the Irish State forestry board) to create economically viable blocks for harvesting was explored by Nieuwenhuis and Tiernan [2] using Mixed Integer Programming. They went on to implement some SFM constraints and compared their economic impact.

However, the concept of SFM requires the optimisation of multiple objectives. The objective of this paper is to describe the development of a multi-criteria Goal Programming (GP) model capable of investigating the long term impact of policy and industry changes on ES provision levels. The industry and policy changes have been determined by a social science team that worked in parallel as part of a larger project known as INTEGRAL [16]. The INTEGRAL project aimed to bridge the gap between European level policy decision makers and the stakeholders within 20 local regions of EU member states. This study will describe the GP decision support system developed and used on one of these regions in Ireland. The GP method is often used in multi-criteria decision making in forest management planning [17,18], with one of the main reasons being that goal constraints (i.e., soft constraints) allow for a model to produce a feasible solution even if the specified constraint is not met. This means that optimal compromises can be achieved between multiple objectives that are not simultaneously achievable.

2. Materials and Methods

The Western Peatland (WP) Case Study Area (CSA) is located in the northwest of Ireland (Figure 1 and Table 1). It is based on one of eight Business Area Units, the method of land division and management used by Coillte. Business Area Unit boundaries follow town land boundaries (as much as possible) and each Business Area Unit was designed to be an independently profitable unit [19]. The location of the WP CSA means that wind exposure is a dominant factor for all land-use options. This study is focused on forestry and hence, the CSA includes only the area that is forest at the beginning of the planning horizon (the planning horizon is the entire number of years that the optimisation algorithm is ran for). Nationally, 44% of state forests are located on peatlands [20] which indicates that peatland forestry is important even at the national level. Approximately 62% of the forests in the CSA have a peatland soil type which is more than the national level and will allow for an analysis of various forest management approaches on Ireland's peatlands.

Table 1. Statistics associated with the Western Peatland case study area.

Descriptor	Western Peatlands
Area (approx. ha)	1,060,000
Forested area (approx. ha)	116,000
Average temperature (°C)	11–12
Typical annual rainfall (mm)	West: 2000 East: 1200–1400

Table 1. *Cont.*

Descriptor	Western Peatlands
Forested land only (as of 2012)	
Forest ownership	
Coillte	64%
Private	36%
Yield potential	
Economically viable forest *	82%
Not economically viable forest	18%
Age class distribution	
0–10 years	15%
11–20 years	34%
21–30 years	27%
31–40 years	15%
41–50 years	6%
51 years or over	3%
Soil type	
Brown earths and brown podzolics	5%
Lithosols	12%
Gleys/peaty gleys and gleyed grey brown podzolics	17%
Flushed blanket peat	48%
Cutaway raised bogs	18%
Elevation	
Less than 200 m	93%
Distance to watercourse	
Less than 200 m	56%
Between 200 and 400 m	26%
400 m or greater	19%

* Forestry that has a Sitka spruce yield class equivalence of 14 or higher is considered to be economically viable [21].

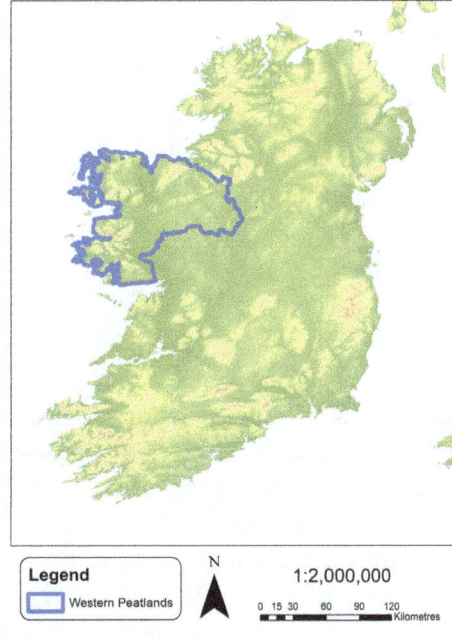

Figure 1. Location and boundaries of the Western Peatland case study area.

2.1. Introduction to the Western Peatland's Ecosystem Services

The landscape's climate and soils enforce limitations for tree species selection. A large proportion of the forest area comprises exotic conifers that have a relatively low level of production on the poor peat soils compared with the levels on more fertile soils located in other areas of Ireland. The timber produced within parts of this area does have viable potential end uses but also sequesters carbon and it is unclear how these forests should be managed.

The WPs is a popular tourism destination. Water based activities such as fishing and leisure cruising are predominant tourism attractions in the area. The mountainous environment and the reasonably undisturbed landscape (when compared to the typical Irish landscape of commercial agriculture) provide also a canvas for many other recreational activities. Excellent water quality for salmonid fish stocks is essential for tourism, and forest management planning also needs to address concerns about the impacts of forest operations on the viability of populations of the threatened freshwater pearl mussel (*Margaritifera margaritifera*), a species native to Ireland.

There is concern amongst some about the dominant proportion of the forested landscape that consists of these exotic conifers, and the issues regarding whether or not these forests should be harvested, and if so, should the land be reforested and with which species, or should the area be turned back to its probable previous native land-use. Landscape restoration and biodiversity enhancement are thought by some to also have a positive effect on tourism in the area.

2.2. Management Approaches

The management interventions have been categorised into seven, mutually exclusive, forest management approaches (Table 2), i.e., Traditional forest management, Continuous Cover Forestry (CCF), Native Woodland Site (NWS), Non-reforested, Buffer zone (areas within certain proximity of road, watercourse or Freshwater Pearl Mussel (FPM) watercourse) and Bog restoration. The characteristics of the forest site determine which approaches are appropriate for implementation. In Remsoft Woodstock, the smallest management unit used is the development type. Each development type is assigned a management approach based on its characteristics and age. An "existing development type" (i.e., a development type that exists before the optimisation matrix is built) is created from polygons in the GIS layer for each unique combination of:

- Species or land-use
- Productivity
- Proportion of a species in a stand
- Environmental zone
- Soil type
- Site preparation type at establishment
- Owner type
- Elevation
- Water risk factor
- Age

The model described in this study is in a linear programming model II formulation [22]. If an existing development type has the appropriate age and attributes, it is eligible for actions which transition a development type (and all of its associated area) into a future development type. For example, if a development type is eligible for a clearfell action, the existing development type and its entire area would transition into a clearfelled future development type. Further actions and transitions can be carried out for a future development type throughout the planning horizon as long as the development type attributes and age make it eligible for such actions to take place. For example, the clearfelled development type might have the option to be reforested, thinned and clearfelled again within the planning horizon and each of these actions will transition the development type. Accordingly, the Western Peatland forestland was classified into 24,300 development type. Transitions over the planning horizon led to a total of 123,800 future development types.

Table 2. Forest management approaches.

Management Approach	Description
Traditional forest management	Forests that are managed under a typical Irish rotation based system, i.e., commercial thinning, clearfelling and artificial restocking.
Continuous Cover Forestry (CCF)	Forest that has a continuous cover. It is thinned at 5-year intervals and can consist of conifers, broadleaves or a mix. It is assumed that stands regenerate naturally.
Native Woodland Site (NWS)	Unique species options at afforestation/reforestation, only available for suitable soil types. Harvesting options are the same as for CCF management.
Non-reforested	Cleared land that has not been reforested after 2 years.
Buffer zone	Areas within certain proximity of a road or watercourse that have undergone a buffer conversion process. There are three types of buffer zone. Within 20 m of a road: 10 m scrubland, 10 m broadleaves; within 10 m of watercourse or FPM watercourse: 10 m scrubland; within 25 m of a watercourse within a 6 km hydrological distance of a live FPM site: 20% native broadleaves and 80% scrubland.
Bog restoration	Area that undergoes a conversion process to bog. This is only permitted on areas with a blanket peat soil type.

2.3. Current Forest Policy and Business As Usual Scenario

Directives outlined by the European Union and enforced nationally by Ireland have designated zones for the purpose of maintaining rare habitats. In these areas, forest operations have been restricted. A list of current forest policies and permissible actions within associated zones were established through a review of literature [21,23–25]. This list was verified and refined by a member of the Forest Service Inspectorate. It was translated into CSA forest management planning rules (Table 3). In addition, the following policies were incorporated in the Business As Usual (BAU) scenario:

(1) All Coillte broadleaf forest must be managed under low impact silvicultural systems (i.e., CCF), as per the company policy decision made in 2005.

(2) A stand must be reforested within 2 years of clearfelling (with certain exceptions, e.g., buffer zones and bog restoration).

(3) Clearfell not permitted of stands still in receipt of forest premiums (i.e., before the age of 20 years on afforested land). Farmers who afforested receive grants that cover the cost of establishment and (as per 2012 regulations) receive premiums annually for the first 20 years of their forest's rotation.

Table 3. Permissible forest operations given current forest policy zones in Ireland.

Relevant Group	Harvesting Options	Reforestation	Fertilisation at Establishment	Tending/Thinning
Special protection areas	Yes	Yes	Yes	Yes
Special areas of conservation	Yes	Native woodland site only	Yes	Yes
proposed National Heritage Area	Yes	Native woodland site only	Yes	Yes
National Heritage Area	Yes	Native woodland site only	Yes	Yes
Old Woodland Site	No	Soil type specific species choice	Yes	No
Native Woodland Site	CCF only	N/A	No	CCF only
Freshwater Pearl Mussel	No CCF	Yes	Yes	Yes
6 KM Freshwater Pearl Mussel zone	No CCF	20 m setback, 5 m broadleaf buffer	No	Yes
Road buffer	Yes	10 m setback then 10 m broadleaf species	Yes	Yes
Non-Freshwater Pearl Mussel buffer zone	No CCF	10 m buffer width	No	Yes

N/A: No clearfelling takes place and therefore reforestation is not an option for these sites.

2.4. GP Mathematical Notation

The GP model used in this study considers the biophysical LP model results described by Corrigan and Nieuwenhuis [26]. Specifically, the objective function value of the maximisation scenario for each ES is used to set that ES goal value (Table 4). Even though Net Present Value (NPV) is not an ES, it is included in this paper in the list of ESs as it is an important factor of the decision making process. In the case of the water sedimentation risk ES, the goal value was rather set as equal to the objective function value of the minimisation scenario (Table 4). Goals are specified as constraints for each Owner Type (OT; i.e., a mutually exclusive landowner type: Coillte, private forest owners, agricultural ruminant production or tillage landowners) (Equation (1) or (2)), and the deviation variables n_{qk} and p_{qk} are used in the objective function (Equation (3)).

$$
\sum_{d=1}^{D} \sum_{i=0}^{H-1} \sum_{j=\min(H,j=i+N_d)}^{\min(H,j=i+M_d)} x_{id,je,k} * c_{id,je,q}
$$
$$
+ \sum_{d=1}^{D} \sum_{i=0}^{H-1} x_{id,Hd,k} * c_{id,Hd,q} + (n_{qk} * NW_{qk}) - t_{qk} \geq 0 \; Q \tag{1}
$$
$$
= 1, 2, \ldots, Q \text{ and } k = 1, 2
$$

$$
\sum_{d=1}^{D} \sum_{i=0}^{H-1} \sum_{j=\min(H,i+N_d)}^{\min(H,i+M_d)} x_{id,je,k} * c_{id,je,q}
$$
$$
+ \sum_{d=1}^{D} \sum_{i=0}^{H-1} x_{id,Hd,k} * c_{id,Hd,q} - (p_{qk} * NW_{qk}) - t_{qk} \geq 0 \; Q \tag{2}
$$
$$
= 1, 2, \ldots, Q \text{ and } k = 1, 2
$$

$$
\text{Min } Z = \sum_{k=1}^{2} \sum_{q=1}^{Q} \left(SF_{qk} * n_{qk} + SF_{qk} * p_{qk} \right) \tag{3}
$$

where $x_{id,je,k}$ is the number of hectares that entered development type (DT) d in year i and left it in year j when they entered DT e, for owner type (OT) k (see Table 5 in the section on normalising ecosystem service goals below); $x_{id,Hd,k}$ = number of hectares that entered DT d in year i stayed in this DT until year H, the end of the planning horizon, for OT k (Table 5); and $c_{id,je,q}$ is the cumulative value for ecosystem service q over the period from year i to year j, in DT d. For the harvest revenue ES, the value is expressed as the NPV; $c_{id,Hd,q}$ is the cumulative value for ecosystem service q over the period from year i to year H, in DT d, for the harvest revenue ES, the value is expressed as the NPV; D is the total number of development types; M_d is the maximum number of years a hectare can be assigned to DT d; N_d is the minimum number of years a hectare can be assigned to DT d; and Q is the total number of ecosystem services.

2.5. Normalising Goals

The scales that quantify ESs are different and therefore their maximum and minimum ranges vary. Therefore, a normalisation procedure must be implemented so that deviations from each goal are comparable. A method was developed to produce normalising weights for the optimisation (Equations (4)–(6)).

$$
ES \; scale_{qk} = \frac{\min(|qk_{max} - qk_{min}|)}{|qk_{max} - qk_{min}|} \tag{4}
$$

$$
NW_{max} = \frac{1}{\min \left(ES \; scale_{qk} \right)} \tag{5}
$$

$$
NW_{xk} = NW_{max} \times ES \; scale_{qk} \tag{6}
$$

where $qk_{max/min}$ is the maximum/minimum biophysical ES supply level for ES q in OT k; $\min(|qk_{max} - qk_{min}|)$ is the ES OT combination with the lowest absolute value of the difference between maximum and minimum values; $ES\ scale_{qk}$ is the scaling factor from 0 to 1; $ES\ scale\ tot_{min}$ is the ES OT combination with the lowest ES normalising value; NW_{max} is the highest NW to create initial "equal" weightings; and NW_{qk} is the NW for ES q in OT k.

2.6. Owner Types and Attitude Changes

The area occupied by each OT was spatially delineated based on a polygon's data source (i.e., Forest Service or Coillte). It was assumed that all members in an OT group will respond to policy changes in a similar manner. OT specific attitude differences were incorporated in the model through the scaling of the normalised weights in the goal programming objective function. A description of how the two types of attitude-based scaling factors were determined is presented below, i.e., the "forced" attitude change (point 1) and the attitude change that varies between OTs (point 2):

(1) This uniform change reflects that all OTs will have to modify their attitude equally. Examples are the stringent policy changes associated with the Water Framework Directive and the assumption that NPV is a factor that must be considered by all OTs when making management decisions.
(2) These relative attitude changes apply to timber, hen harrier, deer forage and cover ESs. The change in attitude is not the same for all OTs and is implemented through OT specific scaling factors. These factors are based on the perceived relative change in attitude of each OT. A summary of these scaling factors and scaling is presented in Table 5.

A summary of the scaled and normalised goals is presented in Table 4.

Table 4. A summary of the scaled and normalising goals process.

Ecosystem Service	Owner Type	Scaling Factor	Goal Value	Normalising Weight (Weight × Scaling Factor *)
Deer cover	Coillte	0.00	27,629,333	0
Deer cover	Private Forest Owners	0.50	14,150,923	53
Deer forage	Coillte	0.00	23,433,803	0
Deer forage	Private Forest Owners	0.50	11,535,438	62
Hen harrier	Coillte	0.00	34,762,208	0
Hen harrier	Private Forest Owners	0.50	17,770,620	20
Water sedimentation risk	Coillte	1.00	12,143	31,622
Water sedimentation risk	Private Forest Owners	1.00	4667	112,882
Timber	Coillte	0.00	49,753,761	0
Timber	Private Forest Owners	0.50	30,263,384	9
Recreation	Coillte	0.00	25,277,976	0
Recreation	Private Forest Owners	0.50	13,542,484	24
NPV	Coillte	1.00	480,548,445	1
NPV	Private Forest Owners	1.00	273,188,452	2

* This is the weight required to normalise the ecosystem service goals multiplied by the scaling factor; hence, if the scaling factor is 0, then the scaled normalising weight will be 0.

Table 5. Scaling factors (and reason) used to represent the relative attitude change of each owner type.

Owner Type	Scaling Factor	Attitude Change
Coillte	N/A	Commercial mandate and fully certified forests mean that the company has changed and is now obliged to manage their forests for all purposes of certification. Hence, Coillte will not change their attitude towards management when new policies are introduced.
Private forest owner	0.50	A diverse group of owners with a range of objectives for their forests. They have committed to having their land in forestry based on the implemented policies at the time they afforested. They will to be influenced by newly implemented policies, however less so than non-forest owning OTs.

Table 5. *Cont.*

Owner Type	Scaling Factor	Attitude Change
OTs currently not owning afforested land were investigated as part of another CSA which focused on afforestation. For transparency reasons, a description of them is included.		
Tillage	0.75	Duesberg et al. [27] found tillage farmers less likely to afforest than the owners of other, less profitable, types of farming systems in Ireland. Forestry is a new potential land-use for them and as a result more farmers in this OT will change attitudes as a result of policy changes than those in the private forest OT.
Ruminant	1.00	This group contains the beef suckling and sheep farming systems which are often financially comparable with forestry. It is expected that the next generation, who will inherit the land in this OT will be more urbanised and open to afforestation and will be influenced more by future forest policy changes than the tillage OT.

2.6.1. Area Constraints

Constraints are in place to ensure that the entire area in year 0 is accounted for over the entire planning horizon (Equations (7) and (8)).

$$x_{0d,je,k} + x_{0d,Hd,k} = A_{dk} \; k = 1, 2, \ldots, d \text{ and } e = 1, 2, \ldots, D \text{ and } j = i, i+1, \ldots, H-1 \qquad (7)$$

$$x_{id,je,k} = \sum_{p=\min(H,j+N_e)}^{\min(H,j+M_e)} x_{je,pf,k} + x_{je,He,k} \; k = 1,2 \text{ and } d,e,f = 1,2,\ldots,D \text{ and } j = 1,2,\ldots,H-1 \qquad (8)$$

where $x_{0d,je,k}$ is the number of hectares assigned to DT d in year 0, the start of the planning horizon, for OT k; A_{dk} is the number of hectares in OT k that are classified as belonging to DT d at the start of the planning horizon. Hectares leaving a DT at year j are not combined with hectares leaving other DTs in the same year, i.e., hectares from different DTs cannot be merged but can only be split. The DT e that a hectare enters has to be compatible with the DT d that it is leaving.

2.6.2. Policy Related Constraints

Constraints, not presented in mathematical notation, ensure that ineligible management prescriptions are not prescribed within politically designated zones. It does this by ensuring that the total area of management interventions not permitted on certain areas or the total area of development type regulations is zero in each year of the entire planning horizon. Another constraint (Equation (8)) ensures that all stands that have been clearfelled are reforested. Equation (9) is not applied for uneconomically viable non-reforested areas when replanting duty is lifted for potential future 2 (Table 7).

$$\sum_{k=1}^{2} \sum_{d=1}^{D} \sum_{i=1}^{H-1} \sum_{j=\min(H,i+N_d)}^{\min(H,i+M_d)} x_{id,je,k} * r_{id,k} = 0 \qquad (9)$$

where $r_{id,k}$ is a binary variable indicating ($r_{id,k} = 1$) the area of development type d in year i for OT k that has not been reforested.

2.6.3. Industry Regulation

Evenness constraints (Equation (10)) were implemented to ensure that there is an even supply of timber for each product assortment category over the planning horizon. The evenness constraint was applied separately to the pulp and stake, and to the small and large sawlog timber assortments.

The constraint ensures that the lowest production level of either assortment in any year of the planning horizon cannot be less than 80% of the level in the year with the highest assortment production.

$$\sum_{k=1}^{2} \sum_{d=1}^{D} x_{id,je,k} * v_{id,a} - (v_{\max a} * 0.80) \geq 0 \; i = 1, 2, \ldots, H \text{ and } a = 1, 2 \qquad (10)$$

where $v_{id,a}$ is the timber assortment (i.e., $a = 1$ for pulp and stake, $a = 2$ for small and large sawlog) volume harvested per hectare of DT d in year i; and $v_{\max a}$ is the maximum harvested volume of assortment a in any one year in the planning horizon.

2.7. Building of Potential Futures

A qualitative scenario building process was carried out by Bonsu et al. [28]. The stakeholder selection process used purposeful maximum variation sampling as described by Patton [29]. A full description of the method that was used to choose stakeholders for their scenario building process is presented in Bonsu et al. [30]. The scenario building process was facilitated by the findings of Corrigan and Nieuwenhuis [26]. The biophysical ranges from this study were described to the stakeholder group in a workshop setting. This provided the stakeholder group with the biophysical limitations of the CSA, and allowed for a participatory process to take place where the views of the local level stakeholders were consolidated. A group decision making module known as "Parmendies Eidos" [31] was used to produce future scenarios that are plausible in the CSA. These scenarios (for the period 2012–2042) will be referred to in this paper as potential futures (PF). These scenarios are based on factors (known as key factors) which are, according to the stakeholders' expert knowledge, most likely to influence forest management within the CSA in the future. The PFs are in the form of consistent combinations of key factor manifestations. These combinations were perceived as likely to happen simultaneously in the same PF. For example, demand for pulpwood increasing and the establishment of a Combined Heat and Power (CHP) plant in the CSA is considered as consistent; however, the establishment of a CHP plant was not considered consistent with a decrease in pulpwood demand. A total of five PFs were developed and six key factors were identified: demand for sawnwood, demand for pulpwood, demand for rural development, water protection, forest policies and regulations, and SFM. These qualitative scenarios have been translated into parameters of the GP model formulation. The PFs were modelled using Remsoft Woodstock [32] using an Intel® Core™ i7-3930K CPU at 3.20 GHz with 32 GB RAM operating Windows 7 Service Pack 1 (64-bit). A BAU PF, as described above, provided a baseline. From this starting point, PFs were developed by changing the relevant baseline parameters and policies in the model. There were four avenues available to implement changes relating to key factor manifestations (Table 6).

Table 6. Summary of methods used to implement policy changes.

	Model Change	Example
1	Alternative management included in the model	Alternative species combination for establishment
2	Changing how ES values are calculated	Removing afforestation premiums from the NPV calculation
3	Changing the spatial arrangements behind the model	Widening buffer zones along watercourses
4	Inclusion of ESs in objective function	Water sedimentation risk was included in the objective function for PFs that include the "more restrictive" manifestation for water sedimentation risk

It should be noted that the same methods of quantifying ESs as developed and presented by Corrigan and Nieuwenhuis [26] were implemented in this model. Regardless of whether ESs are included in the objective function, this model will quantify the provision levels of all ESs. The ESs included in the model were: timber production, carbon sequestered in living trees (the difference between carbon stock from one year to the next is carbon flow, i.e., the ecosystem service), water sedimentation risk, recreation potential, deer forage, deer cover, habitat for red squirrel, hen harrier and species richness of nesting birds ground vegetation. NPV, although not an ecosystem service per se, is calculated and used in the modelling process as one of the main decision making criteria.

Up to five ESs were brought into the GP optimisation to represent the PFs: NPV, timber production, water sedimentation risk, deer forage and deer cover. ESs were chosen for inclusion in the objective function based on the following criteria:

(1)   They represent forest policy-related issues that are currently debated in the local context and will continue to be issues in the future.
(2)   Owner Type demands are diverse and, in some PFs, owners will aim to provide different ESs.

The translation of scenarios into the GP model involved the change of management planning options as listed in methods 1 to 3 (Table 6) and of management policy parameters (Table 7). It involved further the compilation of ES to consider in the GP objective function in each PF (Table 8).

Table 7. A summary of management policy model changes, by key factors, for each potential future.

Potential Future	Demand for Sawnwood	Demand for Pulpwood	Demand for Rural Development	Water Protection	Replanting Requirement	SFM
BAU	Same	Same	No CHP [1] plant	Same	Same	Same
2	Same	Same	No CHP plant	Same	Lifted	Same
3	10% increase in price	10% increase in price	CHP plant built	Same	Same	Same
4	10% increase in price	10% increase in price	CHP plant built	Water measures [2]	Same	SFM measures [3]
5	Same	Same	No CHP plant	Water measures	Same	SFM measures

[1] CHP = combined heat and power plant; [2] Water measures = Water buffer zones doubled: 6 km FPM 25 to 50 m; FPM 10 to 20 m. Increased emphasis on water sedimentation risk ES; [3] SFM measures = Bog restoration an option. Increased emphasis on ecological ESs.

Table 8. Ecosystem service included in the objective function for each potential future.

Potential Future	Deer Cover	Deer Forage	Timber	NPV	Water
BAU				X	
2				X	
3			X	X	
4	X	X	X	X	X
5	X	X		X	X

3. Results

The model outputs contain information on management approaches selected by each owner type (OT). However, as these choices depend to a large extent on the forest composition for each OT, which is quite different at the start of the planning horizon, presenting the results by OT is not very useful within the context of this paper. For this reason, the results section will focus on landscape level trends. The forest age and yield class composition for the BAU PF will be presented first. This initial BAU description will be followed by a comparison of the relative change in selected management approaches and ES provision levels between PFs.

3.1. BAU Potential Future

The age class distributions (Figure 2) show a dip in the 1–10 year age class at the start of the planning horizon (point 1) and this progresses to the 21–30 age class by approximately year 26 (points 1, 2 and 3 in Figure 2). The most recent GIS forest dataset that was available in 2012 for private forests was published in 2007. Since 2007, more non-forested land has been afforested but no update for private forests exists since the 2007 dataset. As a result, there are no privately owned forests that are six years old or less at the beginning of the planning horizon (i.e., the youngest privately owned forests are between six and 10 years old). This accounts for the initial decline in the 1–10 year age class over the first five years of the planning horizon, as the private forests in this age class are all between six and 10 years of age at the start of the planning horizon and moves into the 11–20 year age class by Year 5 of the planning horizon. This dip dissipates between the Years 31 and 36, which corresponds with the average clearfell age of for conifers of 41 years. The age class distribution becomes more smooth at this stage as a proportion of the 31–40 year age class is retained (under the CCF management approach and buffer zones which have been established within the first 10 years of the planning horizon) and will continue to mature, while the areas in this age class that were clearfelled are reforested. This results in the development of a less pronounced dip in the second half of the planning horizon, identified for the 1–10 year age class by Year 48 (point 4 in Figure 2), and continuing to Year 57 (point 6 in Figure 2).

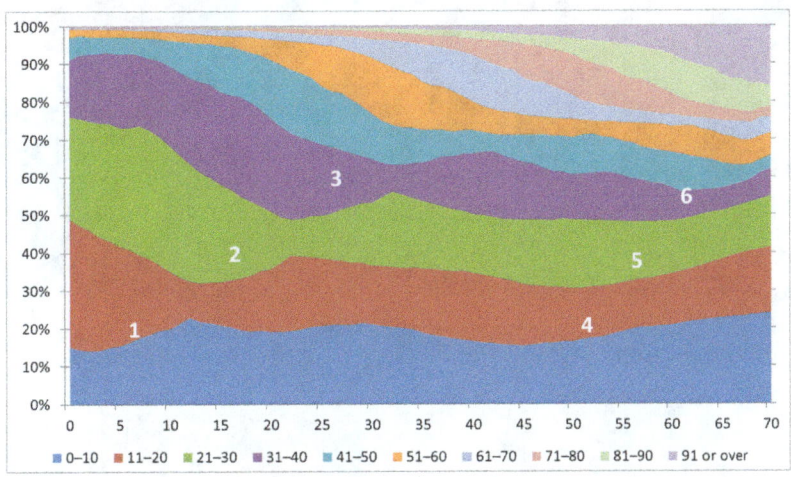

Figure 2. Structure of the forests, in 10-year age classes, for the Business As Usual Potential Future for each year of the planning horizon.

3.1.1. Yield Class Assessment

Initially, 81% of the CSA's forests are productive, i.e., have a yield class equivalent of 14 or higher. For the current rotation, many forests have been established with lodgepole pine (a species suited to marginal sites) with the help of multiple applications of artificial fertiliser. This ensures successful establishment and results in productive forest. Current Irish policy means however that fertilisation is only a once-off option at establishment (on peat sites) to ensure successful re-establishment and this one application does not increase yield class to the same extent, hence the proportion of productive forest decreases to circa 40% over time.

3.1.2. Proportion of the Case Study Area under Each Management Approach

It is assumed that the forest management begins with a traditional rotation based approach. The optimisation selects alternative options for all PFs. This means that the rate of change from traditional forest management can be assessed and compared over the course of the planning horizon (Figure 3).

A relatively large proportion of CCF is established in all PFs (23% of the CSA by year 35 in the BAU PF). One benefit of CCF is that timber can be harvested for the remainder of the planning horizon without the cost of re-establishment (i.e., natural regeneration is assumed). However, the CCF management approach can be selected or sometimes "forced" to happen for a variety of reasons. For instance, CCF stands include all Coillte owned broadleaf forests and many Natura 2000 areas.

The BAU PF and PF 3 are more focused on timber production than the other PFs and as a result, less CCF management is introduced in these PFs compared to PFs 4 or 5. PF 2 has less CCF management as some of the areas that would typically enter CCF are now clearfelled and not reforested (using the non-reforested management approach). The proportion of buffer zone area in the BAU PF in year 35 is 1.8%, while the wider buffer zones and the reduced economic focus in PFs 4 and 5 mean that a higher proportion of the CSA is managed as buffer zone in these scenarios, i.e., 4.19% and 5.29% for PFs 4 and 5 in year 35. PF 2 results in the smallest area of buffer zones; this is due to non-reforestation being an option in this PF. When changing management approach from traditional forest management to CCF or buffer zone, revenue can be generated from timber harvesting without the cost of restocking the area (fully) with trees. The establishment of NWS is not beneficial financially and takes a long time (60 years post establishment) to provide higher habitat ratings, and therefore has little uptake in the PFs.

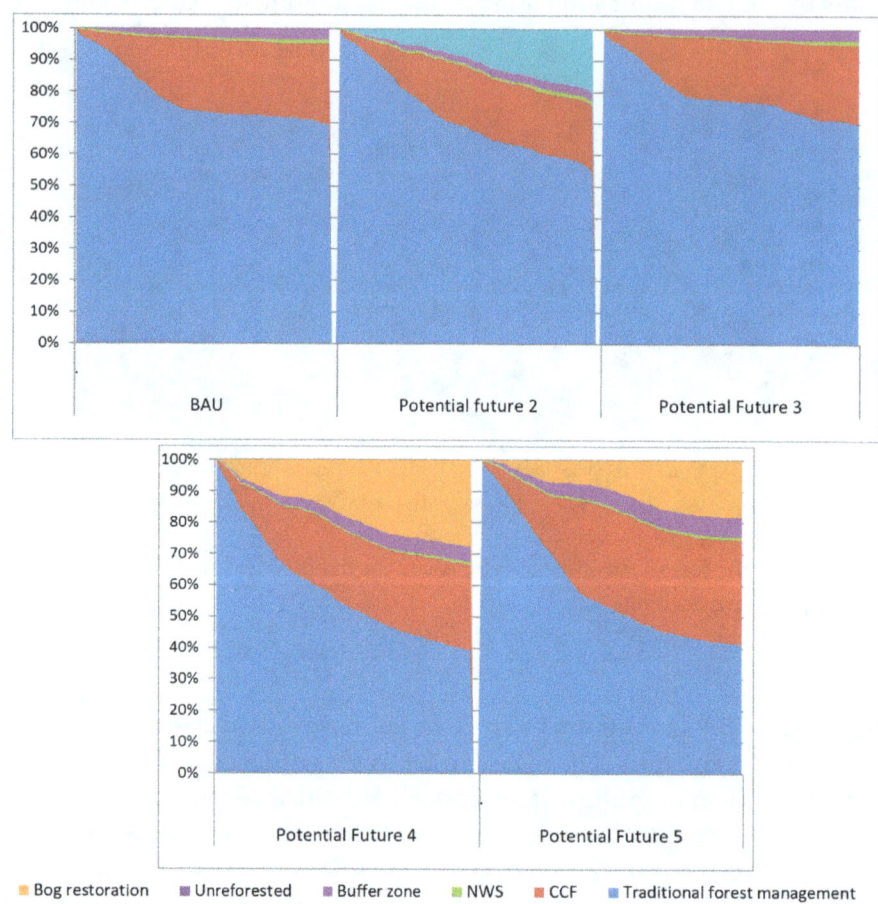

Figure 3. Management approaches as a proportion of the CSA for each Potential Future over the 70-year planning horizon.

The management approaches which are only available when certain key factors are chosen in the model are chosen by the optimisation to reach the ES goals specified in the model. For instance, in PF 2 the key factor change is to allow strategic non-reforestation and circa 20% of the CSA forest is not reforested post-clearfell. It is assumed that a site will be restocked if it is financially viable to do so, however all OTs in PF 2 have the choice not to restock forests if they are unproductive. As there is a much higher proportion of unproductive forest in Coillte owned land, a much higher proportion of Coillte land is non-reforested (13.33% by year 35) than private forest land (0.49% by year 35). In PFs 4 and 5, which are the only PFs where bog restoration is an option, respectively 28% and 18% of the Coillte owned and privately owned CSA forests are restored to bog by the end of the planning horizon.

3.1.3. ES Provisions

Over the first one third of the planning horizon for all PFs, many of the mature forests are harvested after which there is a decline in timber production (within the bounds of the volume smoothening constraints), followed by an increase in harvest in the final third of the planning horizon when these harvested and reforested areas mature again. This maturing age class structure in the final two thirds of the planning horizon increases the provision of most biodiversity ESs for the BAU PF (Figure 4), with the exception of hen harrier habitat which is provided by young forest. The increase in harvesting of these mature forests in the final third of the planning horizon means that hen harrier scores increase (as they are attracted to young or open land-use types), while recreation scores decline as high recreational potential is strongly associated with mature forest. Less change is observed for other ESs, as young forest, although not as suited as mature forest to provide these ESs, does provide relatively high levels of ESs, e.g., deer forage and nesting birds.

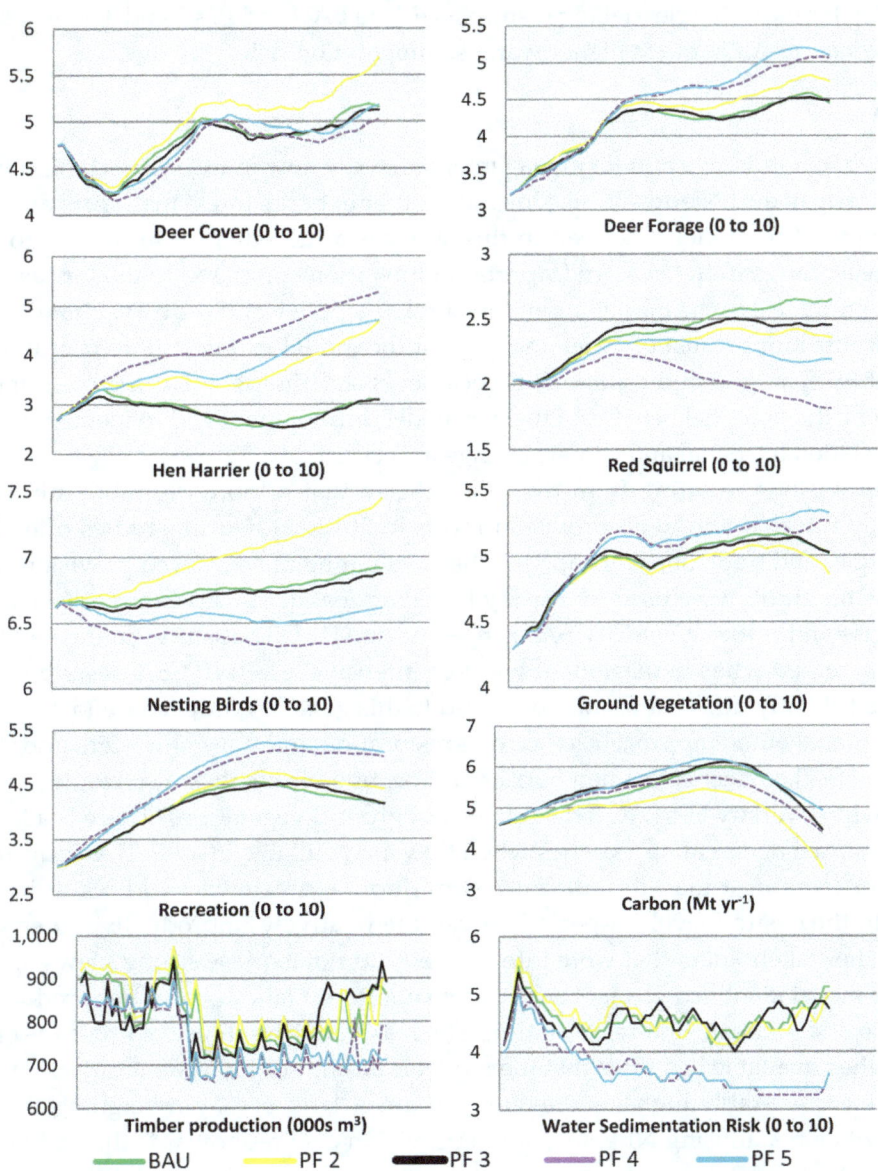

Figure 4. ES provisions for all Potential Futures (PF) over a 70-year planning horizon (*x*-axis). The *y*-axis values are specific to each ecosystem service provision level achieved, while the maximum and minimum values of each ES scale are specified on each graph (except for timber and carbon).

A majority (38 out of 50) of the 10 ES provision levels in all five PFs had increased by the end of the planning horizon compared to their values at the beginning of the planning horizon (and the water sedimentation risk had decreased in most PFs). This gives an indication that the dominant management approach at the start of the planning horizon, traditional forest management (the proportion of which declines in all PFs), is focused on maximising only NPV over time. The CSA's composition changes to a more multifunctional environment through the use of other management approaches in all PFs and hence a more diverse mix of ESs is provided.

PF 4 has the lowest average ES provision level for red squirrel. The larger amount of bog restoration, replacing mature forest in this PF, reduces the provision of this ES considerably compared to the levels in all other PFs. Similarly, habitat provision levels for nesting birds are low for the bog land-use and, as a result, nesting bird habitat declines in PFs 4 and 5.

For PFs in which water protection measures (i.e., wider buffer zones) are implemented (i.e., 4 and 5), the water sedimentation risk is significantly lower over the course of the planning horizon, even accounting for the initial clearfelling disturbance. The BAU and PFs 2 and 3, which do not include wider buffer zones, result in much higher water sedimentation risk.

4. Discussion

The concept of landscape and ES management are very new in Ireland and this study builds on the work of Corrigan and Nieuwenhuis [26] and further exposes forest managers and stakeholders to these concepts. The model described in this study is a first attempt to use GP to model forest related land-use change in Ireland. An important innovation compared to the former Corrigan and Nieuwenhuis study lies in the modelling of forest related policy and industry changes. These make the model outputs more realistic, which means that the model is much more useful for explaining the concept of ES synergies and trade-offs to local level stakeholders. During the interactions with the stakeholders, the potential benefits of further model development and refinement were identified, especially in relation to spatial and climate change aspects.

The ES provisions resulting from the PFs indicate that a more desirable mix of ESs can be achieved in the future than the ES provision levels in 2012. However, studies often indicate that some ES synergies and trade-offs will happen when evaluating the enhanced provisions of others [33]. The results of this study were used in a study by Biber et al. [34] who summarised and compared the ES synergies and trade-offs in 20 CSAs across Europe [16]. That study confirmed that typically biodiversity increased when management was less intensive. The WP CSA described in full in this paper produced slightly different results compared to this general European trend, because while the less intensive management approaches (i.e., bog restoration and non-reforested land) provide high levels of some ESs (i.e., deer cover, hen harrier and nesting birds), they also result in low provision levels of some biodiversity ESs (i.e., red squirrel and ground vegetation). There is a large uptake of these non-forest management approaches when they are available in a PF. This indicates that more diverse and/or less forest than the current composition and extent would result in ES provision levels closer to the desired levels, especially given the relatively unproductive forests in this CSA. Even though the stakeholders that were interviewed did not express strong views against forestry per se [28], the model identified greater benefits from non-forest land-uses on less productive areas, i.e., bog restoration and strategic non-reforestation. The reason for the choice of non-forest management approaches, when available in a PF, is that the model is basing solutions on multiple objectives, some of which have considerably higher scores for non-forest land-uses, outweighing the loss of NPV. For instance, when maximising NPV, recreation provision declines towards the end of the planning horizon (PFs BAU, 2 and 3), meaning that while providing for one objective (NPV) at the highest level, other objectives are negatively affected. When the objective function incorporates two or more ESs that require different types of land-use for their provision, this study indicates that more desired levels of a wider mix of ESs can be provided while only marginally reducing the provision of one ES (in this example, NPV), similar to the findings of other studies [10].

The GP model can accommodate multiple objectives as goals which the GP optimisation aims to achieve simultaneously. The biophysically optimal results from Corrigan and Nieuwenhuis [26] were obtained for one ES at a time without taking into account current policy and industry demands. These were included as hard constraints rather than goals to avoid interfering with the normalised ESs specified as goals (which rely on a balance of calculated weights for their normalisation). None of the goals were reached for any PF. An alternative would be to have the goal levels set at achievable levels by local stakeholders. However, in the case where all goals were achieved, the model's solution could be considered sub-optimal as the landscape has the potential to produce the ESs at higher levels. With this in mind, consulting with an expert panel, as investigated by Hotvedt [35] for sawmill timber supply, could be useful to ensure goals are set at levels appropriate for the goal programming approach. In addition, many of the quantification methods for the wide range of ESs in this study are new and hence stakeholders were less familiar with these methods than the timber experts in the Hotvedt study who were only asked to assign goals for timber supply. With this in mind, the Pareto efficiency method might be useful for comparing relative ES provision levels [36,37]. A transparent method such as this would allow stakeholders to identify the impact of changing the level of one ES provision target on others, with a view towards reaching a consensus on suitable goals. This would be particularly useful when the process involves multidisciplinary, local stakeholder groups.

Caveats

(1)   Some policy decisions are made on a case by case basis. Even if areas share the same characteristics, it is possible for multiple politically enforced management decisions to be made. For example, in Special Areas of Conservation (SACs), reforestation with broadleaves is the typical policy to choose (and the one specified in the model), however in some very special circumstances, conifers may be planted in small areas in SACs for reasons that could not be quantified and included in the model.

(2)   The model described in this study only accommodates spatially related characteristics before the optimisation. This reduces the practical applicability of the resulting PFs. For example it is not possible for the model to determine a suitable habitat for the red squirrel, i.e., a homogenous area of at least 200 to 300 ha of suitable forest is required [38]. Remsoft's heuristic spatial planning software called "Stanley" [39,40] could be used to deal with some of the spatial aspects.

(3)   The method of determining scaling factors for owner types was largely based on qualitative scientific research and the research team's expertise. Although intention to afforest has been linked to farm and social characteristics [27], further research is necessary in Ireland to identify how these characteristics influence forest management and even land-use management in general. The agent-based model developed by Daigneault and Fraser [41] for New Zealand could be a useful approach. Their model, which is focused on agriculture, links various human life stages with the propensity to proceed with certain management intentions.

5. Conclusions

In all PFs, management approaches changed away from the traditional forest management approach that currently dominates in the Western Peatlands. The change of management approach brings with it a change in ES provisions, resulting in a more diverse and also more desirable mixture according to the preferences outlined by the local level stakeholder group. Current forest policy means that many of the financially unproductive forests that are clearfelled must be reforested. The PFs indicate that some deforestation can be beneficial in achieving the stakeholders' desired mix of ESs.

When the findings of this modelling project were described to academic and national level stakeholders, all parties appreciated the quantitative approach, even though the approach only included spatial constraints before the optimisation algorithm was ran. They felt that being able to assess the relative changes in ES provision levels and management approaches in a range of PFs was a useful prompt for discussion, especially for a multidisciplinary stakeholder group with little shared familiarity with such a broad range of ESs. All parties also felt that the approach was useful to investigate policy and industry changes and they appreciated the model's potential to produce estimates of changing ES provision levels and management approaches over time and not just at the start and end of the planning horizon. Coillte, who also use Remsoft, are interested in incorporating several ESs into its management planning procedures.

Acknowledgments: This project has received funding from the European Union's Seventh Program for research, technological development and demonstration under grant agreement No. 282887 (INTEGRAL).

Author Contributions: Edwin Corrigan and Maarten Nieuwenhuis designed the study and interpreted the consolidated results. Edwin Corrigan wrote the paper. Maarten Nieuwenhuis contributed to writing the paper. Both authors contributed substantial data, meta-information, and result interpretation.

Conflicts of Interest: The authors declare no conflict of interest.

Abbreviations

The following abbreviations are used in this manuscript:

BAU	Business As Usual
CCF	Continuous Cover Forestry
CHP	Combined Heat and Power
CSA	Case Study Area
ES	Ecosystem Service
FPM	Freshwater Pearl Mussel
GP	Goal Programming
NPV	Net Present Value
NWS	Native Woodland Site
OT	Owner Type
PF	Potential Future
SFM	Sustainable Forest Management
WP	Western Peatlands

References

1. Barrett, F.; Somers, M.J.; Nieuwenhuis, M. *PractiSFM—An Operational Multi-Resource Inventory Protocol for Sustainable Forest Management*; CABI: Wallingford, UK, 2007; pp. 224–237.

2. Nieuwenhuis, M.; Tiernan, D. The Impact of the Introduction of Sustainable Forest Management Objectives on the Optimisation of PC-based Forest-level Harvest Schedules. *For. Policy Econ.* **2005**, *7*, 689–701. [CrossRef]

3. Turner, B.J.; Chikumbo, O.; Davey, S.M. Optimisation Modelling of Sustainable Forest Management at the Regional Level: An Australian Example. *Ecol. Model.* **2002**, *153*, 157–179. [CrossRef]

4. Food and Agriculture Organization (FAO). *Development of the National-Level Criteria and Indicators for the Sustainable Management of Dry Forest of Asia: Workshop Report*; Asia-Pacific Forestry Commission: Bangkok, Thailand, 2000.

5. Alcamo, J.; Bennett, E.M. *Ecosystems and Human Well-Being: A Framework for Assessment*; Island Press: Washington, DC, USA, 2003.

6. Gómez-Baggethun, E.; de Groot, R.; Lomas, P.L.; Montes, C. The History of Ecosystem Services in Economic Theory and Practice: From Early Notions to Markets and Payment Schemes. *Ecol. Econ.* **2010**, *69*, 1209–1218. [CrossRef]

7. King, R.T. Wildlife and management. *N. Y. Stake Conserv.* **1966**, *20*, 8–11.

8. Westman, W.E. How Much Are Nature's Services Worth? *Science* **1977**, *197*, 960–964. [CrossRef] [PubMed]

9. Millennium Ecosystem Assessment (MEA). Available online: http://www.unep.org/maweb/en/index.aspx (accessed on 5 October 2013).

10. De Groot, R.S.; Alkemade, R.; Braat, L.; Hein, L.; Willemen, L. Challenges in Integrating the Concept of Ecosystem Services and Values in Landscape Planning, Management and Decision Making. *Ecol. Complex.* **2010**, *7*, 260–272. [CrossRef]

11. Fléchard, M.-C.; Carroll, M.S.; Cohn, P.J.; Ní Dhubháin, Á. The Changing Relationships between Forestry and the Local Community in Rural Northwestern Ireland. *Can. J. For. Res.* **2007**, *37*, 1999–2009. [CrossRef]

12. Tiernan, D. Redesigning Afforested Western Peatlands in Ireland. In *After Wise Use—The Future of Peatlands, Proceedings of the 13th International Peat Congress: Peatland Forestry, Tullamore, Ireland, 8–13 June 2008*; Irish Peatland Society: County Kildare, Ireland, 2008.

13. Cregan, M.; Murphy, W. *A Review of Forest Recreation Research Needs in Ireland*; COFORD: Dublin, Ireland, 2006.

14. Ní Dhubháin, Á.; Fléchard, M.-C.; Moloney, R.; O'Connor, D. Stakeholders' Perceptions of Forestry in Rural Areas—Two Case Studies in Ireland. *Land Use Policy* **2009**, *26*, 695–703. [CrossRef]

15. Nieuwenhuis, M.; Williamson, G.P. Harvesting Coillte's Forests: The Right Tree at the Right Time. *Ir. For.* **1993**, *50*, 122–133.

16. INTWEGRAL. Available online: http://www.integral-project.eu/ (accessed on 21 August 2016).

17. Diaz-Balteiro, L.; González-Pachón, J.; Romero, C. Goal Programming in Forest Management: Customising Models for the Decision-maker's Preferences. *Scand. J. For. Res.* **2012**, *28*, 166–173. [CrossRef]

18. Aldea, J.; Martínez-Peña, F.; Romero, C.; Diaz-Balteiro, L. Participatory Goal Programming in Forest Management: An Application Integrating Several Ecosystem Services. *Forests* **2014**, *5*, 3352–3371. [CrossRef]

19. McDonagh, M. (Resource Optimisation Team Leader, Coillte, Oran Town Centre, Oranmore, Galway, Ireland); Corrigan, E. (UCD Forestry, School of Agriculture & Food Science, Dublin, Ireland). Coillte Forest Personal communcation, 2012.

20. Renou, F.; Farrell, E.P. Reclaiming Peatlands for Forestry: The Irish Experience. In *Restoration of Boreal and Temperate Forests*; CRC Press: Boca Raton, FL, USA, 2005.

21. Forest Service. *Irish National Forest Standard*; Magner Communications: Dublin, Ireland, 2000.

22. Johnson, K.N.; Scheurman, H.L. Techniques for Prescribing Optimal Timber Harvest and Investment Under Different Objectives—Discussion and Synthesis. *For. Sci.* **1977**, *23*, a0001–z0001.

23. Forest Service. *Code of Best Forest Practice—Ireland*; Department of the Marine and Natural Resources: Dublin, Ireland, 2000.

24. Forest Service. *Native Woodland Scheme—Establishment*; Department of Agriculture Fisheries and Food: Wexford, Ireland, 2011.

25. Forest Service. *Forest Service Appropriate Assessment Procedure—Appendix B Appropriate Assessment Procedure (AAP) Requirements Regarding Hen Harrier SPAs and Afforestation*; Forest Service: Wexford, Ireland, 2012.

26. Corrigan, E.; Nieuwenhuis, M. A Linear Programming Model to Biophysically Assess Some Ecosystem Service Synergies and Trade-Offs in Two Irish Landscapes. *Forests* **2016**, *7*, 128. [CrossRef]

27. Duesberg, S.; O'Connor, D.; Ní Dhubháin, Á.; Upton, V. Factors Influencing Irish Farmers' Afforestation Intentions. *For. Policy Econ.* **2014**, *39*, 13–20. [CrossRef]

28. Bonsu, N.O.; Dhubháin, Á.N.; O'Connor, D. Evaluating the use of an Integrated Forest Land-Use Planning Approach in addressing Forestry Conflicting Demands: Experience within an Irish Forest Landscape. *Futures* **2016**, in press. [CrossRef]

29. Patton, M.Q. *Qualitative Research & Evaluation Methods*; SAGE Publications: Thousand Oaks, CA, USA, 1990.

30. Bonsu, N.O.; Dhubháin, Á.N.; O'Connor, D. Understanding forest resource conflicts in Ireland: A case study approach. *Land Use Policy* **2015**, in press. [CrossRef]

31. Parmenides. Parmenides Eidos—Visual Reasoning. Available online: https://www.parmenides-eidos.com/eidos9/us/ (accessed on 21 August 2016).

32. Walters, K.R. Design and Development of a Generalised Forest Management System: Woodstock. In Proceedings of the International Symposium on Systems Analysis and Management Decisions in Forestry, Valdivia, Chile, 9–12 march 1993; Remsoft Inc.: Valdivia, Chile, 1993.

33. Pereira, S.; Prieto, A.; Calama, R.; Diaz-Balteiro, L. Optimal Management in *Pinus pinea* L. Stands Combining Silvicultural Schedules for Timber and Cone Production. *Silva Fenn.* **2015**, *49*, 1226. [CrossRef]

34. Biber, P.; Borges, J.G.; Moshammer, R.; Barreiro, S.; Botequim, B.; Brodrechtová, Y.; Brukas, V.; Chirici, G.; Cordero-Debets, R.; Corrigan, E.; et al. How Sensitive Are Ecosystem Services in European Forest Landscapes to Silvicultural Treatment? *Forests* **2015**, *6*, 1666–1695. [CrossRef]

35. Hotvedt, J.E. Application of Linear Goal Programming to Forest Harvest Scheduling. *South. J. Agric. Econ.* **1983**, *15*, 103–108. [CrossRef]

36. Borges, J.G.; Garcia-Gonzalo, J.; Bushenkov, V.; McDill, M.E.; Marques, S.; Oliveira, M.M. Addressing Multicriteria Forest Management with Pareto Frontier Methods: An Application in Portugal. *For. Sci.* **2014**, *60*, 63–72. [CrossRef]

37. Borges, J.G.; Marques, S.; Garcia-Gonzalo, J.; Rahman, A.U.; Bushenkov, V.; Sottomayor, M.; Carvalho, P.O.; Nordström, E.-M. A Multiple Criteria Approach for Negotiating Ecosystem Services Supply Targets and Forest Owners' Programs. *For. Sci.* **2016**. [CrossRef]

38. Pepper, H.; Patterson, G. *Red Squirrel Conservation*; Forestry Commission: Edinburgh, UK, 1998.

39. Walters, K.R.; Feunekes, U.; Cogswell, A.; Cox, E. *A Forest Planning System for Solving Spatial Harvest Scheduling Problems*; Remsoft Inc.: Fredericton, NB, Canada, 1999.

40. Könny, N.; Tóth, S.F.; McDill, M.E.; Rajasekaran, B. Temporal Connectivity of Mature Patches in Forest Planning Models. *For. Sci.* **2014**, *60*, 1089–1099.

41. Daigneault, A.J.; Fraser, M. Estimating Impacts of Climate Change Policy on Land-use: An Agent Based Modeling Approach. In Proceedings of the Agricultural & Applied Economics Association's 2012 Annual Meeting, Seattle, WA, USA, 12–14 August 2012.

PERMISSIONS

LIST OF CONTRIBUTORS

Oscar José Rover
Department of Zootechny and Rural Development, Federal University of Santa Catarina, Florianópolis-SC 88034-001, Brazil

Bernardo Corrado de Gennaro and Luigi Roselli
Department of Agricultural and Environmental Science, University of Bari Aldo Moro, 70126 Bari, Italy

Yan Ning, Yadi Li and Shuangshuang Yang
Department of Construction and Real Estate, Southeast University, Nanjing 210096, China

Chuanjing Ju
Department of Business Administration, Southeast University, Nanjing 210096, China

Brian Deal
Department of Landscape Architecture, University of Illinois, Champaign, IL 61820, USA

Haozhi Pan
Department of Urban and Regional Planning, University of Illinois, Champaign, IL 61820, USA

Li Cui and Kuo-Jui Wu
School of Business, Dalian University of Technology, Panjin 124221, China

Ming-Lang Tseng
Department of Business Administration, Lunghwa University of Science and Technology, Taoyuan 33306, Taiwan

Jiefang Dong
State Key Laboratory of Desert and Oasis Ecology, Xinjiang Institute of Ecology and Geography, Chinese Academy of Sciences, Urumqi 830011, China
Department of Economics and Management, Yuncheng University, Yuncheng 044000, China

Jieyu Huang
Department of Economics and Management, Yuncheng University, Yuncheng 044000, China

Chun Deng
Department of Economics and Management, Yuncheng University, Yuncheng 044000, China
School of Economics & Management, Northwest University, Xi'an 710127, China

Rongrong Li
School of Economic & Management, China University of Petroleum (Huadong), Qingdao 266580, China

Hailin Mu
Key Laboratory of Ocean Energy Utilization and Energy Conservation of Ministry of Education, School of Energy and Power Engineering, Dalian University of Technology, Linggong Road 2, Dalian 116024, Liaoning, China

Shuai Shao
Key Laboratory of Ocean Energy Utilization and Energy Conservation of Ministry of Education, School of Energy and Power Engineering, Dalian University of Technology, Linggong Road 2, Dalian 116024, Liaoning, China
School of Innovation and Entrepreneurship, Dalian University of Technology, Linggong Road 2, Dalian 116024, Liaoning, China
School of Environmental Science and Technology, Dalian University of Technology, Linggong Road 2, Dalian 116024, Liaoning, China

Fenglin Yang, Yun Zhang and Jinhua Li
School of Environmental Science and Technology, Dalian University of Technology, Linggong Road 2, Dalian 116024, Liaoning, China

Giovanna Gavana
Department of Economics, University of Insubria, 21100 Varese VA, Italy

Pietro Gottardo and Anna Maria Moisello
Department of Economics and Management, University of Pavia, 27100 Pavia PV, Italy

Martha Chaves
Sociology of Development and Change Group, Wageningen University, P.O. Box 8130, 6706 KN Wageningen, The Netherlands

Thomas Macintyre
MINGAS in Transition Research Group, Calle 8 # 16-218 Rozo, Palmira, Colombia

Gerard Verschoor
Sociology of Development and Change Group, Wageningen University, Hollandseweg 1, 6706 KN Wageningen, The Netherlands

Arjen E. J. Wals
Education and Competence Studies Group (ECS), Wageningen University, Hollandseweg 1, 6706 KN Wageningen, The Netherlands

Yujie Xiao
Jiangsu Key Laboratory of Modern Logistics, School of Marketing and Logistic Management, Nanjing University of Finance and Economics, Nanjing 210046, China
Business School, Nanjing University, Nanjing 210093, China

Shuai Yang
School of Economics and Management, Changshu Institute of Technology, Changshu 215500, China

J. A. M. Hufen
QA+ Research and Consultancy, Boddens Hosangweg 83, 2481KX Woubrugge, The Netherlands

Andreas Kiesel, Moritz Wagner and Iris Lewandowski
Department Biobased Products and Energy Crops, Institute of Crop Science, University of Hohenheim, Fruwirthstrasse 23, 70599 Stuttgart, Germany

Raffaella Cagliano
Department of Management, Economics and Industrial Engineering, Polytechnic University of Milan, 20133 Milano, Italy

Sarah Behnam
Department of Management, Economics and Industrial Engineering, Polytechnic University of Milan, 20133 Milano, Italy
Department of Industrial Engineering, Business Administration and Statistics, Universidad Politécnica de Madrid, 28006 Madrid, Spain

Edwin Corrigan and Maarten Nieuwenhuis
UCD Forestry, School of Agriculture & Food Science, UCD, Belfield, Dublin 4, Ireland

Index